FUNDAMENTAL SOLUTIONS IN ELASTODYNAMICS

This work is a compilation of fundamental solutions (or Green's functions) for classical or canonical problems in elastodynamics, presented with a common format and notation. These formulas describe the displacements and stresses elicited by transient and harmonic sources in solid elastic media such as full spaces, half-spaces, and strata and plates in both two and three dimensions, using the three major coordinate systems (Cartesian, cylindrical, and spherical). Such formulas are useful for numerical methods and practical application to problems of wave propagation in elasticity, soil dynamics, earthquake engineering, mechanical vibration, or geophysics. Together with the plots of the response functions, this work should serve as a valuable reference to engineers and scientists alike. These formulas were heretofore found only scattered throughout the literature. The solutions are tabulated without proof, but giving reference to appropriate modern papers and books containing full derivations. Most formulas in the book have been programmed and tested within the MATLAB environment, and the programs thus developed are both listed and available for free download.

Eduardo Kausel earned his first professional degree in 1967, graduating as a civil engineer from the University of Chile, and then worked at Chile's National Electricity Company. In 1969 he carried out postgraduate studies at the Technical University in Darmstadt. He earned his Master of Science (1972) and Doctor of Science (1974) degrees from MIT. Following graduation, Dr. Kausel worked at Stone and Webster Engineering Corporation in Boston, and then joined the MIT faculty in 1978, where he has remained since. He is a registered professional engineer in the State of Massachusetts, is a senior member of various professional organizations (ASCE, SSA, EERI, IACMG), and has extensive experience as a consulting engineer.

Among the honors he has received are a 1989 Japanese Government Research Award for Foreign Specialists from the Science and Technology Agency, a 1992 Honorary Faculty Membership in Epsilon Chi, the 1994 Konrad Zuse Guest Professorship at the University of Hamburg in Germany, the Humboldt Prize from the German Government in 2000, and the 2001 MIT-CEE Award for Conspicuously Effective Teaching.

Dr. Kausel is best known for his work on dynamic soil–structure interaction, and for his very successful Green's functions (fundamental solutions) for the dynamic analysis of layered media, which are incorporated in a now widely used program. Dr. Kausel is the author of more than 150 technical papers and reports in the areas of structural dynamics, earthquake engineering, and computational mechanics.

Fundamental Solutions in Elastodynamics

A Compendium

EDUARDO KAUSEL

Massachusetts Institute of Technology

CAMBRIDGE
UNIVERSITY PRESS

CAMBRIDGE UNIVERSITY PRESS
Cambridge, New York, Melbourne, Madrid, Cape Town,
Singapore, São Paulo, Delhi, Tokyo, Mexico City

Cambridge University Press
32 Avenue of the Americas, New York, NY 10013-2473, USA

www.cambridge.org
Information on this title: www.cambridge.org/9780521375993

First published 2006
First paperback edition 2011

A catalog record for this publication is available from the British Library

Library of Congress Cataloging in Publication data

Kausel, E.
Fundamental solutions in elastodynamics : a compendium / Eduardo Kausel.
   p.   cm.
Includes bibliographical references.
ISBN-13: 978-0-521-85570-9
ISBN-10: 0-521-85570-5
1. Elasticity.   2. Dynamics.   3. Green's functions.   I. Title.
QA931.K375   2006
531′.382 – dc22   2005020866

ISBN 978-0-521-85570-9 Hardback
ISBN 978-0-521-37599-3 Paperback

Contents

Preface

We present in this work a collection of fundamental solutions, or so-called *Green's functions*, for some classical or canonical problems in elastodynamics. Such formulas provide the dynamic response functions for transient point sources acting within isotropic, elastic media, in both the frequency domain and the time domain, and in both two and three dimensions. The bodies considered are full spaces, half-spaces, and plates of infinite lateral extent, while the sources range from point and line forces to torques, seismic moments, and pressure pulses. By appropriate convolutions, these solutions can be extended to spatially distributed sources and/or sources with an arbitrary variation in time.

These fundamental solutions, as their name implies, constitute invaluable tools for a large class of numerical solution techniques for wave propagation problems in elasticity, soil dynamics, earthquake engineering, or geophysics. Examples are the Boundary Integral (or element) Method (BIM), which is often used to obtain the solution to wave propagation problems in finite bodies of irregular shape, even while working with the Green's functions for a full space.

The solutions included herein are found scattered throughout the literature, and no single book was found to deal with them all in one place. In addition, each author, paper, or book uses sign conventions and symbols that differ from one another, or they include only partial results, say only the solution in the frequency domain or for some particular value of Poisson's ratio. Sometimes, published results are also displayed in unconventional manners, for example, taking forces to be positive down, but displacements up, or scaling the displays in unusual ways or using too small a scale, and so forth. Thus, it was felt that a compendium of the known solutions in a common format would serve a useful purpose. With this in mind, we use throughout a consistent notation, coordinate systems, and sign convention, which should greatly facilitate the application of these fundamental solutions. Also, while we anguished initially at the choice of symbols for the angles in spherical coordinates, we decided in the end to use θ for the azimuth and ϕ for the polar angle. Although this contravenes the common notation, it provides consistency between spherical and cylindrical coordinates and eases the transition between one and the other system.

We tabulate these solutions herein without proof, giving reference to appropriate modern papers and books containing full derivations, but making no effort at establishing the original sources of the derivations or, for that matter, providing a historical account of these solutions. In some cases, we give no references, in which case we have developed the formulae ourselves using established methods, either because an appropriate reference was not known to us, not readily available, or for purely pragmatic reasons. Yet, recognizing that these are all classical problems, we do not claim to have discovered new formulas. Also, the tables may not necessarily be complete in that solutions for some additional classical problems, or important extensions to these, may exist of which we may be unaware. If and when these are brought to our attention, we shall be happy to consider them with proper credit when preparing a revised version of this work.

Finally, we have programmed most formulas within the MATLAB or other programming environment, and provide plots of response functions that could be used to verify the correctness of a particular implementation. Also, we have made every effort at checking the formulas themselves for correctness and dimensional consistency. Nonetheless, the possibility always exists that errors may remain undetected in some of these formulas. If the reader should find any such errors, we shall be thankful if they are brought to our attention.

Eduardo Kausel   Cambridge, September 2005

1 Fundamentals

1.1 Notation and table of symbols

Except where noted, the following symbols will be used consistently throughout this work:

$a = \beta / \alpha$	Ratio of S- and P-wave velocities
$\mathbf{b} = b_j \hat{\mathbf{e}}_j$	Body load vector
C_R	Rayleigh-wave velocity
$\mathbf{C}_n$	3×3 Bessel matrix, cylindrical coordinates and flat layers (see Table 10.2)
$\mathbf{g}_j = g_{ij}\hat{\mathbf{e}}_i$	Green's function vector for the *frequency domain* response due to a unit load in direction j
$\mathbf{F}_n$	3×3 traction matrix, cylindrical coordinates (see Table 10.4)
$\mathbf{F}_n^{(1)}, \mathbf{F}_n^{(2)}$	As $\mathbf{F}_n$ above, assembled with first and second Hankel functions
$\mathcal{F}$	Fourier transform operator
$\mathcal{F}^{-1}$	Inverse Fourier transform operator
g_{ij}	Green's function for the *frequency domain* response in direction i due to a unit load in direction j
$g_{ij,k} = \partial g_{ij}/\partial x_k$	Derivative with respect to the *receiver* location
$g'_{ij,k} = \partial g_{ij}/\partial x'_k$	Derivative with respect to the *source* location
G_{ij}	Green's function for the *frequency domain* response due to a dipole
$h_n^{(1)}(kR), h_n^{(2)}(kR)$	First and second spherical Hankel functions of order n
$H_n^{(1)}(kr), H_n^{(2)}(kr)$	First and second Hankel functions of order n (Bessel functions of the third kind)
$\mathbf{H}_n$	3×3 displacement matrix, cylindrical coordinates (see Table 10.3)
$\mathbf{H}_n^{(1)}, \mathbf{H}_n^{(2)}$	As $\mathbf{H}_n$ above, assembled with first and second Hankel functions
$\mathbf{H}_m$	3×3 spherical Bessel matrix, spherical coordinates (see Table 10.7)

1

$$\mathcal{H}(t - t_\alpha) = \begin{cases} 0 & t < t_\alpha \\ \frac{1}{2} & t = t_\alpha \\ 1 & t \geq t_\alpha \end{cases}$$ Unit step function, or Heaviside function

$\mathrm{i} = \sqrt{-1}$ Imaginary unit (non-italicized)

i, j, k Sub-indices for the numbers 1, 2, 3 or coordinates x, y, z

$\hat{\mathbf{i}}, \hat{\mathbf{j}}, \hat{\mathbf{k}} \equiv \hat{\mathbf{e}}_1, \hat{\mathbf{e}}_2, \hat{\mathbf{e}}_3$ Orthogonal unit basis vectors in Cartesian coordinates

$J_n(kr), Y_n(kr)$ Bessel functions of the first and second kind

$\mathbf{J}$ 3×3 spherical orthogonality condition (Section 9.3, Table 10.6)

k Radial wavenumber

$k_P = \omega/\alpha$ P wavenumber

$k_S = \omega/\beta$ S wavenumber

k_z Vertical wavenumber

$kp = \sqrt{k^2 - k_P^2}$ Vertical wavenumber for P waves, flat layers

$ks = \sqrt{k^2 - k_S^2}$ Vertical wavenumber for S waves, flat layers

$k_\alpha = \sqrt{k_P^2 - k_z^2}$ Radial wavenumber for P waves, cylindrical layers

$k_\beta = \sqrt{k_S^2 - k_z^2}$ Radial wavenumber for S waves, cylindrical layers

$\mathbf{L}_m^n$ 3×3 Spheroidal (co-latitude) matrix (see Tables 10.7, 10.8)

M Intensity of moment, torque, or seismic moment

M_{ij} Displacement in direction i due to seismic moment with axis j.

$\mathbf{p}$ Load vector

$\tilde{\mathbf{p}}$ Load vector in frequency–wavenumber domain

$\mathbf{\mathfrak{p}}$ As above, but vertical component multiplied by $-\mathrm{i} = -\sqrt{-1}$

p Pressure (positive when compressive)

$p = \sqrt{1 - (k_P/k)^2}$ Dimensionless vertical wavenumber for P waves

P Load amplitude

P_m Legendre function (polynomial) of the first kind

P_m^n Associated Legendre function of the first kind

Q_m^n Associated Legendre function of the second kind

R Source–receiver distance in 3-D space

r Source–receiver distance in 2-D space, or range

r, θ, z Cylindrical coordinates (see Fig. 1.2)

$\hat{\mathbf{r}}, \hat{\mathbf{t}}, \hat{\mathbf{k}}$ Orthogonal unit basis vectors in cylindrical coordinates

$\hat{\mathbf{r}}, \hat{\mathbf{s}}, \hat{\mathbf{t}}$ Orthogonal unit basis vectors in spherical coordinates

R, ϕ, θ Spherical coordinates (see Fig. 1.3)

$s = \sqrt{1 - (k_S/k)^2}$ Dimensionless vertical wavenumber for S waves

t Time

$t_P = r/\alpha$ P-wave arrival time

$t_S = r/\beta$ S-wave arrival time

$t_R = r/C_R$	Rayleigh-wave arrival time
T	Torque
$\mathbf{T}_n$	Azimuthal matrix
T_{ij}	Displacement in direction i due to a unit torque with axis j.
$\mathbf{u}$	Displacement vector
$\tilde{\mathbf{u}}$	Displacement vector in frequency-wavenumber domain
u	As above, but vertical component multiplied by $-i = -\sqrt{-1}$
$u_x, u_y, u_z \equiv u_1, u_2, u_3$	Displacement components in Cartesian coordinates
$u_r, u_\theta, u_z \equiv u, v, w$	Displacement components in cylindrical coordinates
u_R, u_ϕ, u_θ	Displacement components in spherical coordinates
u_{xz} or u_{ij}, etc.	Displacement in direction x (or i) due to force in direction z (or j)
U_{ij}, u_{ij}	Green's function for the *time domain* response in direction i due to a unit load in direction j
$x, y, z \equiv x_1, x_2, x_3$	Cartesian coordinates (see Fig. 1.1)
x', y', z'	Cartesian coordinates of the source
$\alpha = \beta\sqrt{2(1-\nu)/(1-2\nu)}$	P-wave velocity
$\beta = \sqrt{\mu/\rho}$	S-wave velocity
γ_i	Direction cosine of R with ith axis (see Cartesian coordinates)
$\delta(t - t_S) = \dfrac{d\mathcal{H}(t - t_S)}{dt} = \begin{cases} 0, & t < t_S \\ \infty, & t = t_S \\ 0, & t \geq t_S \end{cases}$	Dirac-delta singularity function
$\delta_{ij} = \begin{cases} 1, & i = j \\ 0, & i \neq j \end{cases}$	Kronecker delta
λ	Lamé constant
$\lambda + 2\mu = \rho\,\alpha^2$	Constrained modulus
$\mu = \rho\beta^2$	Shear modulus
ν	Poisson's ratio
$\tau = t\beta/r = t/t_S$	Dimensionless time
ρ	Mass density
θ	Azimuth in cylindrical and spherical coordinates
θ_i	Angle between R and the ith axis ($\gamma_i = \cos\theta_i$), $i = 1, 2, 3$
Φ	Dilatational Helmholtz potential
$\chi = \chi(\omega)$	Dimensionless component function of Green's functions (in later chapters, a Helmholtz shear potential)
$X = X(t)$	Inverse Fourier transform of $\chi(\omega)$ (or the convolution of the latter with an arbitrary time function)
$\psi = \psi(\omega)$	Dimensionless component function of Green's functions (in later chapters, a Helmholtz shear potential)
$\Psi = \Psi(t)$	Inverse Fourier transform of $\psi(\omega)$ (or the convolution of the latter with an arbitrary time function)

Ψ Helmholtz vector potential (shear)

ω Frequency (rad/s)

$\Omega_S = \omega r/\beta = \omega t_S$ Dimensionless frequency for S (shear) waves

$\Omega_P = \omega r/\alpha = \omega t_P$ Dimensionless frequency for P (dilatational) waves

1.2 Sign convention

Component of vectors, such as displacements and forces, are always defined positive in the positive coordinate directions, and plots of displacements are always shown upright (i.e., never reversed or upside down).

   Point and line sources will usually – but not always – be located at the origin of coordinates. When this is not the case, it will be indicated explicitly.

   Wave propagation in the frequency–wavenumber domain will assume a dependence of the form $\exp i(\omega t - kx)$, that is, the underlying Fourier transform pairs from frequency–wavenumber domain to the space–time domain are of the form

$$f(\omega, k) = \int_{-\infty}^{+\infty} \int_{-\infty}^{+\infty} f(t, x) e^{-i(\omega t - kx)} dt\, dx = \mathcal{F}[f(t, x)] \tag{1.1}$$

$$f(t, x) = \left(\frac{1}{2\pi}\right)^2 \int_{-\infty}^{+\infty} \int_{-\infty}^{+\infty} f(\omega, k) e^{i(\omega t - kx)} d\omega\, dk = \mathcal{F}^{-1}[f(\omega, k)] \tag{1.2}$$

Important consequences of this transformation convention concern the direction of positive wave propagation and decay, and the location of *poles* for the dynamic system under consideration. These, in turn, relate to the principles of radiation, boundedness at infinity, and causality. Also, this convention calls for the use of second (cylindrical or spherical) Hankel functions when formulating wave propagation problems in infinite media, either in cylindrical or in spherical coordinates, and casting them in the frequency domain.

1.3 Coordinate systems and differential operators

We choose Cartesian coordinates in three-dimensional space forming a *right-handed system*, and we denote these indifferently as either x, y, z or x_1, x_2, x_3. In most cases, we shall assume that $x = x_1$ and $y = y_2$ lie in a horizontal plane, and that $z = x_3$ is up. For two-dimensional (plane strain) problems, the *in-plane* (or SV-P) components will be contained in the vertical plane defined by x and z (i.e., x_1, x_3), and the *anti-plane* (or SH) components will be in the horizontal direction y (or x_2), which is perpendicular to the plane of wave propagation. On the one hand, this convention facilitates the conversion between Cartesian and either cylindrical or spherical coordinates; on the other, it provides a convenient x–y reference system when working in horizontal planes (i.e., in a bird's-eye view). Nonetheless, you may rotate these systems to suit your convenience.

1.3.1 Cartesian coordinates

a) Three-dimensional space (Fig. 1.1a)

   Source–receiver distance $R = \sqrt{x^2 + y^2 + z^2}$ \hfill (1.3a)

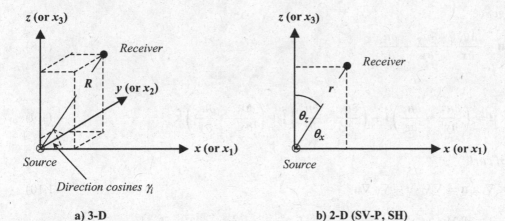

a) 3-D   **b) 2-D (SV-P, SH)**

Figure 1.1: Cartesian coordinates.

Direction cosines of R

$$\gamma_i = \cos\theta_i = \frac{\partial R}{\partial x_i} = \frac{x_i}{R} \tag{1.3b}$$

First derivatives of direction cosines

$$\gamma_{i,j} = \frac{\partial \gamma_i}{\partial x_j} = \frac{1}{R}(\delta_{ij} - \gamma_i \gamma_j) \tag{1.3c}$$

Second derivatives of direction cosines

$$\gamma_{i,jk} = \frac{1}{R^2}(3\gamma_i \gamma_j \gamma_k - \gamma_i \delta_{jk} - \gamma_j \delta_{ki} - \gamma_k \delta_{ij}) \tag{1.3d}$$

Implied summations

$$\delta_{ii} = \delta_{11} + \delta_{22} + \delta_{33} = 3, \tag{1.3e}$$
$$\gamma_i \gamma_i = \gamma_1^2 + \gamma_2^2 + \gamma_3^2 = 1$$

Nabla operator

$$\nabla = \hat{\mathbf{i}}\frac{\partial}{\partial x} + \hat{\mathbf{j}}\frac{\partial}{\partial y} + \hat{\mathbf{k}}\frac{\partial}{\partial z} \tag{1.4}$$

Gradient of vector

$$\nabla\mathbf{u} = \frac{\partial u_x}{\partial x}\hat{\mathbf{i}}\hat{\mathbf{i}} + \frac{\partial u_y}{\partial x}\hat{\mathbf{i}}\hat{\mathbf{j}} + \frac{\partial u_z}{\partial x}\hat{\mathbf{i}}\hat{\mathbf{k}} + \frac{\partial u_x}{\partial y}\hat{\mathbf{j}}\hat{\mathbf{i}} + \frac{\partial u_y}{\partial y}\hat{\mathbf{j}}\hat{\mathbf{j}} + \frac{\partial u_z}{\partial y}\hat{\mathbf{j}}\hat{\mathbf{k}} + \frac{\partial u_x}{\partial z}\hat{\mathbf{k}}\hat{\mathbf{i}} + \frac{\partial u_y}{\partial z}\hat{\mathbf{k}}\hat{\mathbf{j}} + \frac{\partial u_z}{\partial z}\hat{\mathbf{k}}\hat{\mathbf{k}} \tag{1.5}$$

where the products of the form $\hat{\mathbf{i}}\hat{\mathbf{i}}$, etc., are tensor bases, or *dyads*, that is, $\nabla\mathbf{u}$ is a tensor. For example, two distinct projections of this tensor are

$$\hat{\mathbf{k}}\cdot\nabla\mathbf{u} = \frac{\partial u_x}{\partial z}\hat{\mathbf{i}} + \frac{\partial u_y}{\partial z}\hat{\mathbf{j}} + \frac{\partial u_z}{\partial z}\hat{\mathbf{k}} = \frac{\partial\mathbf{u}}{\partial z} \tag{1.6}$$

and

$$(\nabla\mathbf{u})\cdot\hat{\mathbf{k}} = \frac{\partial u_z}{\partial x}\hat{\mathbf{i}} + \frac{\partial u_z}{\partial y}\hat{\mathbf{j}} + \frac{\partial u_z}{\partial z}\hat{\mathbf{k}} = \nabla u_z \tag{1.7}$$

Divergence

$$\nabla \cdot \mathbf{u} = \frac{\partial u_x}{\partial x} + \frac{\partial u_y}{\partial y} + \frac{\partial u_z}{\partial z} \tag{1.8}$$

Curl

$$\nabla \times \mathbf{u} = \left(\frac{\partial u_z}{\partial y} - \frac{\partial u_y}{\partial z} \right) \hat{\mathbf{i}} + \left(\frac{\partial u_x}{\partial z} - \frac{\partial u_z}{\partial x} \right) \hat{\mathbf{j}} + \left(\frac{\partial u_y}{\partial x} - \frac{\partial u_x}{\partial y} \right) \hat{\mathbf{k}} \tag{1.9}$$

Curl of curl

$$\nabla \times \nabla \times \mathbf{u} = \nabla\nabla \cdot \mathbf{u} - \nabla \cdot \nabla \mathbf{u} \tag{1.10}$$

$$\nabla\nabla \cdot \mathbf{u} = \left[\frac{\partial^2 u_x}{\partial x^2} + \frac{\partial^2 u_y}{\partial x\,\partial y} + \frac{\partial^2 u_z}{\partial x\,\partial z} \right] \hat{\mathbf{i}} + \left[\frac{\partial^2 u_x}{\partial y\,\partial x} + \frac{\partial^2 u_y}{\partial y^2} + \frac{\partial^2 u_z}{\partial y\,\partial z} \right] \hat{\mathbf{j}}$$

$$+ \left[\frac{\partial^2 u_x}{\partial z\,\partial x} + \frac{\partial^2 u_y}{\partial z\,\partial y} + \frac{\partial^2 u_z}{\partial z^2} \right] \hat{\mathbf{k}} \tag{1.11}$$

Laplacian

$$\nabla^2 \mathbf{u} = \nabla \cdot \nabla \mathbf{u} = \nabla^2 u_x \hat{\mathbf{i}} + \nabla^2 u_y \hat{\mathbf{j}} + \nabla^2 u_z \hat{\mathbf{k}} = \left(\frac{\partial^2}{\partial x^2} + \frac{\partial^2}{\partial y^2} + \frac{\partial^2}{\partial z^2} \right) \mathbf{u} \tag{1.12}$$

Wave equation

$$(\lambda + \mu)\nabla\nabla \cdot \mathbf{u} + \mu\nabla^2 \mathbf{u} + \mathbf{b} = \rho\ddot{\mathbf{u}} \tag{1.13}$$

b) *Two-dimensional space (x–z), (Fig. 1.1b)*

Source–receiver distance $\qquad r = \sqrt{x^2 + z^2}$ $\hfill$ (1.14a)

Direction cosines of r $\qquad \gamma_1 = \cos\theta_1 = \sin\theta_z,$ $\hfill$ (1.14b)

$\qquad\qquad\qquad\qquad\qquad \gamma_2 = 0,$ $\hfill$ (1.14c)

$\qquad\qquad\qquad\qquad\qquad \gamma_3 = \cos\theta_3 = \cos\theta_z$ $\hfill$ (1.14d)

Implied summations $\qquad \delta_{ii} = \delta_{11} + \delta_{33} = 2,$ $\hfill$ (1.14e)

$\qquad\qquad\qquad\qquad\qquad \gamma_i \gamma_i = \gamma_1^2 + \gamma_3^2 = 1$ $\hfill$ (1.14f)

1.3.2 Cylindrical coordinates

Source–receiver distance $\qquad R = \sqrt{x^2 + y^2 + z^2}$ $\hfill$ (1.15a)

Range $\qquad\qquad\qquad\qquad r = \sqrt{x^2 + y^2}$ $\hfill$ (1.15b)

Azimuth $\qquad\qquad\qquad\quad \tan\theta = y/x$ $\hfill$ (1.15c)

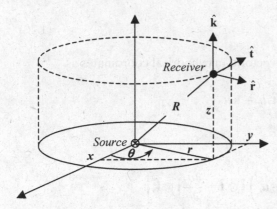

Figure 1.2: Cylindrical coordinates.

Direction cosines

$$\gamma_1 = \frac{r \cos \theta}{\sqrt{r^2 + z^2}} = \frac{r \cos \theta}{R} \tag{1.15d}$$

$$\gamma_2 = \frac{r \sin \theta}{\sqrt{r^2 + z^2}} = \frac{r \sin \theta}{R} \tag{1.15c}$$

$$\gamma_3 = \frac{z}{\sqrt{r^2 + z^2}} = \frac{z}{R} \tag{1.15f}$$

Basis vectors

$$\hat{\mathbf{r}} = \hat{\mathbf{i}} \, \cos \theta + \hat{\mathbf{j}} \, \sin \theta \tag{1.15g}$$

$$\hat{\mathbf{t}} = -\hat{\mathbf{i}} \, \sin \theta + \hat{\mathbf{j}} \, \cos \theta \tag{1.15h}$$

$$\hat{\mathbf{k}} = \hat{\mathbf{k}} \tag{1.15i}$$

Conversion between cylindrical and Cartesian coordinates

$$\mathbf{u} = u_x \hat{\mathbf{i}} + u_y \hat{\mathbf{j}} + u_z \hat{\mathbf{k}}$$

$$= u_r \hat{\mathbf{r}} + u_\theta \hat{\mathbf{t}} + u_z \hat{\mathbf{k}} \tag{1.16}$$

$$\begin{bmatrix} u_r \\ u_\theta \\ u_z \end{bmatrix} = \begin{bmatrix} \cos \theta & \sin \theta & 0 \\ -\sin \theta & \cos \theta & 0 \\ 0 & 0 & 1 \end{bmatrix} \begin{bmatrix} u_x \\ u_y \\ u_z \end{bmatrix} \tag{1.17}$$

$$\begin{bmatrix} u_x \\ u_y \\ u_z \end{bmatrix} = \begin{bmatrix} \cos \theta & -\sin \theta & 0 \\ \sin \theta & \cos \theta & 0 \\ 0 & 0 & 1 \end{bmatrix} \begin{bmatrix} u_r \\ u_\theta \\ u_z \end{bmatrix} \tag{1.18}$$

Nabla operator

$$\nabla = \hat{\mathbf{r}} \frac{\partial}{\partial r} + \hat{\mathbf{t}} \frac{1}{r} \frac{\partial}{\partial \theta} + \hat{\mathbf{k}} \frac{\partial}{\partial z} \tag{1.19}$$

Allowing the symbol $\otimes$ to stand for the scalar product, the dot product, or the cross product, and considering that

$$\frac{\partial\hat{\mathbf{r}}}{\partial\theta}=\hat{\mathbf{t}},\quad\frac{\partial\hat{\mathbf{t}}}{\partial\theta}=-\hat{\mathbf{r}},\quad\frac{\partial\hat{\mathbf{k}}}{\partial\theta}=\mathbf{0}\tag{1.20}$$

we can write a generic nabla operation on a vector $\mathbf{u}$ in cylindrical coordinates as

$$\nabla\otimes\mathbf{u}=\left(\hat{\mathbf{r}}\frac{\partial}{\partial r}+\hat{\mathbf{t}}\frac{1}{r}\frac{\partial}{\partial\theta}+\hat{\mathbf{k}}\frac{\partial}{\partial z}\right)\otimes\left(\hat{\mathbf{r}}\,u_r+\hat{\mathbf{t}}\,u_\theta+\hat{\mathbf{k}}\,u_z\right)$$

$$=\frac{\partial u_r}{\partial r}\hat{\mathbf{r}}\otimes\hat{\mathbf{r}}+\frac{\partial u_\theta}{\partial r}\hat{\mathbf{r}}\otimes\hat{\mathbf{t}}+\frac{\partial u_z}{\partial r}\hat{\mathbf{r}}\otimes\hat{\mathbf{k}}$$

$$+\frac{1}{r}\left[\left(\frac{\partial u_r}{\partial\theta}-u_\theta\right)\hat{\mathbf{t}}\otimes\hat{\mathbf{r}}+\left(\frac{\partial u_\theta}{\partial\theta}+u_r\right)\hat{\mathbf{t}}\otimes\hat{\mathbf{t}}+\frac{\partial u_z}{\partial\theta}\hat{\mathbf{t}}\otimes\hat{\mathbf{k}}\right]$$

$$+\frac{\partial u_r}{\partial z}\hat{\mathbf{k}}\otimes\hat{\mathbf{r}}+\frac{\partial u_\theta}{\partial z}\hat{\mathbf{k}}\otimes\hat{\mathbf{t}}+\frac{\partial u_z}{\partial z}\hat{\mathbf{k}}\otimes\hat{\mathbf{k}}\tag{1.21}$$

Specializing this expression to the scalar, dot, and cross products, we obtain

Gradient

$$\nabla\mathbf{u}=\frac{\partial u_r}{\partial r}\hat{\mathbf{r}}\hat{\mathbf{r}}+\frac{\partial u_\theta}{\partial r}\hat{\mathbf{r}}\hat{\mathbf{t}}+\frac{\partial u_z}{\partial r}\hat{\mathbf{r}}\hat{\mathbf{k}}$$

$$+\frac{1}{r}\left[\left(\frac{\partial u_r}{\partial\theta}-u_\theta\right)\hat{\mathbf{t}}\hat{\mathbf{r}}+\left(\frac{\partial u_\theta}{\partial\theta}+u_r\right)\hat{\mathbf{t}}\hat{\mathbf{t}}+\frac{\partial u_z}{\partial\theta}\hat{\mathbf{t}}\hat{\mathbf{k}}\right]$$

$$+\frac{\partial u_r}{\partial z}\hat{\mathbf{k}}\hat{\mathbf{r}}+\frac{\partial u_\theta}{\partial z}\hat{\mathbf{k}}\hat{\mathbf{t}}+\frac{\partial u_z}{\partial z}\hat{\mathbf{k}}\hat{\mathbf{k}}\tag{1.22}$$

where the products of the form $\hat{\mathbf{r}}\hat{\mathbf{r}}$, etc., are tensor bases, or *dyads*.

Divergence

$$\nabla\cdot\mathbf{u}=\frac{\partial u_r}{\partial r}+\frac{1}{r}\left(\frac{\partial u_\theta}{\partial\theta}+u_r\right)+\frac{\partial u_z}{\partial z}=\frac{1}{r}\left[\frac{\partial(ru_r)}{\partial r}+\frac{\partial u_\theta}{\partial\theta}+\frac{\partial(ru_z)}{\partial z}\right]\tag{1.23}$$

Curl

$$\nabla\times\mathbf{u}=\left(\frac{1}{r}\frac{\partial u_z}{\partial\theta}-\frac{\partial u_\theta}{\partial z}\right)\hat{\mathbf{r}}+\left(\frac{\partial u_r}{\partial z}-\frac{\partial u_z}{\partial r}\right)\hat{\mathbf{t}}+\frac{1}{r}\left(\frac{\partial(ru_\theta)}{\partial r}-\frac{\partial u_r}{\partial\theta}\right)\hat{\mathbf{k}}\tag{1.24}$$

Curl of curl

$$\nabla\times\nabla\times\mathbf{u}=\nabla\nabla\cdot\mathbf{u}-\nabla\cdot\nabla\mathbf{u}\tag{1.25}$$

$$\nabla\nabla\cdot\mathbf{u}=\hat{\mathbf{r}}\frac{\partial}{\partial r}\left[\frac{1}{r}\frac{\partial(ru_r)}{\partial r}+\frac{1}{r}\frac{\partial u_\theta}{\partial\theta}+\frac{\partial u_z}{\partial z}\right]+\hat{\mathbf{t}}\frac{1}{r}\frac{\partial}{\partial\theta}\left[\frac{1}{r}\frac{\partial(ru_r)}{\partial r}+\frac{1}{r}\frac{\partial u_\theta}{\partial\theta}+\frac{\partial u_z}{\partial z}\right]$$

$$+\hat{\mathbf{k}}\frac{\partial}{\partial z}\left[\frac{1}{r}\frac{\partial(ru_r)}{\partial r}+\frac{1}{r}\frac{\partial u_\theta}{\partial\theta}+\frac{\partial u_z}{\partial z}\right]\tag{1.26}$$

Laplacian

$$\nabla^2 = \nabla \cdot \nabla = \frac{\partial^2}{\partial r^2} + \frac{1}{r}\frac{\partial}{\partial r} + \frac{1}{r^2}\frac{\partial^2}{\partial \theta^2} + \frac{\partial^2}{\partial z^2}$$

$$\nabla^2 \mathbf{u} = \nabla \cdot \nabla \mathbf{u} = \hat{\mathbf{r}}\left[\frac{\partial^2 u_r}{\partial r^2} + \frac{1}{r}\frac{\partial u_r}{\partial r} + \frac{1}{r^2}\left(\frac{\partial^2 u_r}{\partial \theta^2} - u_r - 2\frac{\partial u_\theta}{\partial \theta}\right) + \frac{\partial^2 u_r}{\partial z^2}\right]$$

$$+ \hat{\mathbf{t}}\left[\frac{\partial^2 u_\theta}{\partial r^2} + \frac{1}{r}\frac{\partial u_\theta}{\partial r} + \frac{1}{r^2}\left(\frac{\partial^2 u_\theta}{\partial \theta^2} - u_\theta + 2\frac{\partial u_r}{\partial \theta}\right) + \frac{\partial^2 u_\theta}{\partial z^2}\right]$$

$$+ \hat{\mathbf{k}}\left[\frac{\partial^2 u_z}{\partial r^2} + \frac{1}{r}\frac{\partial u_z}{\partial r} + \frac{1}{r^2}\frac{\partial^2 u_z}{\partial \theta^2} + \frac{\partial^2 u_z}{\partial z^2}\right] \tag{1.27}$$

(Note: $\partial^2/\partial\theta^2$ in ∇^2 acts both on the components of $\mathbf{u}$ and on the basis vectors $\hat{\mathbf{r}}$, $\hat{\mathbf{t}}$.)

Wave equation

$$(\lambda + \mu)\nabla\nabla \cdot \mathbf{u} + \mu\nabla \cdot \nabla\mathbf{u} + \mathbf{b} = \rho\ddot{\mathbf{u}} \tag{1.28}$$

Expansion of a vector in Fourier series in the azimuth

$$u = \sum_{n=0}^{\infty} u_n \begin{pmatrix} \cos n\theta \\ \sin n\theta \end{pmatrix} \tag{1.29a}$$

$$v = \sum_{n=0}^{\infty} v_n \begin{pmatrix} -\sin n\theta \\ \cos n\theta \end{pmatrix} \tag{1.29b}$$

$$w = \sum_{n=0}^{\infty} w_n \begin{pmatrix} \cos n\theta \\ \sin n\theta \end{pmatrix}. \tag{1.29c}$$

in which $u \equiv u_r$, $v \equiv u_\theta$, $w \equiv u_z$, and either the lower or the upper element in the parentheses must be used, as may be necessary. Also, u_n, v_n, w_n are the coefficients of the Fourier series, which do not depend on θ, but only on r and z, that is, $u_n = u_n(r, z)$, and so forth.

1.3.3 Spherical coordinates

Source–receiver distance	$R = \sqrt{x^2 + y^2 + z^2}$	(1.30a)
Range	$r = \sqrt{x^2 + y^2}$	(1.30b)
Azimuth	$\tan\theta = y/x, \quad 0 \le \theta \le 2\pi$	(1.30c)
Polar angle	$\phi = \arccos(z/R), \quad 0 \le \phi \le \pi$	(1.30d)
Direction cosines	$\gamma_1 = \sin\phi\cos\theta \quad\Rightarrow\quad \cos\theta = \dfrac{\gamma_1}{\sqrt{1-\gamma_3^2}}$	(1.30e)
	$\gamma_2 = \sin\phi\sin\theta \quad\Rightarrow\quad \sin\theta = \dfrac{\gamma_2}{\sqrt{1-\gamma_3^2}}$	(1.30f)
	$\gamma_3 = \cos\phi \quad\Rightarrow\quad \sin\phi = \sqrt{1-\gamma_3^2}$	(1.30g)

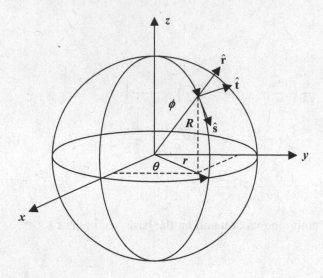

Figure 1.3: Spherical coordinates.

Basis vectors $\hat{\mathbf{r}} = \hat{\mathbf{i}}\sin\phi\,\cos\theta + \hat{\mathbf{j}}\,\sin\phi\,\sin\theta + \hat{\mathbf{k}}\cos\phi$ Radial   (1.31a)

$\hat{\mathbf{t}} = -\hat{\mathbf{i}}\sin\theta + \hat{\mathbf{j}}\cos\theta$ Tangential   (1.31b)

$\hat{\mathbf{s}} = \hat{\mathbf{i}}\cos\phi\,\cos\theta + \hat{\mathbf{j}}\cos\phi\,\sin\theta - \hat{\mathbf{k}}\sin\phi$ Meridional   (1.31c)

Conversion between spherical and Cartesian coordinates

$$\mathbf{u} = u_x\,\hat{\mathbf{i}} + u_y\,\hat{\mathbf{j}} + u_z\,\hat{\mathbf{k}}$$
$$= u_R\,\hat{\mathbf{r}} + u_\theta\,\hat{\mathbf{t}} + u_\phi\,\hat{\mathbf{s}} \tag{1.32}$$

$$\begin{bmatrix} u_R \\ u_\phi \\ u_\theta \end{bmatrix} = \begin{bmatrix} \sin\phi\,\cos\theta & \sin\phi\,\sin\theta & \cos\phi \\ \cos\phi\,\cos\theta & \cos\phi\,\sin\theta & -\sin\phi \\ -\sin\theta & \cos\theta & 0 \end{bmatrix} \begin{bmatrix} u_x \\ u_y \\ u_z \end{bmatrix} \tag{1.33}$$

$$\begin{bmatrix} u_x \\ u_y \\ u_z \end{bmatrix} = \begin{bmatrix} \sin\phi\,\cos\theta & \cos\phi\,\cos\theta & -\sin\theta \\ \sin\phi\,\sin\theta & \cos\phi\,\sin\theta & \cos\theta \\ \cos\phi & -\sin\phi & 0 \end{bmatrix} \begin{bmatrix} u_R \\ u_\phi \\ u_\theta \end{bmatrix} \tag{1.34}$$

Nabla operator

$$\nabla = \hat{\mathbf{r}}\frac{\partial}{\partial R} + \hat{\mathbf{s}}\frac{1}{R}\frac{\partial}{\partial\phi} + \hat{\mathbf{t}}\frac{1}{R\sin\phi}\frac{\partial}{\partial\theta} \tag{1.35}$$

Allowing the symbol $\otimes$ to stand for the scalar product, the dot product, or the cross product, and considering that

$$\frac{\partial\hat{\mathbf{r}}}{\partial\phi} = \hat{\mathbf{s}}, \qquad \frac{\partial\hat{\mathbf{s}}}{\partial\phi} = -\hat{\mathbf{r}}, \qquad \frac{\partial\hat{\mathbf{t}}}{\partial\phi} = \mathbf{0}, \tag{1.36a}$$

$$\frac{\partial\hat{\mathbf{r}}}{\partial\theta} = \sin\phi\,\hat{\mathbf{t}}, \qquad \frac{\partial\hat{\mathbf{s}}}{\partial\theta} = \cos\phi\,\hat{\mathbf{t}}, \qquad \frac{\partial\hat{\mathbf{t}}}{\partial\theta} = -(\sin\phi\,\hat{\mathbf{r}} + \cos\phi\,\hat{\mathbf{s}}) \tag{1.36b}$$

we can write a generic nabla operation on a vector **u** in spherical coordinates as

$$
\nabla \otimes \mathbf{u} = \left(\hat{\mathbf{r}} \frac{\partial}{\partial R} + \hat{\mathbf{s}} \frac{1}{R} \frac{\partial}{\partial \phi} + \hat{\mathbf{t}} \frac{1}{R \sin \phi} \frac{\partial}{\partial \theta} \right) \otimes (\hat{\mathbf{r}} u_R + \hat{\mathbf{s}} u_\phi + \hat{\mathbf{t}} u_\theta)
$$

$$
= \frac{\partial u_R}{\partial R} \hat{\mathbf{r}} \otimes \hat{\mathbf{r}} + \frac{\partial u_\phi}{\partial R} \hat{\mathbf{r}} \otimes \hat{\mathbf{s}} + \frac{\partial u_\theta}{\partial R} \hat{\mathbf{r}} \otimes \hat{\mathbf{t}}
$$

$$
+ \frac{1}{R} \left[\left(\frac{\partial u_R}{\partial \phi} - u_\phi \right) \hat{\mathbf{s}} \otimes \hat{\mathbf{r}} + \left(\frac{\partial u_\phi}{\partial \phi} + u_R \right) \hat{\mathbf{s}} \otimes \hat{\mathbf{s}} + \frac{\partial u_\theta}{\partial \phi} \hat{\mathbf{s}} \otimes \hat{\mathbf{t}} \right]
$$

$$
+ \frac{1}{R \sin \phi} \left[\left(\frac{\partial u_R}{\partial \theta} - u_\theta \sin \phi \right) \hat{\mathbf{t}} \otimes \hat{\mathbf{r}} + \left(\frac{\partial u_\phi}{\partial \theta} - u_\theta \cos \phi \right) \hat{\mathbf{t}} \otimes \hat{\mathbf{s}} \right.
$$

$$
\left. + \left(\frac{\partial u_\theta}{\partial \theta} + u_R \sin \phi + u_\phi \cos \phi \right) \hat{\mathbf{t}} \otimes \hat{\mathbf{t}} \right] \tag{1.37}
$$

Specializing this expression to the scalar, dot, and cross products, we obtain

Gradient

$$
\nabla \mathbf{u} = \frac{\partial u_R}{\partial R} \hat{\mathbf{r}} \hat{\mathbf{r}} + \frac{\partial u_\phi}{\partial R} \hat{\mathbf{r}} \hat{\mathbf{s}} + \frac{\partial u_\theta}{\partial R} \hat{\mathbf{r}} \hat{\mathbf{t}}
$$

$$
+ \frac{1}{R} \left[\left(\frac{\partial u_R}{\partial \phi} - u_\phi \right) \hat{\mathbf{s}} \hat{\mathbf{r}} + \left(\frac{\partial u_\phi}{\partial \phi} + u_r \right) \hat{\mathbf{s}} \hat{\mathbf{s}} + \frac{\partial u_\theta}{\partial \phi} \hat{\mathbf{s}} \hat{\mathbf{t}} \right]
$$

$$
+ \frac{1}{R \sin \phi} \left[\left(\frac{\partial u_R}{\partial \theta} - u_\theta \sin \phi \right) \hat{\mathbf{t}} \hat{\mathbf{r}} + \left(\frac{\partial u_\phi}{\partial \theta} - u_0 \cos \phi \right) \hat{\mathbf{t}} \hat{\mathbf{s}} \right.
$$

$$
\left. + \left(\frac{\partial u_\theta}{\partial \theta} + u_R \sin \phi + u_\phi \cos \phi \right) \hat{\mathbf{t}} \hat{\mathbf{t}} \right] \tag{1.38}
$$

where the products of the form $\hat{\mathbf{r}} \hat{\mathbf{r}}$, etc., are tensor bases, or *dyads*.

Divergence

$$
\nabla \cdot \mathbf{u} = \frac{\partial u_R}{\partial R} + \frac{2}{R} u_R + \frac{1}{R} \frac{\partial u_\phi}{\partial \phi} + \frac{u_\phi}{R} \cot \phi + \frac{1}{R \sin \phi} \frac{\partial u_\theta}{\partial \theta} \tag{1.39}
$$

Curl

$$
\nabla \times \mathbf{u} = \left[\frac{1}{R} \frac{\partial u_\theta}{\partial \phi} - \frac{1}{R \sin \phi} \frac{\partial u_\phi}{\partial \theta} + \frac{u_\theta}{R} \cot \phi \right] \hat{\mathbf{r}}
$$

$$
+ \left[\frac{1}{R \sin \phi} \frac{\partial u_R}{\partial \theta} - \frac{\partial u_\theta}{\partial R} - \frac{u_\theta}{R} \right] \hat{\mathbf{s}}
$$

$$
+ \left[\frac{\partial u_\phi}{\partial R} + \frac{u_\phi}{R} - \frac{1}{R} \frac{\partial u_R}{\partial \phi} \right] \hat{\mathbf{t}} \tag{1.40}
$$

Curl of curl

$$\nabla \times \nabla \times \mathbf{u} = \nabla \nabla \cdot \mathbf{u} - \nabla \cdot \nabla \mathbf{u} \tag{1.41}$$

$$
\begin{aligned}
\nabla \nabla \cdot \mathbf{u} = & \left[\frac{\partial^2 u_R}{\partial R^2} + \frac{2}{R}\left(\frac{\partial u_R}{\partial R} - \frac{u_R}{R} \right) \right. \\
& + \frac{1}{R}\left(\frac{\partial}{\partial \phi}\left(\frac{\partial u_\phi}{\partial R} - \frac{u_\phi}{R} \right) + \cot\phi \left(\frac{\partial u_\phi}{\partial R} - \frac{u_\phi}{R} \right) \right) \\
& \left. + \frac{1}{R \sin\phi} \frac{\partial}{\partial \theta}\left(\frac{\partial u_\theta}{\partial R} - \frac{u_\theta}{R} \right) \right] \hat{\mathbf{r}} \\
& + \frac{1}{R}\left[\frac{\partial}{\partial \phi}\left(\frac{\partial u_R}{\partial R} + \frac{2u_R}{R} + \frac{1}{R}\frac{\partial u_\phi}{\partial \phi} \right) + \frac{1}{R \sin\phi}\left(\frac{\partial u_\phi}{\partial \phi}\cos\phi - \frac{u_\phi}{\sin\phi} \right) \right. \\
& \left. + \frac{1}{R \sin\phi} \frac{\partial}{\partial \theta}\left(\frac{\partial u_\theta}{\partial \phi} - u_\theta \cot\phi \right) \right] \hat{\mathbf{s}} \\
& + \frac{1}{R \sin\phi} \frac{\partial}{\partial \theta}\left[\frac{\partial u_R}{\partial R} + \frac{2u_R}{R} + \frac{1}{R}\frac{\partial u_\phi}{\partial \phi} + \frac{u_\phi}{R}\cot\phi + \frac{1}{R \sin\phi}\frac{\partial u_\theta}{\partial \theta} \right] \hat{\mathbf{t}} \tag{1.42}
\end{aligned}
$$

Laplacian

$$\nabla^2 = \frac{\partial^2}{\partial R^2} + \frac{2}{R}\frac{\partial}{\partial R} + \frac{1}{R^2}\frac{\partial^2}{\partial \phi^2} + \frac{\cot\phi}{R^2}\frac{\partial}{\partial \phi} + \frac{1}{R^2 \sin^2\phi}\frac{\partial^2}{\partial \theta^2} \tag{1.43}$$

$$
\begin{aligned}
\nabla \cdot \nabla \mathbf{u} = & \left[\frac{\partial^2 u_R}{\partial R^2} + \frac{2}{R}\frac{\partial u_R}{\partial r} + \frac{1}{R^2}\left(\frac{\partial^2 u_R}{\partial \phi^2} - 2\frac{\partial u_\phi}{\partial \phi} - 2u_R \right. \right. \\
& \left. \left. + \cot\phi\left(\frac{\partial u_R}{\partial \phi} - 2u_\phi \right) + \frac{1}{\sin^2\phi}\frac{\partial^2 u_R}{\partial \theta^2} - \frac{2}{\sin\phi}\frac{\partial u_\theta}{\partial \theta} \right) \right] \hat{\mathbf{r}} \\
& + \left[\frac{\partial^2 u_\phi}{\partial R^2} + \frac{2}{R}\frac{\partial u_\phi}{\partial R} + \frac{1}{R^2}\left(\frac{\partial^2 u_\phi}{\partial \phi^2} + 2\frac{\partial u_R}{\partial \phi} + \cot\phi\frac{\partial u_\phi}{\partial \phi} \right. \right. \\
& \left. \left. + \frac{1}{\sin^2\phi}\left(\frac{\partial^2 u_\phi}{\partial \theta^2} - 2\cos\phi\frac{\partial u_\theta}{\partial \theta} - u_\phi \right) \right) \right] \hat{\mathbf{s}} \\
& + \left[\frac{\partial^2 u_\theta}{\partial R^2} + \frac{2}{R}\frac{\partial u_\theta}{\partial R} + \frac{1}{R^2}\left(\frac{\partial^2 u_\theta}{\partial \phi^2} + \cot\phi\frac{\partial u_\theta}{\partial \phi} + \frac{1}{\sin^2\phi}\left(\frac{\partial^2 u_\theta}{\partial \theta^2} - u_\theta \right) \right. \right. \\
& \left. \left. + \frac{2}{\sin\phi}\frac{\partial}{\partial \theta}(u_R + u_\phi \cot\phi) \right) \right] \hat{\mathbf{t}} \tag{1.44}
\end{aligned}
$$

(Note: $\partial/\partial\phi$, $\partial/\partial\theta$ in ∇^2 act both on the components of $\mathbf{u}$ and on the basis vectors.)

Wave equation

$$(\lambda + \mu)\nabla\nabla \cdot \mathbf{u} + \mu\nabla \cdot \nabla \mathbf{u} = \rho\ddot{\mathbf{u}} \tag{1.45}$$

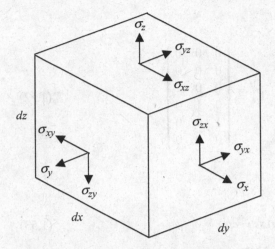

Figure 1.4: Stresses (and strains) in Cartesian coordinates.

1.4 Strains, stresses, and the elastic wave equation

1.4.1 Cartesian coordinates

Strains

$$\varepsilon_x = \frac{\partial u_x}{\partial x}, \qquad \varepsilon_y = \frac{\partial u_y}{\partial y}, \qquad \varepsilon_z = \frac{\partial u_z}{\partial z}$$

$$\varepsilon_{yz} = \frac{\partial u_z}{\partial y} + \frac{\partial u_y}{\partial z}, \qquad \varepsilon_{zx} = \frac{\partial u_x}{\partial z} + \frac{\partial u_z}{\partial x}, \qquad \varepsilon_{xy} = \frac{\partial u_y}{\partial x} + \frac{\partial u_x}{\partial y}$$

$$\varepsilon_{zy} = \varepsilon_{yz}, \qquad \varepsilon_{zx} = \varepsilon_{xz}, \qquad \varepsilon_{yx} = \varepsilon_{xy}$$

$$(1.46)$$

In matrix notation, this can be written as

$$\mathbf{u} = \begin{bmatrix} u_x & u_y & u_z \end{bmatrix}^T \qquad = displacement\ vector \qquad (1.47)$$

$$\boldsymbol{\varepsilon} = \begin{bmatrix} \varepsilon_x & \varepsilon_y & \varepsilon_z & \varepsilon_{yz} & \varepsilon_{xz} & \varepsilon_{xy} \end{bmatrix}^T \qquad = strain\ vector \qquad (1.48)$$

$$\boldsymbol{\varepsilon} = \mathbf{L}\mathbf{u} \qquad = strain\text{--}displacement\ relation \qquad (1.49)$$

$$\mathbf{L}^T = \begin{Bmatrix} \dfrac{\partial}{\partial x} & 0 & 0 & 0 & \dfrac{\partial}{\partial z} & \dfrac{\partial}{\partial y} \\[2mm] 0 & \dfrac{\partial}{\partial y} & 0 & \dfrac{\partial}{\partial z} & 0 & \dfrac{\partial}{\partial x} \\[2mm] 0 & 0 & \dfrac{\partial}{\partial z} & \dfrac{\partial}{\partial y} & \dfrac{\partial}{\partial x} & 0 \end{Bmatrix} \qquad (1.50)$$

which can be abbreviated as

$$\mathbf{L} = \mathbf{L}_x \frac{\partial}{\partial x} + \mathbf{L}_y \frac{\partial}{\partial y} + \mathbf{L}_z \frac{\partial}{\partial z} \qquad (1.51)$$

where by inspection

$$\mathbf{L}_x = \begin{Bmatrix} 1 & 0 & 0 \\ 0 & 0 & 0 \\ 0 & 0 & 0 \\ 0 & 0 & 0 \\ 0 & 0 & 1 \\ 0 & 1 & 0 \end{Bmatrix}, \quad \mathbf{L}_y = \begin{Bmatrix} 0 & 0 & 0 \\ 0 & 1 & 0 \\ 0 & 0 & 0 \\ 0 & 0 & 1 \\ 0 & 0 & 0 \\ 1 & 0 & 0 \end{Bmatrix}, \quad \mathbf{L}_z = \begin{Bmatrix} 0 & 0 & 0 \\ 0 & 0 & 0 \\ 0 & 0 & 1 \\ 0 & 1 & 0 \\ 1 & 0 & 0 \\ 0 & 0 & 0 \end{Bmatrix} \qquad (1.52)$$

Stresses

$$\sigma = [\sigma_x \quad \sigma_y \quad \sigma_z \quad \sigma_{yz} \quad \sigma_{xz} \quad \sigma_{xy}]^T \quad = \textit{stress vector} \qquad (1.53)$$

$$\sigma = \mathbf{D}\varepsilon \quad\qquad\qquad\qquad = \textit{constitutive law} \qquad (1.54)$$

For a fully anisotropic medium, the symmetric constitutive matrix is

$$\mathbf{D} = \begin{Bmatrix} d_{11} & d_{12} & d_{13} & d_{14} & d_{15} & d_{16} \\ d_{12} & d_{22} & d_{23} & d_{24} & d_{25} & d_{26} \\ d_{13} & d_{23} & d_{33} & d_{34} & d_{35} & d_{36} \\ d_{14} & d_{24} & d_{34} & d_{44} & d_{45} & d_{46} \\ d_{15} & d_{25} & d_{35} & d_{45} & d_{55} & d_{56} \\ d_{16} & d_{26} & d_{36} & d_{46} & d_{56} & d_{66} \end{Bmatrix} \qquad (1.55)$$

whereas for an isotropic medium, this matrix is

$$\mathbf{D} = \begin{Bmatrix} \lambda + 2\mu & \lambda & \lambda & 0 & 0 & 0 \\ \lambda & \lambda + 2\mu & \lambda & 0 & 0 & 0 \\ \lambda & \lambda & \lambda + 2\mu & 0 & 0 & 0 \\ 0 & 0 & 0 & \mu & 0 & 0 \\ 0 & 0 & 0 & 0 & \mu & 0 \\ 0 & 0 & 0 & 0 & 0 & \mu \end{Bmatrix} \qquad (1.56)$$

in which $\lambda = $ Lamé constant, and $\mu = $ shear modulus. In this case, the stress–strain relationship is

$$\sigma_j = 2\mu \varepsilon_j + \lambda \varepsilon_{vol}, \qquad \varepsilon_{vol} = \varepsilon_x + \varepsilon_y + \varepsilon_z$$

$$\sigma_{yz} = \mu \varepsilon_{yz}, \qquad\qquad \sigma_{zx} = \mu \varepsilon_{zx}, \qquad\qquad \sigma_{xy} = \mu \varepsilon_{xy}$$

$$\sigma_{zy} = \sigma_{yz}, \qquad\qquad \sigma_{zx} = \sigma_{xz}, \qquad\qquad \sigma_{yx} = \sigma_{xy}$$

$$(1.57)$$

The stresses, inertial loads, and body loads satisfy the dynamic equilibrium equation

$$\mathbf{b} - \rho \ddot{\mathbf{u}} + \mathbf{L}^T \sigma = \mathbf{0} \qquad (1.58)$$

Stresses in principal surfaces

The stresses (or tractions) acting in planes perpendicular to the x, y, z axes, which are needed in the formulation for layered media, can be expressed symbolically as

$$\mathbf{s}_x = [\sigma_x \quad \sigma_{yx} \quad \sigma_{zx}]^T = \mathbf{L}_x^T \boldsymbol{\sigma} = \mathbf{L}_x^T \mathbf{D} \mathbf{L} \mathbf{u} = \mathbf{D}_{xx} \frac{\partial \mathbf{u}}{\partial x} + \mathbf{D}_{xy} \frac{\partial \mathbf{u}}{\partial y} + \mathbf{D}_{xz} \frac{\partial \mathbf{u}}{\partial z}$$

$$\mathbf{s}_y = [\sigma_{xy} \quad \sigma_y \quad \sigma_{zy}]^T = \mathbf{L}_y^T \boldsymbol{\sigma} = \mathbf{L}_y^T \mathbf{D} \mathbf{L} \mathbf{u} = \mathbf{D}_{yx} \frac{\partial \mathbf{u}}{\partial x} + \mathbf{D}_{yy} \frac{\partial \mathbf{u}}{\partial y} + \mathbf{D}_{yz} \frac{\partial \mathbf{u}}{\partial z}$$

$$\mathbf{s}_z = [\sigma_{xz} \quad \sigma_{yz} \quad \sigma_z]^T = \mathbf{L}_z^T \boldsymbol{\sigma} = \mathbf{L}_z^T \mathbf{D} \mathbf{L} \mathbf{u} = \mathbf{D}_{zx} \frac{\partial \mathbf{u}}{\partial x} + \mathbf{D}_{zy} \frac{\partial \mathbf{u}}{\partial y} + \mathbf{D}_{zz} \frac{\partial \mathbf{u}}{\partial z}$$

(1.59)

which involve the 3×3 material matrices

$$\mathbf{D}_{ij} = \mathbf{L}_i^T \mathbf{D} \mathbf{L}_j, \quad i, j = x, y, z$$

(1.60)

These matrices are not symmetric, but $\mathbf{D}_{ji} = \mathbf{D}_{ij}^T$. For a fully anisotropic medium, they are

$$\mathbf{D}_{xx} = \begin{Bmatrix} d_{11} & d_{16} & d_{15} \\ d_{16} & d_{66} & d_{56} \\ d_{15} & d_{56} & d_{55} \end{Bmatrix}, \quad \mathbf{D}_{yy} = \begin{Bmatrix} d_{66} & d_{26} & d_{46} \\ d_{26} & d_{22} & d_{24} \\ d_{46} & d_{24} & d_{44} \end{Bmatrix}, \quad \mathbf{D}_{zz} - \begin{Bmatrix} d_{55} & d_{45} & d_{35} \\ d_{45} & d_{44} & d_{34} \\ d_{35} & d_{34} & d_{33} \end{Bmatrix}$$

(1.61a)

$$\mathbf{D}_{xy} = \begin{Bmatrix} d_{16} & d_{12} & d_{14} \\ d_{66} & d_{26} & d_{46} \\ d_{56} & d_{25} & d_{45} \end{Bmatrix}, \quad \mathbf{D}_{xz} = \begin{Bmatrix} d_{15} & d_{14} & d_{13} \\ d_{56} & d_{46} & d_{36} \\ d_{55} & d_{45} & d_{35} \end{Bmatrix}, \quad \mathbf{D}_{yz} = \begin{Bmatrix} d_{56} & d_{46} & d_{36} \\ d_{25} & d_{24} & d_{23} \\ d_{45} & d_{44} & d_{34} \end{Bmatrix}$$

(1.61b)

For an isotropic medium, they are

$$\mathbf{D}_{xx} = \begin{Bmatrix} \lambda + 2\mu & 0 & 0 \\ 0 & \mu & 0 \\ 0 & 0 & \mu \end{Bmatrix}, \quad \mathbf{D}_{yy} = \begin{Bmatrix} \mu & 0 & 0 \\ 0 & \lambda + 2\mu & 0 \\ 0 & 0 & \mu \end{Bmatrix}, \quad \mathbf{D}_{zz} = \begin{Bmatrix} \mu & 0 & 0 \\ 0 & \mu & 0 \\ 0 & 0 & \lambda + 2\mu \end{Bmatrix}$$

(1.62a)

$$\mathbf{D}_{xy} = \mathbf{D}_{yx}^T = \begin{Bmatrix} 0 & \lambda & 0 \\ \mu & 0 & 0 \\ 0 & 0 & 0 \end{Bmatrix}, \quad \mathbf{D}_{xz} = \mathbf{D}_{zx}^T = \begin{Bmatrix} 0 & 0 & \lambda \\ 0 & 0 & 0 \\ \mu & 0 & 0 \end{Bmatrix}, \quad \mathbf{D}_{yz} = \mathbf{D}_{zy}^T = \begin{Bmatrix} 0 & 0 & 0 \\ 0 & 0 & \lambda \\ 0 & \mu & 0 \end{Bmatrix}$$

(1.62b)

Wave equation

Combining the equations of equilibrium and stress–strain relations, we obtain immediately

$$\mathbf{b} - \rho\ddot{\mathbf{u}} + \mathbf{L}^T\mathbf{D}\mathbf{L}\mathbf{u} = \mathbf{0}, \qquad \mathbf{b} = \{b_x \quad b_y \quad b_z\}^T \tag{1.63}$$

where $\mathbf{b} = \mathbf{b}(x, y, z, t)$ is the body load vector. Expanding the last term, we obtain

$$\mathbf{b} + \mathbf{D}_{xx}\frac{\partial^2\mathbf{u}}{\partial x^2} + \mathbf{D}_{yy}\frac{\partial^2\mathbf{u}}{\partial y^2} + \mathbf{D}_{zz}\frac{\partial^2\mathbf{u}}{\partial z^2} + (\mathbf{D}_{xy} + \mathbf{D}_{yx})\frac{\partial^2\mathbf{u}}{\partial x\,\partial y} + (\mathbf{D}_{yz} + \mathbf{D}_{zy})\frac{\partial^2\mathbf{u}}{\partial y\,\partial z}$$

$$+ (\mathbf{D}_{xz} + \mathbf{D}_{zx})\frac{\partial^2\mathbf{u}}{\partial x\,\partial z} = \rho\ddot{\mathbf{u}}$$

$$\tag{1.64}$$

For an isotropic medium, the wave equation in Cartesian coordinates is

$$\rho\ddot{\mathbf{u}} = \mathbf{b} + \begin{Bmatrix} \lambda + 2\mu & 0 & 0 \\ 0 & \mu & 0 \\ 0 & 0 & \mu \end{Bmatrix}\frac{\partial^2\mathbf{u}}{\partial x^2} + \begin{Bmatrix} \mu & 0 & 0 \\ 0 & \lambda + 2\mu & 0 \\ 0 & 0 & \mu \end{Bmatrix}\frac{\partial^2\mathbf{u}}{\partial y^2}$$

$$+ \begin{Bmatrix} \mu & 0 & 0 \\ 0 & \mu & 0 \\ 0 & 0 & \lambda + 2\mu \end{Bmatrix}\frac{\partial^2\mathbf{u}}{\partial z^2} + \begin{Bmatrix} 0 & \lambda + \mu & 0 \\ \lambda + \mu & 0 & 0 \\ 0 & 0 & 0 \end{Bmatrix}\frac{\partial^2\mathbf{u}}{\partial x\,\partial y}$$

$$+ \begin{Bmatrix} 0 & 0 & 0 \\ 0 & 0 & \lambda + \mu \\ 0 & \lambda + \mu & 0 \end{Bmatrix}\frac{\partial^2\mathbf{u}}{\partial y\,\partial z} + \begin{Bmatrix} 0 & 0 & \lambda + \mu \\ 0 & 0 & 0 \\ \lambda + \mu & 0 & 0 \end{Bmatrix}\frac{\partial^2\mathbf{u}}{\partial x\,\partial z} \tag{1.65}$$

The equations for plane strain problems are obtained from the above by discarding all partial derivatives with respect to (say) y. As a result, the equations decompose into two *uncoupled* problems, the SV-P problem (involving u_x and u_z) and the SH problem (involving u_y).

1.4.2 Cylindrical coordinates

Strains

$$\varepsilon_r = \frac{\partial u_r}{\partial r}, \qquad \varepsilon_\theta = \frac{u_r}{r} + \frac{\partial u_\theta}{r\,\partial\theta}, \qquad \varepsilon_z = \frac{\partial u_z}{\partial z}$$

$$\varepsilon_{\theta z} = \frac{\partial u_\theta}{\partial z} + \frac{\partial u_z}{r\,\partial\theta}, \qquad \varepsilon_{rz} = \frac{\partial u_r}{\partial z} + \frac{\partial u_z}{\partial r}, \qquad \varepsilon_{r\theta} = \frac{\partial u_r}{r\,\partial\theta} + \frac{\partial u_\theta}{\partial r} - \frac{u_\theta}{r}$$

$$\tag{1.66}$$

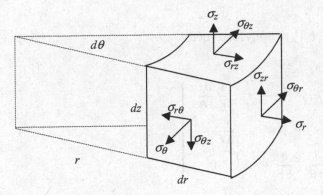

Figure 1.5: Stresses (and strains) in cylindrical coordinates.

$$\mathbf{u} = [u_r \quad u_\theta \quad u_z] \qquad = \textit{displacement vector} \qquad (1.67)$$

$$\boldsymbol{\varepsilon} = [\varepsilon_r \quad \varepsilon_\theta \quad \varepsilon_z \quad \varepsilon_{\theta z} \quad \varepsilon_{rz} \quad \varepsilon_{r\theta}]^T \quad = \textit{strain vector} \qquad (1.68)$$

$$\boldsymbol{\varepsilon} = \mathbf{L}_\varepsilon \mathbf{u} \qquad = \textit{strain–displacement relation} \qquad (1.69)$$

$$
\mathbf{L}_\varepsilon^T =
\left\{
\begin{array}{cccccc}
\dfrac{\partial}{\partial r} & \dfrac{1}{r} & 0 & 0 & \dfrac{\partial}{\partial z} & \dfrac{1}{r}\dfrac{\partial}{\partial \theta} \\[2ex]
0 & \dfrac{1}{r}\dfrac{\partial}{\partial \theta} & 0 & \dfrac{\partial}{\partial z} & 0 & \dfrac{\partial}{\partial r} - \dfrac{1}{r} \\[2ex]
0 & 0 & \dfrac{\partial}{\partial z} & \dfrac{1}{r}\dfrac{\partial}{\partial \theta} & \dfrac{\partial}{\partial r} & 0
\end{array}
\right\}
\qquad (1.70)
$$

which can be abbreviated as

$$\mathbf{L}_\varepsilon = \mathbf{L}_r \frac{\partial}{\partial r} + \mathbf{L}_\theta \frac{1}{r}\frac{\partial}{\partial \theta} + \mathbf{L}_z \frac{\partial}{\partial z} + \mathbf{L}_1 \frac{1}{r} \qquad (1.71)$$

Stresses

$$\boldsymbol{\sigma} = [\sigma_r \quad \sigma_\theta \quad \sigma_z \quad \sigma_{\theta z} \quad \sigma_{rz} \quad \sigma_{r\theta}]^T \quad = \textit{stress vector} \qquad (1.72)$$

$$\boldsymbol{\sigma} = \mathbf{D}\boldsymbol{\varepsilon} \qquad = \textit{constitutive law} \qquad (1.73)$$

$\mathbf{D}$ is the same as for Cartesian coordinates, although its physical meaning is somewhat different. For an anisotropic medium, we assume here that the elements d_{ij} of $\mathbf{D}$ are *independent* of the azimuth θ for any range r, so this defines a cylindrical type of anisotropy. For isotropic media, we have

$$
\begin{aligned}
&\sigma_j = 2\mu\,\varepsilon_j + \lambda\,\varepsilon_{vol}, &&j = r, \theta, z, &&\varepsilon_{vol} = \varepsilon_r + \varepsilon_\theta + \varepsilon_z \\
&\sigma_{\theta z} = \mu\,\varepsilon_{\theta z}, &&\sigma_{rz} = \mu\,\varepsilon_{rz}, &&\sigma_{r\theta} = \mu\,\varepsilon_{r\theta} \\
&\sigma_{z\theta} = \sigma_{\theta z}, &&\sigma_{zr} = \sigma_{rz}, &&\sigma_{\theta r} = \sigma_{r\theta}
\end{aligned}
$$

$$(1.74)$$

The dynamic equilibrium equation is

$$\mathbf{b} - \rho\ddot{\mathbf{u}} + \mathbf{L}_\sigma^T \boldsymbol{\sigma} = \mathbf{0}, \quad \mathbf{b}^T = [b_r \ \ b_\theta \ \ b_z]^T \doteq \text{body load vector} \tag{1.75}$$

with

$$\mathbf{L}_\sigma^T = \begin{Bmatrix} \dfrac{\partial}{\partial r} + \dfrac{1}{r} & -\dfrac{1}{r} & 0 & 0 & \dfrac{\partial}{\partial z} & \dfrac{1}{r}\dfrac{\partial}{\partial \theta} \\[2ex] 0 & \dfrac{1}{r}\dfrac{\partial}{\partial \theta} & 0 & \dfrac{\partial}{\partial z} & 0 & \dfrac{\partial}{\partial r} + \dfrac{2}{r} \\[2ex] 0 & 0 & \dfrac{\partial}{\partial z} & \dfrac{1}{r}\dfrac{\partial}{\partial \theta} & \dfrac{\partial}{\partial r} + \dfrac{1}{r} & 0 \end{Bmatrix} \tag{1.76}$$

which can be written compactly as

$$\mathbf{L}_\sigma = \mathbf{L}_r \frac{\partial}{\partial r} + \mathbf{L}_\theta \frac{1}{r}\frac{\partial}{\partial \theta} + \mathbf{L}_z \frac{\partial}{\partial z} + (\mathbf{L}_r - \mathbf{L}_1)\frac{1}{r} \tag{1.77}$$

In full, the operator matrices are

$$\mathbf{L}_r = \begin{Bmatrix} 1 & 0 & 0 \\ 0 & 0 & 0 \\ 0 & 0 & 0 \\ 0 & 0 & 0 \\ 0 & 0 & 1 \\ 0 & 1 & 0 \end{Bmatrix}, \ \mathbf{L}_\theta = \begin{Bmatrix} 0 & 0 & 0 \\ 0 & 1 & 0 \\ 0 & 0 & 0 \\ 0 & 0 & 1 \\ 0 & 0 & 0 \\ 1 & 0 & 0 \end{Bmatrix}, \ \mathbf{L}_z = \begin{Bmatrix} 0 & 0 & 0 \\ 0 & 0 & 0 \\ 0 & 0 & 1 \\ 0 & 1 & 0 \\ 1 & 0 & 0 \\ 0 & 0 & 0 \end{Bmatrix}, \ \mathbf{L}_1 = \begin{Bmatrix} 0 & 0 & 0 \\ 1 & 0 & 0 \\ 0 & 0 & 0 \\ 0 & 0 & 0 \\ 0 & 0 & 0 \\ 0 & -1 & 0 \end{Bmatrix}$$

$$\tag{1.78}$$

Notice that $\mathbf{L}_r \equiv \mathbf{L}_x$, $\mathbf{L}_\theta \equiv \mathbf{L}_y$, and $\mathbf{L}_z \equiv \mathbf{L}_z$ are the same as in Cartesian coordinates, while $\mathbf{L}_1$ is new. Also, observe that $\mathbf{L}_\sigma \neq \mathbf{L}_\varepsilon$.

Stresses in principal surfaces

The stresses in radial, azimuthal and vertical surfaces are

$$\mathbf{s}_r = [\sigma_r \ \ \sigma_{r\theta} \ \ \sigma_{rz}]^T = \mathbf{L}_r^T \boldsymbol{\sigma}$$

$$= \mathbf{D}_{rr}\frac{\partial \mathbf{u}}{\partial r} + \mathbf{D}_{r\theta}\frac{1}{r}\frac{\partial \mathbf{u}}{\partial \theta} + \mathbf{D}_{rz}\frac{\partial \mathbf{u}}{\partial z} + \mathbf{D}_{r1}\frac{\mathbf{u}}{r}$$

$$\mathbf{s}_\theta = [\sigma_{r\theta} \ \ \sigma_\theta \ \ \sigma_{z\theta}]^T = \mathbf{L}_\theta^T \boldsymbol{\sigma}$$

$$= \mathbf{D}_{\theta r}\frac{\partial \mathbf{u}}{\partial r} + \mathbf{D}_{\theta\theta}\frac{1}{r}\frac{\partial \mathbf{u}}{\partial \theta} + \mathbf{D}_{\theta z}\frac{\partial \mathbf{u}}{\partial z} + \mathbf{D}_{\theta 1}\frac{\mathbf{u}}{r}$$

$$\mathbf{s}_z = [\sigma_{rz} \ \ \sigma_{\theta z} \ \ \sigma_z]^T = \mathbf{L}_z^T \boldsymbol{\sigma}$$

$$= \mathbf{D}_{zr}\frac{\partial \mathbf{u}}{\partial r} + \mathbf{D}_{z\theta}\frac{1}{r}\frac{\partial \mathbf{u}}{\partial \theta} + \mathbf{D}_{zz}\frac{\partial \mathbf{u}}{\partial z} + \mathbf{D}_{z1}\frac{\mathbf{u}}{r}$$

$$\tag{1.79}$$

For an isotropic medium, the matrices $\mathbf{D}_{rr}$ etc. are identical to those in Cartesian coordinates, with the replacement $x \to r$, $y \to \theta$, $z \to z$. In addition, we have the three new matrices $\mathbf{D}_{r1}$, $\mathbf{D}_{\theta 1}$, $\mathbf{D}_{z1}$ and their transposes. For a cylindrically anisotropic medium, these matrices are

$$\mathbf{D}_{rr} = \begin{Bmatrix} d_{11} & d_{16} & d_{15} \\ d_{16} & d_{66} & d_{56} \\ d_{15} & d_{56} & d_{55} \end{Bmatrix}, \quad \mathbf{D}_{\theta\theta} = \begin{Bmatrix} d_{66} & d_{26} & d_{46} \\ d_{26} & d_{22} & d_{24} \\ d_{46} & d_{24} & d_{44} \end{Bmatrix}, \quad \mathbf{D}_{zz} = \begin{Bmatrix} d_{55} & d_{45} & d_{35} \\ d_{45} & d_{44} & d_{34} \\ d_{35} & d_{34} & d_{33} \end{Bmatrix}$$

$$(1.80a)$$

$$\mathbf{D}_{r\theta} = \begin{Bmatrix} d_{16} & d_{12} & d_{14} \\ d_{66} & d_{26} & d_{46} \\ d_{56} & d_{25} & d_{45} \end{Bmatrix}, \quad \mathbf{D}_{rz} = \begin{Bmatrix} d_{15} & d_{14} & d_{13} \\ d_{56} & d_{46} & d_{36} \\ d_{55} & d_{45} & d_{35} \end{Bmatrix}, \quad \mathbf{D}_{r1} = \begin{Bmatrix} d_{12} & -d_{16} & 0 \\ d_{26} & -d_{66} & 0 \\ d_{25} & -d_{56} & 0 \end{Bmatrix}$$

$$(1.80b)$$

$$\mathbf{D}_{\theta z} = \begin{Bmatrix} d_{56} & d_{46} & d_{36} \\ d_{25} & d_{24} & d_{23} \\ d_{45} & d_{44} & d_{34} \end{Bmatrix}, \quad \mathbf{D}_{\theta 1} = \begin{Bmatrix} d_{26} & -d_{66} & 0 \\ d_{22} & -d_{26} & 0 \\ d_{24} & -d_{46} & 0 \end{Bmatrix}, \quad \mathbf{D}_{z1} = \begin{Bmatrix} d_{25} & -d_{56} & 0 \\ d_{24} & -d_{46} & 0 \\ d_{23} & -d_{36} & 0 \end{Bmatrix}$$

$$(1.80c)$$

$$\mathbf{D}_{11} = \begin{Bmatrix} d_{22} & -d_{26} & 0 \\ -d_{26} & d_{66} & 0 \\ 0 & 0 & 0 \end{Bmatrix}$$

$$(1.80d)$$

Observe that these matrices are generally not symmetric, but $\mathbf{D}_{\beta\alpha} = \mathbf{D}_{\alpha\beta}^{\mathrm{T}}$. For a cylindrically isotropic material, the matrices are

$$\mathbf{D}_{rr} = \begin{Bmatrix} \lambda + 2\mu & 0 & 0 \\ 0 & \mu & 0 \\ 0 & 0 & \mu \end{Bmatrix}, \quad \mathbf{D}_{\theta\theta} = \begin{Bmatrix} \mu & 0 & 0 \\ 0 & \lambda + 2\mu & 0 \\ 0 & 0 & \mu \end{Bmatrix}, \quad \mathbf{D}_{zz} = \begin{Bmatrix} \mu & 0 & 0 \\ 0 & \mu & 0 \\ 0 & 0 & \lambda + 2\mu \end{Bmatrix}$$

$$(1.81a)$$

$$\mathbf{D}_{r\theta} = \mathbf{D}_{\theta r}^{T} = \begin{Bmatrix} 0 & \lambda & 0 \\ \mu & 0 & 0 \\ 0 & 0 & 0 \end{Bmatrix}, \quad \mathbf{D}_{rz} = \mathbf{D}_{zr}^{T} = \begin{Bmatrix} 0 & 0 & \lambda \\ 0 & 0 & 0 \\ \mu & 0 & 0 \end{Bmatrix}, \quad \mathbf{D}_{\theta z} = \mathbf{D}_{z\theta}^{T} = \begin{Bmatrix} 0 & 0 & 0 \\ 0 & 0 & \lambda \\ 0 & \mu & 0 \end{Bmatrix}$$

$$(1.81b)$$

$$\mathbf{D}_{r1} = \mathbf{D}_{1r} = \begin{Bmatrix} \lambda & 0 & 0 \\ 0 & -\mu & 0 \\ 0 & 0 & 0 \end{Bmatrix}, \quad \mathbf{D}_{\theta 1} = \mathbf{D}_{1\theta}^{T} = \begin{Bmatrix} 0 & -\mu & 0 \\ \lambda + 2\mu & 0 & 0 \\ 0 & 0 & 0 \end{Bmatrix}$$

$$(1.81c)$$

$$\mathbf{D}_{z1} = \mathbf{D}_{1z}^T = \left\{ \begin{array}{ccc} 0 & 0 & 0 \\ 0 & 0 & 0 \\ \lambda & 0 & 0 \end{array} \right\}, \quad \mathbf{D}_{11} = \left\{ \begin{array}{ccc} \lambda+2\mu & 0 & 0 \\ 0 & \mu & 0 \\ 0 & 0 & 0 \end{array} \right\} \tag{1.81d}$$

Wave equation

$$\mathbf{b} - \rho\ddot{\mathbf{u}} + \mathbf{L}_\sigma^T \mathbf{D} \mathbf{L}_\varepsilon \mathbf{u} = \mathbf{0}, \qquad \mathbf{b}^T = [b_r \quad b_\theta \quad b_z]^T \tag{1.82}$$

in which **b** is the vector of body loads. Expansion of the second term yields

$$\mathbf{b} - \rho\ddot{\mathbf{u}} + \mathbf{D}_{rr}\frac{\partial^2 \mathbf{u}}{\partial r^2} + \mathbf{D}_{\theta\theta}\frac{1}{r^2}\frac{\partial^2 \mathbf{u}}{\partial \theta^2} + \mathbf{D}_{zz}\frac{\partial^2 \mathbf{u}}{\partial z^2} + (\mathbf{D}_{r\theta} + \mathbf{D}_{\theta r})\frac{1}{r}\frac{\partial^2 \mathbf{u}}{\partial r \partial \theta}$$

$$+ (\mathbf{D}_{rz} + \mathbf{D}_{zr})\frac{\partial^2 \mathbf{u}}{\partial r \partial z} + (\mathbf{D}_{\theta z} + \mathbf{D}_{z\theta})\frac{1}{r}\frac{\partial^2 \mathbf{u}}{\partial z \partial \theta} + \mathbf{D}_{rr}\frac{1}{r}\frac{\partial \mathbf{u}}{\partial r} + (\mathbf{D}_{\theta 1} - \mathbf{D}_{1\theta})\frac{1}{r^2}\frac{\partial \mathbf{u}}{\partial \theta}$$

$$+ (\mathbf{D}_{rz} + \mathbf{D}_{z1} - \mathbf{D}_{1z})\frac{1}{r}\frac{\partial \mathbf{u}}{\partial z} - \mathbf{D}_{11}\frac{\mathbf{u}}{r^2} = \mathbf{0}$$

$$\tag{1.83}$$

For an isotropic medium, this results in the wave equation in cylindrical coordinates:

$$\rho\ddot{\mathbf{u}} = \mathbf{b} + \left\{ \begin{array}{ccc} \lambda+2\mu & 0 & 0 \\ 0 & \mu & 0 \\ 0 & 0 & \mu \end{array} \right\} \left(\frac{\partial^2 \mathbf{u}}{\partial r^2} + \frac{1}{r}\frac{\partial \mathbf{u}}{\partial r}\right) + \left\{ \begin{array}{ccc} \mu & 0 & 0 \\ 0 & \lambda+2\mu & 0 \\ 0 & 0 & \mu \end{array} \right\} \frac{1}{r^2}\frac{\partial^2 \mathbf{u}}{\partial \theta^2}$$

$$+ \left\{ \begin{array}{ccc} \mu & 0 & 0 \\ 0 & \mu & 0 \\ 0 & 0 & \lambda+2\mu \end{array} \right\} \frac{\partial^2 \mathbf{u}}{\partial z^2} + \left\{ \begin{array}{ccc} 0 & \lambda+\mu & 0 \\ \lambda+\mu & 0 & 0 \\ 0 & 0 & 0 \end{array} \right\} \frac{1}{r}\frac{\partial^2 \mathbf{u}}{\partial r \partial \theta}$$

$$+ \left\{ \begin{array}{ccc} 0 & 0 & \lambda+\mu \\ 0 & 0 & 0 \\ \lambda+\mu & 0 & 0 \end{array} \right\} \frac{\partial^2 \mathbf{u}}{\partial r \partial z} + \left\{ \begin{array}{ccc} 0 & 0 & 0 \\ 0 & 0 & \lambda+\mu \\ 0 & \lambda+\mu & 0 \end{array} \right\} \frac{1}{r}\frac{\partial^2 \mathbf{u}}{\partial z \partial \theta}$$

$$+ \left\{ \begin{array}{ccc} 0 & -(\lambda+3\mu) & 0 \\ \lambda+3\mu & 0 & 0 \\ 0 & 0 & 0 \end{array} \right\} \frac{1}{r^2}\frac{\partial \mathbf{u}}{\partial \theta} + \left\{ \begin{array}{ccc} 0 & 0 & 0 \\ 0 & 0 & 0 \\ \lambda+\mu & 0 & 0 \end{array} \right\} \frac{1}{r}\frac{\partial \mathbf{u}}{\partial z}$$

$$- \left\{ \begin{array}{ccc} \lambda+2\mu & 0 & 0 \\ 0 & \mu & 0 \\ 0 & 0 & 0 \end{array} \right\} \frac{\mathbf{u}}{r^2} \tag{1.84}$$

1.4.3 Spherical coordinates

Strains and stresses

$$\varepsilon_R = \frac{\partial u_R}{\partial R},$$

$$\varepsilon_\phi = \frac{u_R}{R} + \frac{\partial u_\phi}{R\partial\phi}$$

$$\varepsilon_\theta = \frac{u_R}{R} + \cot\phi\frac{u_\phi}{R} + \frac{1}{\sin\phi}\frac{\partial u_\theta}{R\partial\theta},$$

$$\varepsilon_{\phi\theta} = \frac{1}{\sin\phi}\frac{\partial u_\phi}{R\partial\theta} + \frac{\partial u_\theta}{R\partial\phi} - \cot\phi\frac{u_\theta}{R}$$

$$\varepsilon_{R\theta} = \frac{1}{\sin\phi}\frac{\partial u_R}{R\partial\theta} + \frac{\partial u_\theta}{\partial R} - \frac{u_\theta}{R},$$

$$\varepsilon_{R\phi} = \frac{\partial u_R}{R\partial\phi} + \frac{\partial u_\phi}{\partial R} - \frac{u_\phi}{R}$$

$$(1.85)$$

$$\sigma_j = 2\mu\,\varepsilon_j + \lambda\,\varepsilon_{vol}, \qquad j = R,\phi,\theta, \qquad \varepsilon_{vol} = \varepsilon_R + \varepsilon_\phi + \varepsilon_\theta$$

$$\sigma_{\phi\theta} = \mu\,\varepsilon_{\phi\theta}, \qquad \sigma_{R\theta} = \mu\,\varepsilon_{R\theta}, \qquad \sigma_{R\phi} = \mu\,\varepsilon_{R\phi}$$

$$\sigma_{\theta\phi} = \sigma_{\phi\theta}, \qquad \sigma_{\theta R} = \sigma_{R\theta}, \qquad \sigma_{\phi R} = \sigma_{R\phi}$$

$$(1.86)$$

$\mathbf{u} = [u_R \quad u_\phi \quad u_\theta]^T$	*Displacement vector*	(1.87)
$\mathbf{b} = [b_R \quad b_\phi \quad b_\theta]^T$	*Body load vector*	(1.88)
$\boldsymbol{\varepsilon} = [\varepsilon_R \quad \varepsilon_\phi\varepsilon_\theta \quad \varepsilon_{\phi\theta} \quad \varepsilon_{R\theta} \quad \varepsilon_{R\phi}]^T$	*Strain vector*	(1.89)
$\boldsymbol{\sigma} = [\sigma_R \quad \sigma_\phi \quad \sigma_\theta \quad \sigma_{\phi\theta} \quad \sigma_{R\theta} \quad \sigma_{R\phi}]^T$	*Stress vector*	(1.90)
$\boldsymbol{\varepsilon} = \mathbf{L}_\varepsilon\mathbf{u}$	*Strain–displacement relation*	(1.91)
$\boldsymbol{\sigma} = \mathbf{D}\boldsymbol{\varepsilon}$	*Constitutive law*	(1.92)
$\mathbf{L}_\sigma^\mathbf{T}\boldsymbol{\sigma} + \mathbf{b} = \rho\,\ddot{\mathbf{u}}$	*Dynamic equilibrium equation*	(1.93)

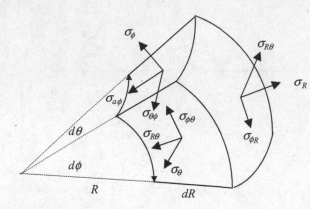

Figure 1.6: Stresses (and strains) in spherical coordinates.

where

$$
\mathbf{L}_\varepsilon^T = \left\{
\begin{array}{ccccccc}
\dfrac{\partial}{\partial R} & \dfrac{1}{R} & \dfrac{1}{R} & 0 & \dfrac{1}{R\sin\phi}\dfrac{\partial}{\partial\theta} & \dfrac{1}{R}\dfrac{\partial}{\partial\phi} \\[2ex]
0 & \dfrac{1}{R}\dfrac{\partial}{\partial\phi} & \dfrac{\cot\phi}{R} & \dfrac{1}{R\sin\phi}\dfrac{\partial}{\partial\theta} & 0 & \dfrac{\partial}{\partial R}-\dfrac{1}{R} \\[2ex]
0 & 0 & \dfrac{1}{R\sin\phi}\dfrac{\partial}{\partial\theta} & \dfrac{1}{R}\dfrac{\partial}{\partial\phi}-\dfrac{\cot\phi}{R} & \dfrac{\partial}{\partial R}-\dfrac{1}{R} & 0
\end{array}
\right\}
\tag{1.94}
$$

$$
\mathbf{L}_\sigma^T = \left\{
\begin{array}{ccccccc}
\dfrac{\partial}{\partial R}+\dfrac{2}{R} & -\dfrac{1}{R} & -\dfrac{1}{R} & 0 & \dfrac{1}{R\sin\phi}\dfrac{\partial}{\partial\theta} & \dfrac{1}{R}\dfrac{\partial}{\partial\phi}+\dfrac{\cot\phi}{R} \\[2ex]
0 & \dfrac{1}{R}\dfrac{\partial}{\partial\phi}+\dfrac{\cot\phi}{R} & -\dfrac{\cot\phi}{R} & \dfrac{1}{R\sin\phi}\dfrac{\partial}{\partial\theta} & 0 & \dfrac{\partial}{\partial R}+\dfrac{3}{R} \\[2ex]
0 & 0 & \dfrac{1}{R\sin\phi}\dfrac{\partial}{\partial\theta} & \dfrac{1}{R}\dfrac{\partial}{\partial\phi}+\dfrac{2\cot\phi}{R} & \dfrac{\partial}{\partial R}+\dfrac{3}{R} & 0
\end{array}
\right\}
$$
$$\tag{1.95}$$

The two differential operators $\mathbf{L}_\varepsilon$ and $\mathbf{L}_\sigma$ can be written as

$$
\mathbf{L}_\varepsilon = \mathbf{L}_R\frac{\partial}{\partial R} + \mathbf{L}_\phi\frac{1}{R}\frac{\partial}{\partial\phi} + \mathbf{L}_\theta\frac{1}{R\sin\phi}\frac{\partial}{\partial\theta} + \mathbf{L}_1\frac{1}{R} + \mathbf{L}_2\frac{\cot\phi}{R}
\tag{1.96}
$$

$$
\mathbf{L}_\sigma = \mathbf{L}_R\frac{\partial}{\partial R} + \mathbf{L}_\phi\frac{1}{R}\frac{\partial}{\partial\phi} + \mathbf{L}_\theta\frac{1}{R\sin\phi}\frac{\partial}{\partial\theta} + (2\mathbf{L}_R-\mathbf{L}_1)\frac{1}{R} + (\mathbf{L}_\phi-\mathbf{L}_2)\frac{\cot\phi}{R}
\tag{1.97}
$$

with matrices

$$
\mathbf{L}_R = \left\{
\begin{array}{ccc}
1&0&0\\0&0&0\\0&0&0\\0&0&0\\0&0&1\\0&1&0
\end{array}
\right\},\quad
\mathbf{L}_\phi = \left\{
\begin{array}{ccc}
0&0&0\\0&1&0\\0&0&0\\0&0&1\\0&0&0\\1&0&0
\end{array}
\right\},\quad
\mathbf{L}_\theta = \left\{
\begin{array}{ccc}
0&0&0\\0&0&0\\0&0&1\\0&1&0\\1&0&0\\0&0&0
\end{array}
\right\}
\tag{1.98a}
$$

$$\mathbf{L}_1 = \begin{Bmatrix} 0 & 0 & 0 \\ 1 & 0 & 0 \\ 1 & 0 & 0 \\ 0 & 0 & 0 \\ 0 & 0 & -1 \\ 0 & -1 & 0 \end{Bmatrix}, \quad \mathbf{L}_2 = \begin{Bmatrix} 0 & 0 & 0 \\ 0 & 0 & 0 \\ 0 & 1 & 0 \\ 0 & 0 & -1 \\ 0 & 0 & 0 \\ 0 & 0 & 0 \end{Bmatrix} \tag{1.98b}$$

Stresses in principal surfaces

$$\mathbf{s}_R = [\sigma_R \quad \sigma_{R\phi} \quad \sigma_{R\theta}]^T = \mathbf{L}_R^T \boldsymbol{\sigma} = \mathbf{L}_R^T \mathbf{D} \mathbf{L}_\varepsilon \mathbf{u}$$

$$= \left[\mathbf{D}_{RR} \frac{\partial}{\partial R} + \mathbf{D}_{R\phi} \frac{1}{R} \frac{\partial}{\partial \phi} + \mathbf{D}_{R\theta} \frac{1}{R \sin\phi} \frac{\partial}{\partial \theta} \right.$$

$$\left. + \mathbf{D}_{R1} \frac{1}{R} + \mathbf{D}_{R2} \frac{\cot\phi}{R} \right] \mathbf{u}$$

$$\mathbf{s}_\phi = [\sigma_{\phi R} \quad \sigma_\phi \quad \sigma_{\phi\theta}]^T = \mathbf{L}_\phi^T \boldsymbol{\sigma} = \mathbf{L}_\phi^T \mathbf{D} \mathbf{L}_\varepsilon \mathbf{u}$$

$$= \left[\mathbf{D}_{\phi R} \frac{\partial}{\partial R} + \mathbf{D}_{\phi\phi} \frac{1}{R} \frac{\partial}{\partial \phi} + \mathbf{D}_{\phi\theta} \frac{1}{R \sin\phi} \frac{\partial}{\partial \theta} \right.$$

$$\left. + \mathbf{D}_{\phi 1} \frac{1}{R} + \mathbf{D}_{\phi 2} \frac{\cot\phi}{R} \right] \mathbf{u}$$

$$\mathbf{s}_\theta = [\sigma_{\theta R} \quad \sigma_{\theta\phi} \quad \sigma_\theta]^T = \mathbf{L}_\theta^T \boldsymbol{\sigma} = \mathbf{L}_\theta^T \mathbf{D} \mathbf{L}_\varepsilon \mathbf{u}$$

$$= \left[\mathbf{D}_{\theta R} \frac{\partial}{\partial R} + \mathbf{D}_{\theta\phi} \frac{1}{R} \frac{\partial}{\partial \phi} + \mathbf{D}_{\theta\theta} \frac{1}{R \sin\phi} \frac{\partial}{\partial \theta} \right.$$

$$\left. + \mathbf{D}_{\theta 1} \frac{1}{R} + \mathbf{D}_{\theta 2} \frac{\cot\phi}{R} \right] \mathbf{u}$$

$$\tag{1.99}$$

which uses the definition $\mathbf{D}_{\alpha\beta} = \mathbf{L}_\alpha^T \mathbf{D} \mathbf{L}\beta$, with $\alpha, \beta = R, \phi, \theta, 1, 2$. Observe that these matrices are generally not symmetric, but $\mathbf{D}_{\beta\alpha} = \mathbf{D}_{\alpha\beta}^T$. For a spherically anisotropic material, the matrices are

$$\mathbf{D}_{RR} = \begin{Bmatrix} d_{11} & d_{16} & d_{15} \\ d_{16} & d_{66} & d_{56} \\ d_{15} & d_{56} & d_{55} \end{Bmatrix}, \quad \mathbf{D}_{\phi\phi} = \begin{Bmatrix} d_{66} & d_{26} & d_{46} \\ d_{26} & d_{22} & d_{24} \\ d_{46} & d_{24} & d_{44} \end{Bmatrix} \tag{1.100a}$$

$$\mathbf{D}_{\theta\theta} = \begin{Bmatrix} d_{55} & d_{45} & d_{35} \\ d_{45} & d_{44} & d_{34} \\ d_{35} & d_{34} & d_{33} \end{Bmatrix}, \quad \mathbf{D}_{R\phi} = \begin{Bmatrix} d_{16} & d_{12} & d_{14} \\ d_{66} & d_{26} & d_{46} \\ d_{56} & d_{25} & d_{45} \end{Bmatrix} \tag{1.100b}$$

$$\mathbf{D}_{R\theta} = \begin{Bmatrix} d_{15} & d_{14} & d_{13} \\ d_{56} & d_{46} & d_{36} \\ d_{55} & d_{45} & d_{35} \end{Bmatrix}, \qquad \mathbf{D}_{\phi\theta} = \begin{Bmatrix} d_{56} & d_{46} & d_{36} \\ d_{25} & d_{24} & d_{23} \\ d_{45} & d_{44} & d_{34} \end{Bmatrix} \tag{1.100c}$$

$$\mathbf{D}_{11} = \begin{Bmatrix} d_{22} + 2d_{23} + d_{33} & -(d_{26} + d_{36}) & -(d_{25} + d_{35}) \\ -(d_{26} + d_{36}) & d_{66} & d_{56} \\ -(d_{25} + d_{35}) & d_{56} & d_{55} \end{Bmatrix}, \qquad \mathbf{D}_{22} = \begin{Bmatrix} 0 & 0 & 0 \\ 0 & d_{33} & -d_{34} \\ 0 & -d_{34} & d_{44} \end{Bmatrix} \tag{1.100d}$$

$$\mathbf{D}_{12} = \begin{Bmatrix} 0 & d_{23} + d_{33} & -(d_{24} + d_{34}) \\ 0 & -d_{36} & d_{46} \\ 0 & -d_{35} & d_{45} \end{Bmatrix}, \qquad \mathbf{D}_{r1} = \begin{Bmatrix} d_{12} + d_{13} & -d_{16} & -d_{15} \\ d_{26} + d_{36} & -d_{66} & -d_{56} \\ d_{25} + d_{35} & -d_{56} & -d_{55} \end{Bmatrix} \tag{1.100e}$$

$$\mathbf{D}_{r2} = \begin{Bmatrix} 0 & d_{13} & -d_{14} \\ 0 & d_{36} & -d_{46} \\ 0 & d_{35} & -d_{45} \end{Bmatrix}, \qquad \mathbf{D}_{\phi1} = \begin{Bmatrix} d_{26} + d_{36} & -d_{66} & -d_{56} \\ d_{22} + d_{23} & -d_{26} & -d_{25} \\ d_{24} + d_{34} & -d_{46} & -d_{45} \end{Bmatrix} \tag{1.100f}$$

$$\mathbf{D}_{\phi2} = \begin{Bmatrix} 0 & d_{36} & -d_{46} \\ 0 & d_{23} & -d_{24} \\ 0 & d_{34} & -d_{44} \end{Bmatrix}, \qquad \mathbf{D}_{\theta1} = \begin{Bmatrix} d_{25} + d_{35} & -d_{56} & -d_{55} \\ d_{24} + d_{34} & -d_{46} & -d_{45} \\ d_{23} + d_{33} & -d_{36} & -d_{35} \end{Bmatrix} \tag{1.100g}$$

$$\mathbf{D}_{\theta2} = \begin{Bmatrix} 0 & d_{35} & -d_{45} \\ 0 & d_{34} & -d_{44} \\ 0 & d_{33} & -d_{34} \end{Bmatrix} \tag{1.100h}$$

For an isotropic medium, the material matrices are

$$\mathbf{D}_{RR} = \begin{Bmatrix} \lambda + 2\mu & 0 & 0 \\ 0 & \mu & 0 \\ 0 & 0 & \mu \end{Bmatrix}, \quad \mathbf{D}_{\phi\phi} = \begin{Bmatrix} \mu & 0 & 0 \\ 0 & \lambda + 2\mu & 0 \\ 0 & 0 & \mu \end{Bmatrix}, \quad \mathbf{D}_{\theta\theta} = \begin{Bmatrix} \mu & 0 & 0 \\ 0 & \mu & 0 \\ 0 & 0 & \lambda + 2\mu \end{Bmatrix} \tag{1.101a}$$

$$\mathbf{D}_{R\phi} = \begin{Bmatrix} 0 & \lambda & 0 \\ \mu & 0 & 0 \\ 0 & 0 & 0 \end{Bmatrix}, \quad \mathbf{D}_{\phi\theta} = \begin{Bmatrix} 0 & 0 & 0 \\ 0 & 0 & \lambda \\ 0 & \mu & 0 \end{Bmatrix}, \quad \mathbf{D}_{R\theta} = \begin{Bmatrix} 0 & 0 & \lambda \\ 0 & 0 & 0 \\ \mu & 0 & 0 \end{Bmatrix} \tag{1.101b}$$

$$\mathbf{D}_{R1} = \begin{Bmatrix} 2\lambda & 0 & 0 \\ 0 & -\mu & 0 \\ 0 & 0 & -\mu \end{Bmatrix}, \quad \mathbf{D}_{\phi1} = \begin{Bmatrix} 0 & -\mu & 0 \\ 2(\lambda + \mu) & 0 & 0 \\ 0 & 0 & 0 \end{Bmatrix}, \quad \mathbf{D}_{\theta1} = \begin{Bmatrix} 0 & 0 & -\mu \\ 0 & 0 & 0 \\ 2(\lambda + \mu) & 0 & 0 \end{Bmatrix} \tag{1.101c}$$

$$\mathbf{D}_{R2} = \begin{Bmatrix} 0 & \lambda & 0 \\ 0 & 0 & 0 \\ 0 & 0 & 0 \end{Bmatrix}, \quad \mathbf{D}_{\phi 2} = \begin{Bmatrix} 0 & 0 & 0 \\ 0 & \lambda & 0 \\ 0 & 0 & -\mu \end{Bmatrix}, \quad \mathbf{D}_{\theta 2} = \begin{Bmatrix} 0 & 0 & 0 \\ 0 & 0 & -\mu \\ 0 & \lambda + 2\mu & 0 \end{Bmatrix} \tag{1.101d}$$

$$\mathbf{D}_{11} = \begin{Bmatrix} 4(\lambda + \mu) & 0 & 0 \\ 0 & \mu & 0 \\ 0 & 0 & \mu \end{Bmatrix}, \quad \mathbf{D}_{12} = \begin{Bmatrix} 0 & 2(\lambda + \mu) & 0 \\ 0 & 0 & 0 \\ 0 & 0 & 0 \end{Bmatrix}, \quad \mathbf{D}_{22} = \begin{Bmatrix} 0 & 0 & 0 \\ 0 & \lambda + 2\mu & 0 \\ 0 & 0 & \mu \end{Bmatrix}$$
$$\tag{1.101e}$$

Wave equation

The dynamic equilibrium equation can be written as

$$\mathbf{L}_\sigma^T \mathbf{D} \mathbf{L}_\varepsilon \mathbf{u} + \mathbf{b} = \rho \ddot{\mathbf{u}} \tag{1.102}$$

For any arbitrary, spherically anisotropic material with $\mathbf{D}_{\alpha\beta} = \mathbf{L}_\alpha^T \mathbf{D} \mathbf{L}_\beta$, the wave equation is

$$
\begin{aligned}
\rho \ddot{\mathbf{u}} = {} & \mathbf{b} + \mathbf{D}_{RR} \frac{\partial^2}{\partial R^2} + \mathbf{D}_{\phi\phi} \frac{1}{R^2} \frac{\partial^2}{\partial \phi^2} + \mathbf{D}_{\theta\theta} \frac{1}{R^2 \sin^2 \phi} \frac{\partial^2}{\partial \theta^2} + (\mathbf{D}_{R\phi} + \mathbf{D}_{\phi R}) \frac{1}{R} \frac{\partial^2}{\partial R \partial \phi} \\[2mm]
& + (\mathbf{D}_{R\theta} + \mathbf{D}_{\theta R}) \frac{1}{R \sin \phi} \frac{\partial^2}{\partial R \partial \theta} + (\mathbf{D}_{\phi\theta} + \mathbf{D}_{\theta\phi}) \frac{1}{R^2 \sin \phi} \frac{\partial^2}{\partial \phi \partial \theta} \\[2mm]
& + (2\mathbf{D}_{RR} + \mathbf{D}_{R1} - \mathbf{D}_{1R}) \frac{1}{R} \frac{\partial}{\partial R} + (\mathbf{D}_{\phi R} + \mathbf{D}_{R2} - \mathbf{D}_{2R}) \frac{\cot \phi}{R} \frac{\partial}{\partial R} \\[2mm]
& + (\mathbf{D}_{R\phi} + \mathbf{D}_{\phi 1} - \mathbf{D}_{1\phi}) \frac{1}{R^2} \frac{\partial}{\partial \phi} + (\mathbf{D}_{\phi\phi} + \mathbf{D}_{\phi 2} - \mathbf{D}_{2\phi}) \frac{\cot \phi}{R^2} \frac{\partial}{\partial \phi} \\[2mm]
& + (\mathbf{D}_{R\theta} + \mathbf{D}_{\theta 1} - \mathbf{D}_{1\theta}) \frac{1}{R^2 \sin \phi} \frac{\partial}{\partial \theta} + (\mathbf{D}_{\theta 2} - \mathbf{D}_{2\theta}) \frac{\cot \phi}{R^2 \sin \phi} \frac{\partial}{\partial \theta} \\[2mm]
& + (\mathbf{D}_{R1} - \mathbf{D}_{11}) \frac{1}{R^2} + (\mathbf{D}_{R2} + \mathbf{D}_{\phi 1} - \mathbf{D}_{12} - \mathbf{D}_{21}) \frac{\cot \phi}{R^2} \\[2mm]
& - \mathbf{D}_{\phi 2} \frac{1}{R^2 \sin^2 \phi} + (\mathbf{D}_{\phi 2} - \mathbf{D}_{22}) \frac{\cot^2 \phi}{R^2}
\end{aligned}
$$

$$\tag{1.103}$$

For an isotropic material, this results in the elastic wave equation in spherical coordinates:

$$
\rho\ddot{\mathbf{u}} = \mathbf{b} + \left\{ \begin{bmatrix} \lambda+2\mu & 0 & 0 \\ 0 & \mu & 0 \\ 0 & 0 & \mu \end{bmatrix} \left(\frac{\partial^2}{\partial R^2} + \frac{2}{R}\frac{\partial}{\partial R} - \frac{2}{R^2} \right) \right.
$$

$$
+ \begin{bmatrix} \mu & 0 & 0 \\ 0 & \lambda+2\mu & 0 \\ 0 & 0 & \mu \end{bmatrix} \left(\frac{1}{R^2}\frac{\partial^2}{\partial\phi^2} + \frac{\cot\phi}{R^2}\frac{\partial}{\partial\phi} \right)
$$

$$
+ \begin{bmatrix} \mu & 0 & 0 \\ 0 & \mu & 0 \\ 0 & 0 & \lambda+2\mu \end{bmatrix} \frac{1}{R^2\sin^2\phi}\frac{\partial^2}{\partial\theta^2} + \begin{bmatrix} 0 & \lambda+\mu & 0 \\ \lambda+\mu & 0 & 0 \\ 0 & 0 & 0 \end{bmatrix} \frac{1}{R}\frac{\partial^2}{\partial R\partial\phi}
$$

$$
+ \begin{bmatrix} 0 & 0 & \lambda+\mu \\ 0 & 0 & 0 \\ \lambda+\mu & 0 & 0 \end{bmatrix} \frac{1}{R\sin\phi}\frac{\partial^2}{\partial R\partial\theta} + \begin{bmatrix} 0 & 0 & 0 \\ 0 & 0 & \lambda+\mu \\ 0 & \lambda+\mu & 0 \end{bmatrix} \frac{1}{R^2\sin\phi}\frac{\partial^2}{\partial\phi\,\partial\theta}
$$

$$
+ \begin{bmatrix} 0 & \lambda+\mu & 0 \\ 0 & 0 & 0 \\ 0 & 0 & 0 \end{bmatrix} \frac{\cot\phi}{R}\frac{\partial}{\partial R} + \begin{bmatrix} 0 & -(\lambda+3\mu) & 0 \\ 2(\lambda+2\mu) & 0 & 0 \\ 0 & 0 & 0 \end{bmatrix} \frac{1}{R^2}\frac{\partial}{\partial\phi}
$$

$$
+ \begin{bmatrix} 0 & 0 & -(\lambda+3\mu) \\ 0 & 0 & \cdot \\ 2(\lambda+2\mu) & 0 & 0 \end{bmatrix} \frac{1}{R^2\sin\phi}\frac{\partial}{\partial\theta}
$$

$$
+ \begin{bmatrix} 0 & 0 & 0 \\ 0 & 0 & -(\lambda+3\mu) \\ 0 & \lambda+3\mu & 0 \end{bmatrix} \frac{\cot\phi}{R^2\sin\phi}\frac{\partial}{\partial\theta} + \begin{bmatrix} 0 & -(\lambda+3\mu) & 0 \\ 0 & 0 & 0 \\ 0 & 0 & 0 \end{bmatrix} \frac{\cot\phi}{R^2}
$$

$$
\left. + \begin{bmatrix} 0 & 0 & 0 \\ 0 & -\lambda & 0 \\ 0 & 0 & \mu \end{bmatrix} \frac{1}{R^2\sin^2\phi} + \begin{bmatrix} 0 & 0 & 0 \\ 0 & -2\mu & 0 \\ 0 & 0 & -2\mu \end{bmatrix} \frac{\cot^2\phi}{R^2} \right\}\mathbf{u} \tag{1.104}
$$

2 Dipoles

In most cases, we provide solutions for point forces only. Additional solutions for single couples, (moments or torques), tensile crack sources, double couples, and bimoments (moments of moments) can be obtained by differentiation. This is done as described in the following.

2.1 Point dipoles or doublets: single couples and tensile crack sources

Let $g_j(\mathbf{r}, \mathbf{r}') = g_j(\mathbf{r}, \mathbf{r}', t)$ or $g_j(\mathbf{r}, \mathbf{r}') = g_j(\mathbf{r}, \mathbf{r}', \omega)$ be the Green's function for a *receiver* placed at location $\mathbf{r}$ due to an impulsive or harmonic *source*, a point load $P = 1$ acting at location $\mathbf{r}'$ in some principal direction j. Next, apply two such loads in opposite directions at some small distance $2a$ apart in the neighborhood of $\mathbf{r}'$, as shown in Figure 2.1. Define $\mathbf{a} = a\hat{\mathbf{a}}$ to be a vector connecting the midpoint between the points of application of these forces to one of these, and let $\hat{\mathbf{a}}$ be a unit vector in that direction. Together, this pair of forces causes a displacement field

$$\mathbf{u} = P[g_j(\mathbf{r}, \mathbf{r}' + \mathbf{a}) - g_j(\mathbf{r}, \mathbf{r}' - \mathbf{a})] \tag{2.1}$$

Writing $\mathbf{a} = a\,\hat{\mathbf{a}} = \Delta x_1'\,\hat{\mathbf{e}}_1 + \Delta x_2'\,\hat{\mathbf{e}}_2 + \Delta x_3'\,\hat{\mathbf{e}}_3$ and expanding in Taylor series, we obtain

$$\begin{aligned}
g_j(\mathbf{r}, \mathbf{r}' \pm \mathbf{a}) &= g_j(\mathbf{r}, x_1' \pm \Delta x_1', x_2' \pm \Delta x_2', x_3' \pm \Delta x_3') \\
&= g_j(\mathbf{r}, x_1', x_2', x_3') \pm \left[\Delta x_1' \frac{\partial g_j}{\partial x_1'} + \Delta x_2' \frac{\partial g_j}{\partial x_2'} + \Delta x_3' \frac{\partial g_j}{\partial x_3'} \right] + \cdots \\
&= g_j(\mathbf{r}, \mathbf{r}') \pm \left[\frac{\partial g_j}{\partial x_1'} \hat{\mathbf{e}}_1 + \frac{\partial g_j}{\partial x_2'} \hat{\mathbf{e}}_2 + \frac{\partial g_j}{\partial x_3'} \hat{\mathbf{e}}_3 \right] \cdot [\Delta x_1' \hat{\mathbf{e}}_1 + \Delta x_2' \hat{\mathbf{e}}_2 + \Delta x_3' \hat{\mathbf{e}}_3] + \cdots \\
&= g_j(\mathbf{r}, \mathbf{r}') \pm (\nabla' g_j) \cdot \mathbf{a} + O(a^2) \tag{2.2}
\end{aligned}$$

so

$$\mathbf{u}(\mathbf{r}, \mathbf{r}', \mathbf{a}) = 2Pa\, \nabla' g_j \cdot \hat{\mathbf{a}} + O(a^2) \tag{2.3}$$

in which the primes indicates that the gradient must be taken with respect to the source point, not the receiver, and j identifies the direction of the force. Applying the previous

27

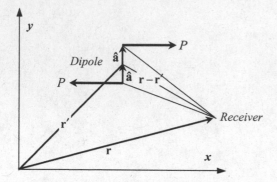

Figure 2.1: From point forces to point dipoles.

equation to the two neighboring sources, and defining the strength M of the dipole to be $M = 2Pa$, we obtain in the limit $a \to 0$ the Green's function for simple point dipoles, or *doublets*, as

$$\mathbf{G} = M \nabla' \mathbf{g}_j \cdot \hat{\mathbf{a}} = M \left(g_{ij,k} \hat{\mathbf{e}}_i \, \hat{\mathbf{e}}_k \right) \cdot \left(\alpha_m \hat{\mathbf{e}}_m \right) = M \, \alpha_k \, g_{ij,k} \hat{\mathbf{e}}_i \, \hat{\mathbf{e}}_k \tag{2.4}$$

that is, the Green's functions for point dipoles are obtained from the gradient of the Green's functions for point forces *with respect to the location of the source* (a tensor with nine components), projected along the direction of the moment arm. Choosing the direction of the moment arm to coincide in turn with the three principal directions (i.e., $\hat{\mathbf{a}} = \hat{\mathbf{e}}_k$) and setting the strength of the dipoles to unity (i.e., $M = 1$), we obtain the Green's functions for the nine possible simple force dipoles as

$$\mathbf{G}_{jk} = \nabla' \mathbf{g}_j \cdot \hat{\mathbf{e}}_k = \left[\frac{\partial \mathbf{g}_j}{\partial x_1'} \hat{\mathbf{e}}_1 + \frac{\partial \mathbf{g}_j}{\partial x_2'} \hat{\mathbf{e}}_2 + \frac{\partial \mathbf{g}_j}{\partial x_3'} \hat{\mathbf{e}}_3 \right] \cdot \hat{\mathbf{e}}_k = \frac{\partial \mathbf{g}_j}{\partial x_k'}$$

$$= \frac{\partial}{\partial x_k'} \left[g_{1j} \hat{\mathbf{e}}_1 + g_{2j} \hat{\mathbf{e}}_2 + g_{3j} \hat{\mathbf{e}}_3 \right] \equiv g_{ij,k}' \hat{\mathbf{e}}_i = G_{ijk} \hat{\mathbf{e}}_i \tag{2.5}$$

Observe that the dipoles constitute a third order tensor G_{ijk}. The three dipoles with equal indices $\mathbf{G}_{(jj)}$ (or $\mathbf{G}_{xx}$, etc.) are tensile cracks; the other six are single couples. The indices j, k indicate the directions of the forces and the moment arm. Figure 2.2 shows the nine simple dipoles, chosen so that the forces in the *first* octant are always pointing in the positive direction; the directions of the dipoles are then defined by the right-hand rule. It should be noted that the displacement field elicited by single couples is *not* rotationally symmetric with respect to the axis of the couple, not even in a fully homogeneous space.

Dipoles in a homogeneous space

In the case of a fully homogeneous and unbounded space, the Green's functions depend only on the relative location of the receiver with respect to the source, that is, $\mathbf{g}_j(\mathbf{r}, \mathbf{r}') = \mathbf{g}_j(\mathbf{r} - \mathbf{r}')$. In this situation, the gradient with respect to the source is simply the negative of the gradient with respect to the receiver, in which case we could write

$$\mathbf{G}_{jk} = G_{ijk} \hat{\mathbf{e}}_i = -\nabla \mathbf{g}_j(\mathbf{r}) \cdot \hat{\mathbf{e}}_k = -g_{ij,k} \hat{\mathbf{e}}_i \tag{2.6}$$

By contrast, in a body with physical boundaries or material discontinuities, the location of the source is relevant. For example, in the case of a plate of infinite lateral extent, the depth of the source – but not the lateral position – must be considered. Equivalent forms can then be obtained by recourse to the reciprocity principle, which switches the locations of the source and the receiver.

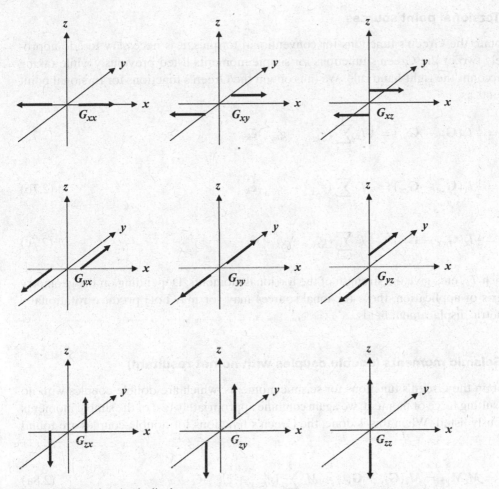

Figure 2.2: The nine simple dipoles.

2.2 Line dipoles

These are dipoles formed by line sources. They are obtained from the preceding equations by setting to zero the partial derivatives with respect to the line direction, and using the Green's functions for line sources in place of those for point forces.

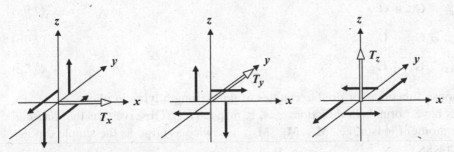

Figure 2.3: Pure torsional moments.

2.3 Torsional point sources

To obtain the Green's functions for conventional torques, it is necessary to add appropriately two of the Green's functions for simple moments listed previously while taking into account the right-hand rule. We thus obtain the Green's functions for torsional point moments as

$$\mathbf{T}_x = \tfrac{1}{2}T_x(\mathbf{G}_{zy} - \mathbf{G}_{yz}) = \tfrac{1}{2}T_x \sum_{k=1}^{3} (g'_{kz,y} - g'_{ky,z})\hat{\mathbf{e}}_k \tag{2.7a}$$

$$\mathbf{T}_y = \tfrac{1}{2}T_y(\mathbf{G}_{xz} - \mathbf{G}_{zx}) = \tfrac{1}{2}T_y \sum_{k=1}^{3} (g'_{kx,z} - g'_{kz,x})\hat{\mathbf{e}}_k \tag{2.7b}$$

$$\mathbf{T}_z = \tfrac{1}{2}T_z(\mathbf{G}_{yx} - \mathbf{G}_{xy}) = \tfrac{1}{2}T_z \sum_{k=1}^{3} (g'_{ky,x} - g'_{kx,y})\hat{\mathbf{e}}_k \tag{2.7c}$$

in which T_x, etc., give the strength of the torsional moments. Depending on the geometry and axis of application, these torsional sources may (or may not) produce rotationally symmetric displacement fields.

2.4 Seismic moments (double couples with no net resultant)

To obtain the Green's functions for seismic moments, which are double couples with no net resulting force or moment, we again combine appropriately two of the simple moments previously listed. When this is done, the Green's functions for double couples are found to be

$$\mathbf{M}_x = M_x\mathbf{M}_{yz} = M_x(\mathbf{G}_{zy} + \mathbf{G}_{yz}) = M_x \sum_{k=1}^{3} (g'_{kz,y} + g'_{ky,z})\hat{\mathbf{e}}_k \tag{2.8a}$$

$$\mathbf{M}_y = M_y\mathbf{M}_{zx} = M_y(\mathbf{G}_{xz} + \mathbf{G}_{zx}) = M_y \sum_{k=1}^{3} (g'_{kx,z} + g'_{kz,x})\hat{\mathbf{e}}_k \tag{2.8b}$$

$$\mathbf{M}_z = M_z\mathbf{M}_{xy} = M_z(\mathbf{G}_{yx} + \mathbf{G}_{xy}) = M_z \sum_{k=1}^{3} (g'_{ky,x} + g'_{kx,y})\hat{\mathbf{e}}_k \tag{2.8c}$$

that is,

$$\mathbf{M}_{yz} \equiv \mathbf{M}_{zy} = \mathbf{G}_{yz} + \mathbf{G}_{zy} \tag{2.9a}$$

$$\mathbf{M}_{zx} \equiv \mathbf{M}_{xz} = \mathbf{G}_{zx} + \mathbf{G}_{xz} \tag{2.9b}$$

$$\mathbf{M}_{xy} \equiv \mathbf{M}_{yx} = \mathbf{G}_{xy} + \mathbf{G}_{yx} \tag{2.9c}$$

are the response functions for double couples of unit strength lying in planes yz, zx, xy. These planes have normals in directions x, y, z, respectively. Observe that the diagonal terms of the moment tensor, i.e., $\mathbf{M}_{xx}$, $\mathbf{M}_{yy}$, $\mathbf{M}_{zz}$, are twice as large as the simple dipoles for tension cracks.

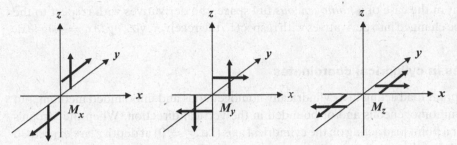

Figure 2.4: Double couples (seismic moments with no net resultant).

2.5 Blast loads (explosive line and point sources)

Linelike and pointlike blast loads represent the limiting cases of cylindrical and spherical cavities of infinitesimal radius subjected to singularly large impulsive or harmonic compressive pressures. The strength of the source is defined so that when the overpressure is integrated over the volume of the cavity, one obtains a unit value. In two and three dimensions, this gives an overpressure

$$p = \frac{\delta(r)}{2\pi r} \rightarrow \iint_A pr\,dr\,d\theta = 1 \qquad \text{2-D} \tag{2.10}$$

$$p = \frac{\delta(r)}{4\pi r^2} \rightarrow \iiint_V pr^2\,dr\,d\theta\,d\phi = 1 \qquad \text{3-D} \tag{2.11}$$

The Green's functions for line and point blast loads can be obtained from the Green's functions for line and point forces by superposition of the three tensile cracks, except that they must be modified by a factor to compensate for the extraction of the infinitesimal particle that fills the cavity[1]:

$$\mathbf{B}_{2D} = \frac{\lambda + 2\mu}{\mu}\left[\frac{\partial\,\mathbf{g}_x}{\partial x'} + \frac{\partial\,\mathbf{g}_z}{\partial z'}\right]$$

$$= \frac{\lambda + 2\mu}{\mu}\left[\left(g'_{xx,x} + g'_{xz,z}\right)\hat{\mathbf{e}}_x + \left(g'_{zx,x} + g'_{zz,z}\right)\hat{\mathbf{e}}_z\right] \qquad \text{Line blast source in 2-D}$$

$$= \left(\frac{\alpha}{\beta}\right)^2 [\mathbf{G}_{xx} + \mathbf{G}_{zz}] \tag{2.12}$$

$$\mathbf{B}_{3D} = \frac{3(\lambda + 2\mu)}{4\mu}\left[\frac{\partial\,\mathbf{g}_x}{\partial x'} + \frac{\partial\,\mathbf{g}_y}{\partial y'} + \frac{\partial\,\mathbf{g}_z}{\partial z'}\right]$$

$$= \frac{3(\lambda + 2\mu)}{4\mu}\sum_{k=1}^{3}\left(g'_{kx,x} + g'_{ky,y} + g'_{kz,z}\right)\hat{\mathbf{e}}_k \qquad \text{Point blast source in 3-D}$$

$$= \frac{3}{4}\left(\frac{\alpha}{\beta}\right)^2 [\mathbf{G}_{xx} + \mathbf{G}_{yy} + \mathbf{G}_{zz}] \tag{2.13}$$

[1] Kausel, E., 1998, Blast loads versus point loads: the missing factor, *Journal of Engineering Mechanics, ASCE*, Vol. 124, No. 2 (Feb.), pp. 243–244.

Observe that in the case of a *homogeneous* full space, the derivatives with respect to the source can be changed into derivatives with respect to the receiver, viz., $\partial g / \partial x_i' = -\partial g / \partial x_i$.

2.6 Dipoles in cylindrical coordinates

Consider a point load acting in a cylindrically homogeneous and unbounded medium, but that can be inhomogeneous and/or bounded in the vertical direction. When the Green's functions for a point load acting on the cylindrical axis (i.e., $r' = 0$) at depth z' are expressed in cylindrical coordinates, they have the general form

$$\mathbf{g} = \sum_{n=0}^{\infty} u_n \begin{pmatrix} c_n \\ s_n \end{pmatrix} \hat{\mathbf{r}} + v_n \begin{pmatrix} -s_n \\ c_n \end{pmatrix} \hat{\mathbf{t}} + w_n \begin{pmatrix} c_n \\ s_n \end{pmatrix} \hat{\mathbf{k}} \qquad (2.14)$$

in which $c_n = \cos n\theta$ and $s_n = \sin n\theta$. In particular, the Green's functions for point loads in the vertical z direction ($n = 0$) and in the horizontal x and y directions (both $n = 1$) are of the form

$$\mathbf{g}_z = U(r, z, z') \hat{\mathbf{r}} + W(r, z, z') \hat{\mathbf{k}} \qquad \qquad \textit{Vertical z load} \qquad (2.15)$$

$$\mathbf{g}_x = u(r, z, z') \cos \theta \, \hat{\mathbf{r}} + v(r, z, z') (-\sin \theta) \hat{\mathbf{t}} + w(r, z, z') \cos \theta \hat{\mathbf{k}} \quad \textit{Horizontal x load} \quad (2.16)$$

$$\mathbf{g}_y = u(r, z, z') \sin \theta \, \hat{\mathbf{r}} + v(r, z, z') \cos \theta \, \hat{\mathbf{t}} + w(r, z, z') \sin \theta \hat{\mathbf{k}} \quad \textit{Horizontal y load} \quad (2.17)$$

The gradient $\nabla \mathbf{g}$ for any of these Green's functions with respect to the *receiver* location can be shown to be given by

$$\nabla \vec{\mathbf{g}} = \hat{\mathbf{r}} \left[\frac{\partial u}{\partial r} \begin{pmatrix} c_n \\ s_n \end{pmatrix} \hat{\mathbf{r}} + \frac{\partial v}{\partial r} \begin{pmatrix} -s_n \\ c_n \end{pmatrix} \hat{\mathbf{t}} + \frac{\partial w}{\partial r} \begin{pmatrix} c_n \\ s_n \end{pmatrix} \hat{\mathbf{k}} \right]$$

$$+ \hat{\mathbf{t}} \left[\frac{nu - v}{r} \begin{pmatrix} -s_n \\ c_n \end{pmatrix} \hat{\mathbf{r}} + \frac{u - nv}{r} \begin{pmatrix} c_n \\ s_n \end{pmatrix} \hat{\mathbf{t}} + \frac{nw}{r} \begin{pmatrix} -s_n \\ c_n \end{pmatrix} \hat{\mathbf{k}} \right]$$

$$+ \hat{\mathbf{k}} \left[\frac{\partial u}{\partial z} \begin{pmatrix} c_n \\ s_n \end{pmatrix} \hat{\mathbf{r}} + \frac{\partial v}{\partial z} \begin{pmatrix} -s_n \\ c_n \end{pmatrix} \hat{\mathbf{t}} + \frac{\partial w}{\partial z} \begin{pmatrix} c_n \\ s_n \end{pmatrix} \hat{\mathbf{k}} \right] \qquad (2.18)$$

in which the products $\hat{\mathbf{r}}\hat{\mathbf{r}}$, $\hat{\mathbf{r}}\hat{\mathbf{t}}$, etc., are tensor bases, or *dyads*.

Now, here we have assumed the medium to be laterally homogeneous, so the dependence of the Green's functions within a given plane is only a function of the relative horizontal position of the receiver and the source. Thus, except for the terms involving the vertical derivatives, the gradient with respect to the receiver is the negative of the gradient with respect to the source. It follows that in the computation of the Green's functions for dipoles, it suffices for us to reverse the signs of the terms other than those for the vertical derivatives. When we substitute the Green's functions for horizontal and vertical loads given previously into the expression for the gradient together with the sign change described, and if we also express the appropriate moment arm direction $\hat{\mathbf{a}}$ in cylindrical coordinates as one of

$$\hat{\mathbf{a}} = \hat{\mathbf{i}} = \hat{\mathbf{r}} \cos \theta - \hat{\mathbf{t}} \sin \theta, \quad \hat{\mathbf{a}} = \hat{\mathbf{j}} = \hat{\mathbf{r}} \sin \theta + \hat{\mathbf{t}} \cos \theta, \quad \hat{\mathbf{a}} = \hat{\mathbf{k}} \qquad (2.19)$$

we obtain after some algebra the Green's functions for unit crack sources (M_{xx}, M_{yy}, $M_{zz} = 1$) and for single couples (M_{xy}, M_{xz}, etc. $= 1$) listed below, which also use the following definitions:

$$u, v, w = \text{components of Green's function for a horizontal point load } (n = 1); \quad (2.20)$$

$$U, W = \text{components of Green's function for a vertical point load } (n = 0) \quad (2.21)$$

$$\mathbf{G}_{xx} = -\frac{1}{2}\left\{\left(\frac{\partial u}{\partial r} + \frac{u-v}{r}\right)\hat{\mathbf{r}} + \left(\frac{\partial w}{\partial r} + \frac{w}{r}\right)\hat{\mathbf{k}}\right.$$

$$\left. + \left(\frac{\partial u}{\partial r} - \frac{u-v}{r}\right)\cos 2\theta\,\hat{\mathbf{r}} - \left(\frac{\partial v}{\partial r} + \frac{u-v}{r}\right)\sin 2\theta\,\hat{\mathbf{t}} + \left(\frac{\partial w}{\partial r} - \frac{w}{r}\right)\cos 2\theta\,\hat{\mathbf{k}}\right\}$$

$$(2.22a)$$

$$\mathbf{G}_{yy} = -\frac{1}{2}\left\{\left(\frac{\partial u}{\partial r} + \frac{u-v}{r}\right)\hat{\mathbf{r}} + \left(\frac{\partial w}{\partial r} + \frac{w}{r}\right)\hat{\mathbf{k}}\right.$$

$$\left. - \left(\frac{\partial u}{\partial r} - \frac{u-v}{r}\right)\cos 2\theta\,\hat{\mathbf{r}} + \left(\frac{\partial v}{\partial r} + \frac{u-v}{r}\right)\sin 2\theta\,\hat{\mathbf{t}} - \left(\frac{\partial w}{\partial r} - \frac{w}{r}\right)\cos 2\theta\,\hat{\mathbf{k}}\right\}$$

$$(2.22b)$$

$$\mathbf{G}_{zz} = +\left\{\frac{\partial U}{\partial z'}\hat{\mathbf{r}} + \frac{\partial W}{\partial z'}\hat{\mathbf{k}}\right\}$$

$$\mathbf{G}_{xy} = -\frac{1}{2}\left\{-\left(\frac{\partial v}{\partial r} - \frac{u-v}{r}\right)\hat{\mathbf{t}} + \left(\frac{\partial u}{\partial r} - \frac{u-v}{r}\right)\hat{\mathbf{r}}\sin 2\theta + \left(\frac{\partial v}{\partial r} + \frac{u-v}{r}\right)\cos 2\theta\,\hat{\mathbf{t}}\right.$$

$$\left. + \left(\frac{\partial w}{\partial r} - \frac{w}{r}\right)\sin 2\theta\hat{\mathbf{k}}\right\} \quad (2.22c)$$

$$\mathbf{G}_{yx} = -\frac{1}{2}\left\{\left(\frac{\partial v}{\partial r} - \frac{u-v}{r}\right)\hat{\mathbf{t}} + \left(\frac{\partial u}{\partial r} - \frac{u-v}{r}\right)\hat{\mathbf{r}}\sin 2\theta + \left(\frac{\partial v}{\partial r} + \frac{u-v}{r}\right)\cos 2\theta\,\hat{\mathbf{t}}\right.$$

$$\left. + \left(\frac{\partial w}{\partial r} - \frac{w}{r}\right)\sin 2\theta\hat{\mathbf{k}}\right\} \quad (2.22d)$$

$$\mathbf{G}_{xz} = +\left\{\frac{\partial u}{\partial z'}\cos\theta\,\hat{\mathbf{r}} - \frac{\partial v}{\partial z'}\sin\theta\,\hat{\mathbf{t}} + \frac{\partial w}{\partial z'}\cos\theta\,\hat{\mathbf{k}}\right\} \quad (2.22e)$$

$$\mathbf{G}_{zx} = -\left\{\frac{\partial U}{\partial r}\cos\theta\,\hat{\mathbf{r}} - \frac{U}{r}\sin\theta\,\hat{\mathbf{t}} + \frac{\partial W}{\partial r}\cos\theta\,\hat{\mathbf{k}}\right\} \quad (2.22f)$$

$$\mathbf{G}_{yz} = +\left\{\frac{\partial u}{\partial z'}\sin\theta\,\hat{\mathbf{r}} + \frac{\partial v}{\partial z'}\cos\theta\,\hat{\mathbf{t}} + \frac{\partial w}{\partial z'}\sin\theta\,\hat{\mathbf{k}}\right\} \quad (2.22g)$$

$$\mathbf{G}_{zy} = -\left\{\frac{\partial U}{\partial r}\sin\theta\,\hat{\mathbf{r}} + \frac{U}{r}\cos\theta\,\hat{\mathbf{t}} + \frac{\partial W}{\partial r}\sin\theta\,\hat{\mathbf{k}}\right\} \quad (2.22h)$$

Observe that the terms with derivatives with respect to the vertical *source* location z' have positive sign. Also, the Green's functions for point torsional moments, seismic double couples, and blast loads in cylindrical coordinates are

$$\mathbf{T}_x = -\frac{1}{2}\left\{\left(\frac{\partial u}{\partial z'} + \frac{\partial U}{\partial r}\right)\sin\theta\,\hat{\mathbf{r}} + \left(\frac{\partial v}{\partial z'} + \frac{U}{r}\right)\cos\theta\,\hat{\mathbf{t}} + \left(\frac{\partial w}{\partial z'} + \frac{\partial W}{\partial r}\right)\sin\theta\,\hat{\mathbf{k}}\right\} \quad (2.23a)$$

$$\mathbf{T}_y = +\frac{1}{2}\left\{\left(\frac{\partial u}{\partial z'} + \frac{\partial U}{\partial r}\right)\cos\theta\,\hat{\mathbf{r}} - \left(\frac{\partial v}{\partial z'} + \frac{U}{r}\right)\sin\theta\,\hat{\mathbf{t}} + \left(\frac{\partial w}{\partial z'} + \frac{\partial W}{\partial r}\right)\cos\theta\,\hat{\mathbf{k}}\right\} \quad (2.23b)$$

$$\mathbf{T}_z = -\frac{1}{2}\left(\frac{\partial v}{\partial r} - \frac{u-v}{r}\right)\hat{\mathbf{t}} \quad (2.23c)$$

$$\mathbf{M}_x = +\left\{\left(\frac{\partial u}{\partial z'} - \frac{\partial U}{\partial r}\right)\sin\theta\,\hat{\mathbf{r}} + \left(\frac{\partial v}{\partial z'} - \frac{U}{r}\right)\cos\theta\,\hat{\mathbf{t}} + \left(\frac{\partial w}{\partial z'} - \frac{\partial W}{\partial r}\right)\sin\theta\,\hat{\mathbf{k}}\right\} \quad (2.23d)$$

$$\mathbf{M}_y = +\left\{\left(\frac{\partial u}{\partial z'} - \frac{\partial U}{\partial r}\right)\cos\theta\,\hat{\mathbf{r}} - \left(\frac{\partial v}{\partial z'} - \frac{U}{r}\right)\sin\theta\,\hat{\mathbf{t}} + \left(\frac{\partial w}{\partial z'} - \frac{\partial W}{\partial r}\right)\cos\theta\,\hat{\mathbf{k}}\right\} \quad (2.23e)$$

$$\mathbf{M}_z = -\left\{\left(\frac{\partial u}{\partial r} - \frac{u-v}{r}\right)\sin 2\theta\,\hat{\mathbf{r}} + \left(\frac{\partial v}{\partial r} + \frac{u-v}{r}\right)\cos 2\theta\,\hat{\mathbf{t}} + \left(\frac{\partial w}{\partial r} - \frac{w}{r}\right)\sin 2\theta\,\hat{\mathbf{k}}\right\}$$

$$(2.23f)$$

$$\mathbf{B}_{3D} = -\frac{3(\lambda + 2\mu)}{4\mu}\left\{\left(\frac{\partial u}{\partial r} + \frac{u-v}{r} - \frac{\partial U}{\partial z'}\right)\hat{\mathbf{r}} + \left(\frac{\partial w}{\partial r} + \frac{w}{r} - \frac{\partial W}{\partial z'}\right)\hat{\mathbf{k}}\right\} \quad (2.23g)$$

in which u, v, w and U, W are again the components of the Green's function in cylindrical coordinates for horizontal and vertical *point loads*, respectively.

Finally, we provide also the expression for one of the several *bimoments* (moments of moments) that can be constructed. Here, we take two equal and opposite torsional moments $\mathbf{T}_z$ in close proximity with moment arm along the y direction, similar in concept to the eighth dipole in Figure 2.2, but replacing the two forces by torsional moments. The reason for including it here is that this dipole of dipoles, together with $\mathbf{g}_x$ (response to horizontal point load) and $\mathbf{G}_{xz}$ (dipole about y), provides a third independent solution for a point source on the axis with azimuthal variation $n = 1$, that is, of the form $[\cos\theta \quad -\sin\theta \quad \cos\theta]$. When considered in combination, these three solutions allow interesting applications, such as in the formulation of transmitting boundaries in cylindrical coordinates via boundary elements. The Green's function for the torsional bimoment is

$$\mathbf{u} = -\left\{\frac{1}{r}\left(\frac{\partial v}{\partial r} - \frac{u-v}{r}\right)\cos\theta\,\hat{\mathbf{r}} - \frac{\partial}{\partial r}\left(\frac{\partial v}{\partial r} - \frac{u-v}{r}\right)\sin\theta\,\hat{\mathbf{t}}\right\} \quad (2.23)$$

3 Two-dimensional problems in full, homogeneous spaces

3.1 Fundamental identities and definitions

$$t_P = \frac{r}{\alpha}, \qquad t_S = \frac{r}{\beta}, \qquad r = \sqrt{x^2 + z^2} \tag{3.1}$$

$$\Omega_P = \frac{\omega r}{\alpha}, \qquad \Omega_S = \frac{\omega r}{\beta}, \qquad a = \frac{\beta}{\alpha} = \sqrt{\frac{1 - 2\nu}{2(1 - \nu)}} \tag{3.2}$$

$$\gamma_i = \cos \theta_i = \frac{x_i}{r}, \qquad \gamma_{i,k} = \frac{\partial \cos \theta_i}{\partial x_k} = \frac{1}{r}(\delta_{ik} - \gamma_i \gamma_k), \qquad \frac{\partial f(r)}{\partial x_k} = \gamma_k \frac{\partial f}{\partial r} \tag{3.3}$$

3.2 Anti-plane line load (SH waves)[1]

A unit impulsive SH line source acts at $x = 0$, $z = 0$ in direction y within an infinite, homogeneous space. The receiver is at x, z. The response is given in Cartesian coordinates; see Fig. 3.2.

Frequency domain

$$g_{yy}(r, \omega) = -\frac{i}{4\mu} H_0^{(2)}(\Omega_S) \tag{3.4}$$

$$\frac{\partial g_{yy}}{\partial r} = \frac{i \Omega_S}{4\mu r} H_1^{(2)}(\Omega_S) \tag{3.5}$$

Time domain, impulse response

$$u_{yy}(r, t) = \frac{1}{2\pi \mu} \frac{\mathcal{H}(t - t_S)}{\sqrt{t^2 - t_S^2}} \tag{3.6}$$

$$\frac{\partial u_{yy}}{\partial r} = \frac{t_S^2}{2\pi \mu r} \frac{\mathcal{H}(t - t_S)}{(t^2 - t_S^2)^{3/2}} \tag{3.7}$$

---

[1] Graff, K. F., 1975, *Wave motion in elastic solids*, Ohio State University Press, pp. 285, 288, eqs. 5.2.21, 5.2.40. However, both formulas are missing a divisor $2\pi\mu$, and eq. 5.2.21 should be conjugated to conform to our sign convention.

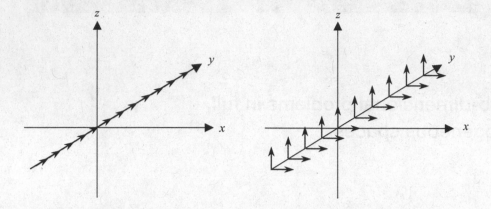

a) Anti-plane (SH) line load   b) In-plane (SV-P) line loads

Figure 3.1: Plane strain loads.

3.3 SH line load in an orthotropic space

Consider an orthotropic elastic space with shear moduli μ_x, μ_z in the plane of wave propagation. The wave equation (including the source term) and the shearing stresses are

$$\rho \frac{\partial^2 u_y}{\partial t^2} - \mu_x \frac{\partial^2 u_y}{\partial x^2} - \mu_z \frac{\partial^2 u_y}{\partial z^2} = b(x, z, t) = P\delta(x)\delta(z)\delta(t) \tag{3.8}$$

$$\tau_{xy} = \mu_x \frac{\partial u_y}{\partial x}, \qquad \tau_{yz} = \mu_z \frac{\partial u_y}{\partial z} \tag{3.9}$$

Using stretching factors λ_x, λ_z, define the scaled coordinates $\tilde{x} = \lambda_x x$ and $\tilde{z} = \lambda_z z$, by means of which the wave equation and stresses can be reduced to the equivalent

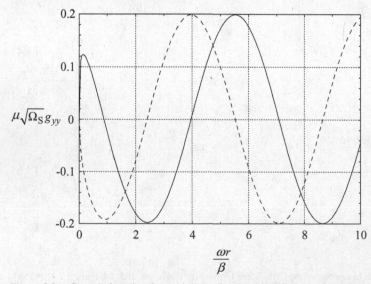

Figure 3.2a: Green's function for anti-plane line load in full space. Solid line, real part; dashed line, imaginary part.

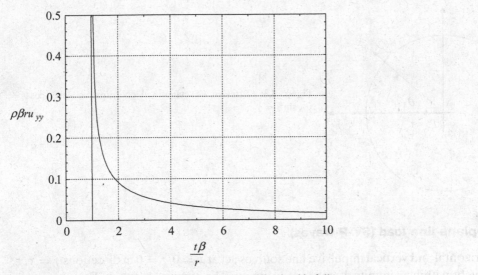

Figure 3.2b: Impulse response function for anti-plane line load in full space.

isotropic form

$$\tilde{\rho}\frac{\partial^2 u_y}{\partial t^2} - \tilde{\mu}\left[\frac{\partial^2 u_y}{\partial \tilde{x}^2} + \frac{\partial^2 u_y}{\partial \tilde{z}^2}\right] = b(\tilde{x}, \tilde{z}, t) = P\delta(\tilde{x})\delta(\tilde{z})\delta(t) \tag{3.10}$$

$$\tilde{\tau}_{xy} = \tilde{\mu}\frac{\partial u_y}{\partial \tilde{x}} = \frac{1}{\lambda_x}\sqrt{\frac{\mu_z}{\mu_x}}\tau_{xy}, \qquad \tilde{\tau}_{yz} = \tilde{\mu}\frac{\partial u_y}{\partial \tilde{z}} = \frac{1}{\lambda_z}\sqrt{\frac{\mu_x}{\mu_z}}\tau_{yz} \tag{3.11}$$

where

$$\tilde{\mu} = \sqrt{\mu_x\mu_z}, \qquad \tilde{\rho} = \frac{\rho}{\lambda_x\lambda_z}, \qquad b = \frac{b}{\lambda_x\lambda_z} \tag{3.12}$$

Choosing $\lambda_x = 1$, this implies $\lambda_z = \sqrt{\mu_x/\mu_z}$, in which case the equivalent parameters are

$$\tilde{x} = x, \qquad \tilde{z} = z\sqrt{\frac{\mu_x}{\mu_z}}, \qquad \tilde{\mu} = \sqrt{\mu_x\mu_z}, \qquad \tilde{\rho} = \rho\sqrt{\frac{\mu_z}{\mu_x}} \tag{3.13}$$

$$\tilde{b} = b\sqrt{\frac{\mu_z}{\mu_x}} = P\delta(x)\delta(\tilde{z})\delta(t) \qquad \left(\text{observe that }\int\delta(\tilde{z})\,d\tilde{z} = \int\delta(z)\,dz = 1\right) \tag{3.14}$$

$$\frac{\tilde{b}}{\tilde{\mu}} = \frac{P}{\mu_x}\delta(x)\,\delta(\tilde{z})\,\delta(t) \quad \rightarrow \quad \text{Green's functions have factor }\frac{1}{\mu_x}\text{ in lieu of }\frac{1}{\mu} \tag{3.15}$$

$$\tilde{r} = \sqrt{x^2 + \frac{\mu_x}{\mu_z}z^2}, \qquad \tilde{\beta} = \sqrt{\frac{\tilde{\mu}}{\tilde{\rho}}} = \sqrt{\frac{\mu_x}{\rho}} = \beta_x, \qquad \tilde{\Omega}_S = \frac{\omega\tilde{r}}{\beta_x}, \qquad \tilde{t}_S = \frac{\tilde{r}}{\beta_x} \tag{3.16}$$

The equations of the previous section for an isotropic medium can now be applied using the equivalent properties defined above. To obtain the stresses, observe that

$$\tau_{xy} = \mu_x\frac{\partial u_y}{\partial x} = \mu_x\frac{\partial \tilde{r}}{\partial x}\frac{\partial u_y}{\partial \tilde{r}} = \mu_x\frac{x}{\tilde{r}}\frac{\partial u_y}{\partial \tilde{r}}, \tag{3.17a}$$

$$\tau_{yz} = \mu_z\frac{\partial u_y}{\partial z} = \mu_z\frac{\partial \tilde{r}}{\partial z}\frac{\partial u_y}{\partial \tilde{r}} = \mu_x\frac{z}{\tilde{r}}\frac{\partial u_y}{\partial \tilde{r}}, \tag{3.17b}$$

where, of course, $u_y = g_{yy}(\tilde{r}, \tilde{\Omega}_S)$ or $u_y = u_{yy}(\tilde{r}, t)$. The factor μ_x in τ_{yz} is not a misprint.

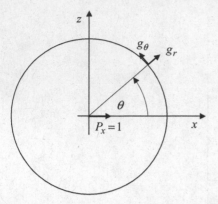

Figure 3.3: Cartesian vs. cylindrical coordinates Here, $\cos\theta = \gamma_x$, $\sin\theta = \gamma_z$.

3.4 In-plane line load (SV-P waves)

Unit horizontal and vertical impulsive line sources act at $x = 0, z = 0$ in directions $j = x, z$, respectively, within an infinite, homogeneous space. The receiver is at x, z. The response is given along the Cartesian directions $i = x, z$ and the cylindrical directions r, θ. *Note*: The pairs of functions ψ, Ψ and χ, X defined below are Fourier transforms of each other, i.e.,

$$\psi = \mathcal{F}(\Psi), \qquad \chi = \mathcal{F}(X) \tag{3.18}$$

Frequency domain[2]

$$g_{ij}(r, \omega) = \frac{1}{\mu}\left\{\psi\,\delta_{ij} + \chi\,\gamma_i\,\gamma_j\right\}, \qquad i,\,j = 1, 3 \text{ or } x, z \tag{3.19}$$

$$\frac{\partial g_{ij}}{\partial x_k} = \frac{1}{\mu}\left\{\gamma_k\left[\frac{\partial\psi}{\partial r}\delta_{ij} + \left(\frac{\partial\psi}{\partial r} - \frac{2\chi}{r}\right)\gamma_i\,\gamma_j\right] + \frac{\chi}{r}\left[\delta_{ik}\gamma_j + \delta_{jk}\gamma_i\right]\right\} \tag{3.20}$$

$$\psi = \frac{i}{4}\left\{\left[\frac{H_1^{(2)}(\Omega_S)}{\Omega_S} - \left(\frac{\beta}{\alpha}\right)^2\frac{H_1^{(2)}(\Omega_P)}{\Omega_P}\right] - H_0^{(2)}(\Omega_S)\right\} \tag{3.21}$$

$$\chi = \frac{i}{4}\left\{\left(\frac{\beta}{\alpha}\right)^2 H_2^{(2)}(\Omega_P) - H_2^{(2)}(\Omega_S)\right\} \tag{3.22}$$

$$\frac{\partial\psi}{\partial r} = \frac{\chi}{r} + \frac{i}{4r}\Omega_S H_1^{(2)}(\Omega_S) \tag{3.23}$$

$$\frac{\partial\chi}{\partial r} = \frac{i}{4r}\left[\left(\frac{\beta}{\alpha}\right)^2\Omega_P H_1^{(2)}(\Omega_P) - \Omega_S H_1^{(2)}(\Omega_S)\right] - 2\frac{\chi}{r} \tag{3.24}$$

Cylindrical components

$$g_{rx} = \frac{1}{\mu}(\psi + \chi)(\cos\theta) \tag{3.25}$$

$$g_{\theta x} = \frac{1}{\mu}\psi(-\sin\theta) \tag{3.26}$$

---

[2] Dominguez, J. and Abascal, R., 1984, On fundamental solutions for the boundary integral equations in static and dynamic elasticity, *Engineering Analysis*, Vol. 1, No. 3, pp. 128–134, eqs. 20, 28.

and more generally, for a load in direction α,

$$g_{r\alpha} = \frac{1}{\mu} (\psi + \chi) (\cos(\theta - \alpha)) \tag{3.27}$$

$$g_{\theta\alpha} = \frac{1}{\mu} \psi (-\sin(\theta - \alpha)) \tag{3.28}$$

Time domain, impulse response[3]

$$u_{ij}(r,t) = \frac{1}{\mu} \left\{ \Psi \, \delta_{ij} + X \gamma_i \, \gamma_j \right\} \tag{3.29}$$

$$\frac{\partial u_{ij}}{\partial x_k} = \frac{1}{\mu} \left\{ \gamma_k \left[\frac{\partial \Psi}{\partial r} \delta_{ij} + \left(\frac{\partial X}{\partial r} - \frac{2X}{r} \right) \gamma_i \, \gamma_j \right] + \frac{X}{r} [\delta_{ik} \gamma_j + \delta_{jk} \gamma_i] \right\} \tag{3.30}$$

$$\Psi = \frac{1}{2\pi} \left\{ \left[\frac{1}{\sqrt{t^2 - t_S^2}} + \frac{\sqrt{t^2 - t_S^2}}{t_S^2} \right] \mathcal{H}(t - t_S) - \left(\frac{\beta}{\alpha} \right)^2 \frac{\sqrt{t^2 - t_P^2}}{t_P^2} \mathcal{H}(t - t_P) \right\} \tag{3.31}$$

$$X = \frac{1}{2\pi} \left\{ \left(\frac{\beta}{\alpha} \right)^2 \left[\frac{1}{\sqrt{t^2 - t_P^2}} + \frac{2\sqrt{t^2 - t_P^2}}{t_P^2} \right] \mathcal{H}(t - t_P) \right.$$

$$\left. - \left[\frac{1}{\sqrt{t^2 - t_S^2}} + \frac{2\sqrt{t^2 - t_S^2}}{t_S^2} \right] \mathcal{H}(t - t_S) \right\} \tag{3.32}$$

$$\frac{\partial \Psi}{\partial r} = \frac{1}{2\pi r} \frac{t_S^2}{(t^2 - t_S^2)^{3/2}} \mathcal{H}(t - t_S) + \frac{X}{r} \tag{3.33}$$

$$\frac{\partial X}{\partial r} = \frac{1}{2\pi r} \left\{ \left(\frac{\beta}{\alpha} \right)^2 \frac{t_P^2}{(t^2 - t_P^2)^{3/2}} \mathcal{H}(t - t_P) - \frac{t_S^2}{(t^2 - t_S^2)^{3/2}} \mathcal{H}(t - t_S) \right\} - \frac{2X}{r} \tag{3.34}$$

Cylindrical components
  Load in direction x

$$u_{rx} = \frac{1}{\mu} (\Psi + X) (\cos \theta) \tag{3.35}$$

$$u_{\theta x} = \frac{1}{\mu} \Psi (-\sin \theta) \tag{3.36}$$

Load in direction z

$$u_{rz} = \frac{1}{\mu} (\Psi + X) (\sin \theta) \tag{3.37}$$

$$u_{\theta z} = \frac{1}{\mu} \Psi (\cos \theta) \tag{3.38}$$

Figures 3.4a–d show the displacement functions, excluding the variation with θ.

[3] Eason, G., Fulton, J., and Sneddon, I. N., 1956, The generation of waves in an infinite elastic solid by a variable body, *Philosophical Transactions of the Royal Society of London*, Vol. 248, pp. 575–607; Graff, K. F., 1975, *Wave motion in elastic solids*, Ohio State University Press, p. 293, eqs. 5.2.76–77.

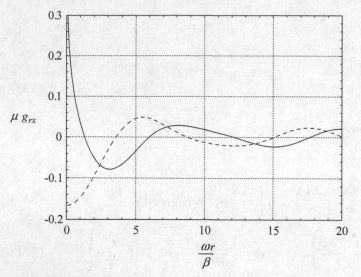

Figure 3.4a: Radial component of Green's function for in-plane line load in full space with $\nu = 0.3$. Varies as $\cos\theta$ with respect to load direction. Solid line: real part. Dashed line: imaginary part.

Strains (and stresses)

Use the expressions given previously for the derivatives $g_{ij,k}$, $u_{ij,k}$.

3.5 Dipoles in plane strain

Note: Solution for SV-P dipoles given in the frequency domain only. To obtain the time domain solution, simply replace the functions ψ, χ by Ψ, X, respectively. These functions, together with their derivatives, are defined in section 3.4. Also, to convert from Cartesian

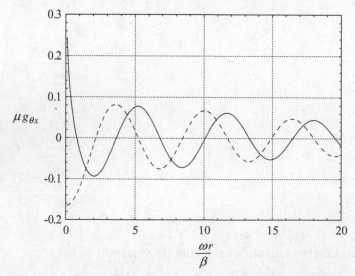

Figure 3.4b: Tangential component of Green's function for in-plane line load in full space with $\nu = 0.3$. Varies as $-\sin\theta$ with respect to load direction. Solid line: real part. Dashed line: imaginary part.

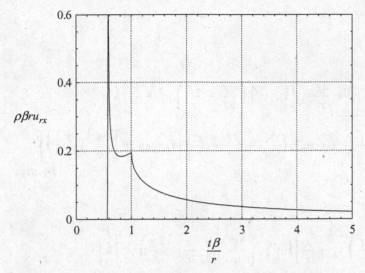

Figure 3.4c: Radial component of impulse response function for in-plane line
load in full space with $\nu = 0.3$. Varies as $\cos\theta$ with respect to load direction.

to polar coordinates we have used $2\,\gamma_x\,\gamma_z = 2\,\sin\theta\,\cos\theta = \sin 2\theta$ and $\gamma_x^2 - \gamma_z^2 = \cos^2\theta - \sin^2\theta = \cos 2\theta$, see Fig. 3.3.

Single dipoles (Fig. 2.2)

Anti-plane dipoles (SH waves)

$$
\mathbf{G}_{yj} = \begin{cases} -\gamma_j \dfrac{\partial g_{yy}}{\partial r}\hat{\mathbf{j}} = -\gamma_j \dfrac{\mathrm{i}\,\Omega_S}{4\mu} H_1^{(2)}(\Omega_S)\hat{\mathbf{j}} \\[3mm] -\gamma_j \dfrac{\partial u_{yy}}{\partial r}\hat{\mathbf{j}} = \dfrac{t_S^2\,\mathcal{H}(t - t_S)}{2\pi\,\mu r\sqrt{t^2 - t_S^2}}\hat{\mathbf{j}} \end{cases} \qquad j = x, z \qquad (3.39)
$$

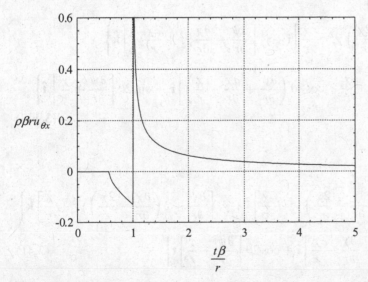

Figure 3.4d: Tangential component of impulse response function for in-plane line
load in full space with $\nu = 0.3$. Varies as $-\sin\theta$ with respect to load direction.

In-plane dipoles (SV-P waves)

$$\mathbf{G}_{xx} = -\left\{ \frac{\partial g_{xx}}{\partial x}\hat{\mathbf{i}} + \frac{\partial g_{zx}}{\partial x}\hat{\mathbf{k}} \right\}$$

$$= -\frac{1}{\mu}\left\{ \gamma_x \left[\frac{\partial \psi}{\partial r} + \frac{\partial \chi}{\partial r}\gamma_x^2 + \frac{2\chi}{r}\gamma_z^2 \right]\hat{\mathbf{i}} + \gamma_z \left[\left(\frac{\partial \chi}{\partial r} - \frac{2\chi}{r} \right)\gamma_x^2 + \frac{\chi}{r} \right]\hat{\mathbf{k}} \right\}$$

$$= -\frac{1}{2\mu}\left\{ \left[\frac{\partial \psi}{\partial r} + \frac{\partial \chi}{\partial r} + \frac{\chi}{r} + \cos 2\theta \left(\frac{\partial \psi}{\partial r} + \frac{\partial \chi}{\partial r} - \frac{\chi}{r} \right) \right]\hat{\mathbf{r}} - \sin 2\theta \left[\frac{\partial \psi}{\partial r} + \frac{\chi}{r} \right]\hat{\mathbf{t}} \right\}$$

$$\text{(3.40a)}$$

$$\mathbf{G}_{zx} = -\left\{ \frac{\partial g_{xz}}{\partial x}\hat{\mathbf{i}} + \frac{\partial g_{zz}}{\partial x}\hat{\mathbf{k}} \right\}$$

$$= -\frac{1}{\mu}\left\{ \gamma_z \left[\left(\frac{\partial \chi}{\partial r} - \frac{2\chi}{r} \right)\gamma_x^2 + \frac{\chi}{r} \right]\hat{\mathbf{i}} + \gamma_x \left[\frac{\partial \psi}{\partial r} + \left(\frac{\partial \chi}{\partial r} - \frac{2\chi}{r} \right)\gamma_z^2 \right]\hat{\mathbf{k}} \right\}$$

$$= -\frac{1}{2\mu}\left\{ \sin 2\theta \left[\frac{\partial \psi}{\partial r} + \frac{\partial \chi}{\partial r} - \frac{\chi}{r} \right]\hat{\mathbf{r}} + \left[\frac{\partial \psi}{\partial r} - \frac{\chi}{r} + \cos 2\theta \left(\frac{\partial \psi}{\partial r} + \frac{\chi}{r} \right) \right]\hat{\mathbf{t}} \right\} \quad \text{(3.40b)}$$

$$\mathbf{G}_{xz} = -\left\{ \frac{\partial g_{xx}}{\partial z}\hat{\mathbf{i}} + \frac{\partial g_{zx}}{\partial z}\hat{\mathbf{k}} \right\}$$

$$= -\frac{1}{\mu}\left\{ \gamma_z \left[\frac{\partial \psi}{\partial r} + \left(\frac{\partial \chi}{\partial r} - \frac{2\chi}{r} \right)\gamma_x^2 \right]\hat{\mathbf{i}} + \gamma_x \left[\left(\frac{\partial \chi}{\partial r} - \frac{2\chi}{r} \right)\gamma_z^2 + \frac{\chi}{r} \right]\hat{\mathbf{k}} \right\}$$

$$= -\frac{1}{2\mu}\left\{ \sin 2\theta \left[\frac{\partial \psi}{\partial r} + \frac{\partial \chi}{\partial r} - \frac{\chi}{r} \right]\hat{\mathbf{r}} + \left[-\left(\frac{\partial \psi}{\partial r} - \frac{\chi}{r} \right) + \cos 2\theta \left(\frac{\partial \psi}{\partial r} + \frac{\chi}{r} \right) \right]\hat{\mathbf{t}} \right\}$$

$$\text{(3.40c)}$$

$$\mathbf{G}_{zz} = -\left\{ \frac{\partial g_{xz}}{\partial z}\hat{\mathbf{i}} + \frac{\partial g_{zz}}{\partial z}\hat{\mathbf{k}} \right\}$$

$$= -\frac{1}{\mu}\left\{ \gamma_x \left[\left(\frac{\partial \chi}{\partial r} - \frac{2\chi}{r} \right)\gamma_z^2 + \frac{\chi}{r} \right]\hat{\mathbf{i}} + \gamma_z \left[\frac{\partial \psi}{\partial r} + \frac{\partial \chi}{\partial r}\gamma_z^2 + \frac{2\chi}{r}\gamma_x^2 \right]\hat{\mathbf{k}} \right\}$$

$$= -\frac{1}{2\mu}\left\{ \left[\frac{\partial \psi}{\partial r} + \frac{\partial \chi}{\partial r} + \frac{\chi}{r} - \cos 2\theta \left(\frac{\partial \psi}{\partial r} + \frac{\partial \chi}{\partial r} - \frac{\chi}{r} \right) \right]\hat{\mathbf{r}} + \sin 2\theta \left[\frac{\partial \psi}{\partial r} + \frac{\chi}{r} \right]\hat{\mathbf{t}} \right\}$$

$$\text{(3.40d)}$$

Double couple (Fig. 2.4)

$$\mathbf{M}_y = \mathbf{G}_{xz} + \mathbf{G}_{zx}$$

$$= -\frac{1}{\mu}\left\{ \gamma_z \left[\frac{\partial \psi}{\partial r} + 2\left(\frac{\partial \chi}{\partial r} - \frac{2\chi}{r} \right)\gamma_x^2 + \frac{\chi}{r} \right]\hat{\mathbf{i}} + \gamma_x \left[\frac{\partial \psi}{\partial r} + 2\left(\frac{\partial \chi}{\partial r} - \frac{2\chi}{r} \right)\gamma_z^2 + \frac{\chi}{r} \right]\hat{\mathbf{k}} \right\}$$

$$= -\frac{1}{2\mu}\left\{ \sin 2\theta \left[\frac{\partial \psi}{\partial r} + \frac{\partial \chi}{\partial r} - \frac{\chi}{r} \right]\hat{\mathbf{r}} + \cos 2\theta \left[\frac{\partial \psi}{\partial r} + \frac{\chi}{r} \right]\hat{\mathbf{t}} \right\} \tag{3.41}$$

Torsional line source (Fig. 2.3)

$$\mathbf{T}_y = \frac{1}{2} \left(\mathbf{G}_{xz} - \mathbf{G}_{zx} \right)$$

$$= -\frac{1}{2\mu} \left\{ \left[\frac{\partial \psi}{\partial r} - \frac{\chi}{r} \right] \left[\gamma_z \hat{\mathbf{i}} - \gamma_x \hat{\mathbf{k}} \right] \right\}$$

$$= -\frac{1}{2\mu} \left\{ -\left[\frac{\partial \psi}{\partial r} - \frac{\chi}{r} \right] \hat{\mathbf{t}} \right\} \tag{3.42}$$

Note: $\mathbf{T}_y$ is a vector along the y direction (i.e., into the paper), which implies a torsional moment that is positive in the clockwise direction, that is, in the $-\hat{\mathbf{t}}$ tangential direction.

Frequency domain

$$\mathbf{T}_y = \frac{i}{8\mu r} \Omega_S H_1^{(2)}(\Omega_S) \hat{\mathbf{t}} \tag{3.43}$$

Time domain, impulse response

$$\mathbf{T}_y = \frac{1}{4\pi \mu r} \frac{t_S^2}{\left(t^2 - t_S^2\right)^{3/2}} \mathcal{H}(t - t_S) \hat{\mathbf{t}} \tag{3.44}$$

Blast load (compressive pressure)

See also the next Section 3.6, and the explanation for the factor $(\alpha/\beta)^2$ in Section 2.5.

$$\mathbf{B}_{2D} = \left(\frac{\alpha}{\beta} \right)^2 \left(\mathbf{G}_{xx} + \mathbf{G}_{zz} \right)$$

$$= -\frac{1}{\mu} \left(\frac{\alpha}{\beta} \right)^2 \left[\frac{\partial \psi}{\partial r} + \frac{\partial \chi}{\partial r} + \frac{\chi}{r} \right] \left[\gamma_x \hat{\mathbf{i}} + \gamma_z \hat{\mathbf{k}} \right]$$

$$= -\frac{1}{\mu} \left(\frac{\alpha}{\beta} \right)^2 \left[\frac{\partial \psi}{\partial r} + \frac{\partial \chi}{\partial r} + \frac{\chi}{r} \right] \hat{\mathbf{r}} \tag{3.45}$$

Frequency domain

$$\mathbf{B}_{2D} = -\frac{i}{4\mu r} \Omega_P H_1^{(2)}(\Omega_P) \hat{\mathbf{r}} \tag{3.46}$$

Time domain, impulse response

$$\mathbf{B}_{2D} = -\frac{1}{2\pi \mu r} \frac{t_P^2}{\left(t^2 - t_P^2\right)^{3/2}} \mathcal{H}(t - t_P) \hat{\mathbf{r}} \tag{3.47}$$

3.6 Line blast source: suddenly applied pressure

A line blast source acts within an infinitesimally small cylindrical cavity. The compressive pressure of the line source has an implied space–time variation $p(r, t) = \mathcal{H}(t)\, \delta(r)/2\pi r$, which, when integrated over the infinitesimal area of the cavity at times $t > 0$, gives a blast source of unit strength. This line blast source produces only radial displacements that consist of cylindrically propagating P waves. The response given is for the radial displacement, positive away from the source.

Frequency domain

The displacement field caused by a harmonically pulsating line of pressure was shown in the previous Section 3.5 to be given by

$$u_r = \frac{\Omega_P}{4 \, i \, \mu r} H_1^{(2)}(\Omega_P), \qquad \Omega_P = \frac{\omega r}{\alpha} \tag{3.48}$$

which is formally the frequency response function that corresponds to an impulsive line of pressure. Observe, however, that in a strict sense, this expression does not have an inverse Fourier transform, because its amplitude grows with frequency, and it thus fails the Dirichlet conditions. Hence, the impulse response given previously is true only in the sense of a distribution. To obtain the frequency response for a suddenly applied pressure (i.e., a step load), we divide by $i\omega$ and obtain

$$G_r(\omega, r) = -\frac{1}{4 \, \mu \, \alpha} H_1^{(2)}(\Omega_P) \tag{3.49}$$

$$\frac{\partial G_r}{\partial r} = \frac{1}{4 \, \mu \, \alpha r} \left[H_1^{(2)}(\Omega_P) - \Omega_P H_0^{(2)}(\Omega_P) \right] \tag{3.50}$$

Time domain, unit step load response

$$U_r(t, R) = \frac{1}{2\pi \mu r} \frac{t}{\sqrt{t^2 - t_P^2}} \mathcal{H}(t - t_P), \qquad t_P = \frac{r}{\alpha} \tag{3.51}$$

$$\frac{\partial U_r}{\partial r} = \frac{t}{2\pi \mu r^2} \left[\frac{t_P^2}{\left(t^2 - t_P^2\right)^{3/2}} - \frac{1}{\sqrt{t^2 - t_P^2}} \right] \mathcal{H}(t - t_P) \tag{3.52}$$

3.7 Cylindrical cavity subjected to pulsating pressure[4]

A cylindrical cavity of radius r_0 is subjected to an internal compressive internal pressure $p(t)$ with Fourier transform $p(\omega) = \int_{-\infty}^{+\infty} p(t) \, e^{-i\omega t} \, dt$, which elicits radial displacements consisting of cylindrically propagating P waves. Response given is for the radial displacement, positive away from the source.

[4] Eringen, A. C. and Suhubi, S. S. (1975), *Elastodynamics*, Academic Press, Vol II, pp. 490–492.

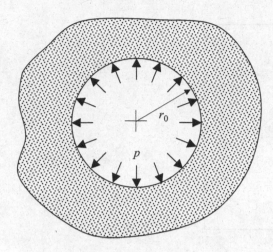

Figure 3.5: Cylindrical cavity subjected to pressure.

Frequency domain

The displacement field caused by a harmonically pulsating pressure $p(\omega)$ at a distance $r \geq r_0$ is

$$g_r(\omega) = \frac{r_0}{\mu} \frac{p(\omega) \, H_1^{(2)}(\Omega_P)}{2 H_1^{(2)}(\Omega_0) - a^{-2} \Omega_0 H_0^{(2)}(\Omega_0)},$$

$$a^{-2} = \frac{\alpha}{\beta}, \qquad \Omega_P = \frac{\omega r}{\alpha}, \qquad \Omega_0 = \frac{\omega r_0}{\alpha} \tag{3.53}$$

which depends on Poisson's ratio because of the factor a.

Observe that if $p(\omega) = 1/\pi r_0^2$, and we consider the limit of a vanishingly small cavity, i.e. $r_0 \to 0$, then $2 H_1^{(2)}(\Omega_0) \to 4i/(\pi\, \Omega_0) = 4i\, r/(\pi r_0)/\Omega_P$, $\Omega_P H_0^{(2)}(\Omega_0) \to 0$. In this case, the above formula converges to that of the line of pressure given in the previous section.

Time domain

No closed-form expression exists. However, a purely numerical solution is still possible by Fourier synthesis:

$$u_r(r, r_0, t) = \frac{1}{2\pi} \int_{-\infty}^{+\infty} g_r(\omega) \, e^{i\omega t} d\omega$$

$$= \frac{r_0}{2\pi\mu} \int_{-\infty}^{+\infty} \frac{p(\omega) \, H_1^{(2)}(\Omega_P) e^{i\omega t}}{2 H_1^{(2)}(\Omega_0) - a^{-2} \Omega_0 H_0^{(2)}(\Omega_0)} d\omega \tag{3.54}$$

but this equation is surprisingly difficult to evaluate numerically, especially when the pressure changes abruptly, because in that case the integrand falls off rather slowly. Indeed, for an impulsive pressure $p(\omega) = 1$, it can be shown that $g_r(\omega)$ behaves asymptotically as

$$g_r(\omega) \, e^{i\omega t} = \frac{\alpha\, a^2}{\mu} \sqrt{\frac{r_0}{r}} \frac{e^{i\omega t'}}{i\omega} \tag{3.55}$$

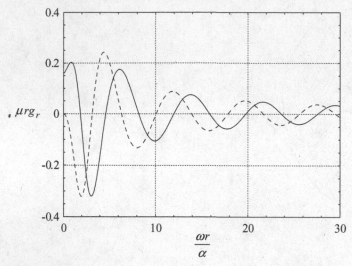

Figure 3.6a: Cylindrical cavity in full space with $\nu = 0.3$: frequency response at $r/r_0 = 5$ for pressure $p\pi r_0^2 = 1$. Solid line, real part; dashed line, imaginary part.

with $t' = t - t_0$, where $t_0 = (r - r_0)/\alpha$ is the time of arrival of P waves at the receiver station. From this asymptotic structure and the properties of Fourier transforms,[5] we infer that the magnitude of the response immediately behind the wave front is (see also Fig. 3.6b)

$$u(r, r_0, t_0^+) = \lim_{\omega \to \infty} \left[i\omega\, g_r(\omega)\, e^{i\omega t_0} \right] = \frac{\alpha\, a^2}{\mu} \sqrt{\frac{r_0}{r}} = \frac{1}{\rho\alpha} \sqrt{\frac{r_0}{r}} \tag{3.56}$$

with t_0^+ the instant immediately *after* passage of the wave front, and ρ the mass density.

A strategy to evaluate the impulse response would be to work in delayed time $t' = t - t_0 \geq 0$, use the asymptotic expansion for the Hankel functions beginning at an appropriately large frequency ω_0 (say, $\Omega_0 > 2\pi$, that is, $\omega_0 = 2\pi\alpha/r_0$), and express the contribution of the two tails in terms of the sine integral, for which accurate and efficient polynomial approximations exist:

$$u_r(r, r_0, t) = \frac{r_0}{2\pi\mu} \int_{-\omega_0}^{+\omega_0} \left[\frac{H_1^{(2)}(\Omega_P)\, e^{i\omega t_0}}{2 H_1^{(2)}(\Omega_0) - a^{-2}\Omega_0 H_0^{(2)}(\Omega_0)} \right] e^{i\omega(t - t_0)}\, d\omega$$

$$+ \frac{1}{2\rho\alpha} \sqrt{\frac{r_0}{r}} \left[1 - \frac{2}{\pi} \mathrm{Si}\left(\omega_0(t - t_0) \right) \right] \tag{3.57}$$

Observe that the kernel of the above integral tends to a constant value at zero frequency. Also, the denominator is never zero for any frequency or Poisson's ratio, which means that the integrand has no real poles. Figures 3.6a,b show the results for an impulsive pressure of strength $p\,\pi r_0^2 = \delta(t)$ at a dimensionless distance $r/r_0 = 5$ in a full space with Poisson's ratio $\nu = 0.30$. This pressure is consistent with that of a line of pressure in Section 3.6, which facilitates comparisons.

[5] Papoulis, A., 1962, *The Fourier integral and its applications*, McGraw-Hill.

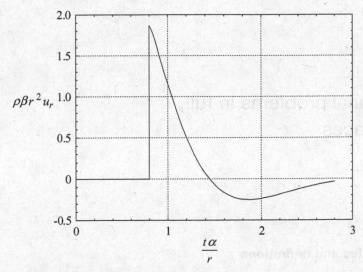

Figure 3.6b: Cylindrical cavity in full space with $\nu = 0.3$, Impulse response at $r/r_0 = 5$ for pressure $p\pi r_0^2 = \delta(t)$.

Observe that the magnitude of the discontinuity in Fig. 3.6b agrees with the formula

$$\frac{\rho\beta r^2}{\pi r_0^2}\frac{1}{\rho\alpha}\sqrt{\frac{r_0}{r}} = \frac{1}{\pi}\frac{\beta}{\alpha}\left(\frac{r}{r_0}\right)^{3/2} = 1.9 \tag{3.58}$$

This expression is obtained by multiplying the displacement scaling factor, the applied pressure, and the jump amplitude behind the wave front obtained previously from the asymptotic properties of Fourier transforms.

4 Three-dimensional problems in full, homogeneous spaces

4.1 Fundamental identities and definitions

$$t_P = \frac{R}{\alpha}, \qquad t_S = \frac{R}{\beta}, \qquad R = \sqrt{x^2 + y^2 + z^2} \tag{4.1}$$

$$\Omega_P = \frac{\omega R}{\alpha}, \qquad \Omega_S = \frac{\omega R}{\beta}, \qquad a = \frac{\beta}{\alpha} = \sqrt{\frac{1 - 2\nu}{2(1 - \nu)}} \tag{4.2}$$

$$\gamma_i = \cos\theta_i = \frac{x_i}{R}, \qquad \gamma_{i,k} = \frac{\partial \cos\theta_i}{\partial x_k} = \frac{1}{R}\left(\delta_{ik} - \gamma_i\,\gamma_k\right), \qquad \frac{\partial f(R)}{\partial x_k} = \gamma_k \frac{\partial f}{\partial R} \tag{4.3}$$

4.2 Point load (Stokes problem)

Unit impulsive point sources act at the origin, in any coordinate direction, within an infinite, homogeneous space. The receiver is at x_1, x_2, x_3.

Frequency domain[1]

$$g_{ij}(R, \omega) = \frac{1}{4\pi\mu R}\{\psi\,\delta_{ij} + \chi\,\gamma_i\,\gamma_j\}, \qquad i, j = 1, 2, 3 \,(\text{or } x, y, z) \tag{4.4}$$

$$\frac{\partial g_{ij}}{\partial x_k} = \frac{1}{4\pi\mu R}\left\{\gamma_k\left[\left(\frac{\partial\psi}{\partial R} - \frac{\psi}{R}\right)\delta_{ij} + \left(\frac{\partial\chi}{\partial R} - \frac{3\chi}{R}\right)\gamma_i\,\gamma_j\right] + \frac{\chi}{R}[\delta_{ik}\gamma_j + \delta_{jk}\gamma_i]\right\} \tag{4.5}$$

$$\psi = e^{-i\Omega_P}\left(\frac{\beta}{\alpha}\right)^2\left\{\frac{i}{\Omega_P} + \frac{1}{\Omega_P^2}\right\} + e^{-i\Omega_S}\left\{1 - \frac{i}{\Omega_S} - \frac{1}{\Omega_S^2}\right\} \tag{4.6}$$

$$\chi = e^{-i\Omega_P}\left(\frac{\beta}{\alpha}\right)^2\left\{1 - \frac{3i}{\Omega_P} - \frac{3}{\Omega_P^2}\right\} - e^{-i\Omega_S}\left\{1 - \frac{3i}{\Omega_S} - \frac{3}{\Omega_S^2}\right\} \tag{4.7}$$

---

[1] Dominguez, J. and Abascal, R., 1984, On fundamental solutions for the boundary integral equations in static and dynamic elasticity, *Engineering Analysis*, Vol. 1, No. 3, pp. 128–134, eqs. 20, 21.

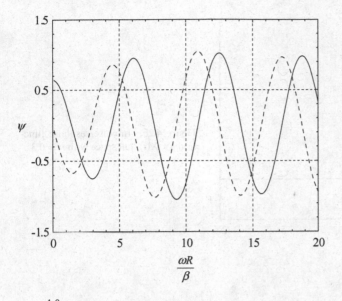

Figure 4.1a: Green's function for $\psi(\omega)$, full space with $\nu = 0.3$. Solid (dashed) line, real (imaginary) part.

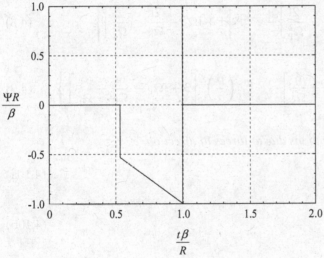

Figure 4.1b: Impulse response function for $\Psi(t)$, full space with $\nu = 0.3$.

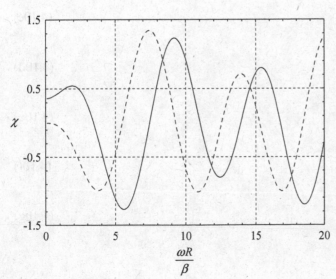

Figure 4.2a: Green's function for $\chi(\omega)$, full space with $\nu = 0.3$. Solid (dashed) line, real (imaginary) part.

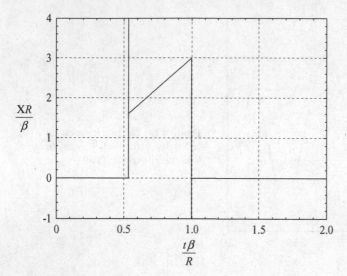

$\dfrac{XR}{\beta}$

$\dfrac{t\beta}{R}$

Figure 4.2b: Impulse response function for $X(t)$, full space with $\nu = 0.3$.

$$\frac{\partial \psi}{\partial R} = \frac{1}{R}\left\{ e^{-i\Omega_P}\left(\frac{\beta}{\alpha}\right)^2\left[1 - \frac{2i}{\Omega_P} - \frac{2}{\Omega_P^2}\right] - e^{-i\Omega_S}\left[1 + i\Omega_S - \frac{2i}{\Omega_S} - \frac{2}{\Omega_S^2}\right]\right\} \tag{4.8}$$

$$\frac{\partial \chi}{\partial R} = \frac{1}{R}\left\{ e^{-i\Omega_S}\left[3 + i\Omega_S - \frac{6i}{\Omega_S} - \frac{6}{\Omega_S^2}\right] - e^{-i\Omega_P}\left(\frac{\beta}{\alpha}\right)^2\left[3 + i\Omega_P - \frac{6i}{\Omega_P} - \frac{6}{\Omega_P^2}\right]\right\} \tag{4.9}$$

Spherical coordinates, Green's functions due to forces in directions x, y, z

$$g_{Rx} = \frac{\psi + \chi}{4\pi\mu R}\sin\phi\cos\theta, \tag{4.10a}$$

$$g_{\phi x} = \frac{\psi}{4\pi\mu R}\cos\phi\cos\theta, \tag{4.10b}$$

$$g_{\theta x} = \frac{\psi}{4\pi\mu R}(-\sin\theta), \tag{4.10c}$$

$$g_{Ry} = \frac{\psi + \chi}{4\pi\mu R}\sin\phi\sin\theta, \tag{4.10d}$$

$$g_{\phi y} = \frac{\psi}{4\pi\mu R}\cos\phi\sin\theta, \tag{4.10e}$$

$$g_{\theta y} = \frac{\psi}{4\pi\mu R}\cos\theta, \tag{4.10f}$$

$$g_{Rz} = \frac{\psi + \chi}{4\pi\mu R}\cos\phi, \tag{4.10g}$$

$$g_{\phi z} = \frac{\psi}{4\pi\mu R}(-\sin\phi), \tag{4.10h}$$

$$g_{\theta z} = 0 \tag{4.10i}$$

Time domain[2]

$$u_{ij} = \frac{1}{4\pi\mu R}\{\Psi\delta_{ij} + X\gamma_i\gamma_j\} \tag{4.11}$$

$$\frac{\partial u_{ij}}{\partial x_k} = \frac{1}{4\pi\mu R}\left\{\gamma_k\left[\left(\frac{\partial\Psi}{\partial R} - \frac{\Psi}{R}\right)\delta_{ij} + \left(\frac{\partial X}{\partial R} - \frac{3X}{R}\right)\gamma_i\gamma_j\right] + \frac{X}{R}[\delta_{ik}\gamma_j + \delta_{jk}\gamma_i]\right\} \tag{4.12}$$

Spherical coordinates, forces in directions x, y, z:

$$u_{Rx} = \frac{\Psi + X}{4\pi\mu R}\sin\phi\cos\theta, \tag{4.13a}$$

$$u_{\phi x} = \frac{\Psi}{4\pi\mu R}\cos\phi\cos\theta, \tag{4.13b}$$

$$u_{\theta x} = \frac{\Psi}{4\pi\mu R}(-\sin\theta), \tag{4.13c}$$

$$u_{Ry} = \frac{\Psi + X}{4\pi\mu R}\sin\phi\sin\theta, \tag{4.13d}$$

$$u_{\phi y} = \frac{\Psi}{4\pi\mu R}\cos\phi\sin\theta, \tag{4.13e}$$

$$u_{\theta y} = \frac{\Psi}{4\pi\mu R}\cos\theta, \tag{4.13f}$$

$$u_{Rz} = \frac{\Psi + X}{4\pi\mu R}\cos\phi, \tag{4.13g}$$

$$u_{\phi z} = \frac{\Psi}{4\pi\mu R}(-\sin\phi), \tag{4.13h}$$

$$u_{\theta z} = 0 \tag{4.13i}$$

Impulse response

$$\Psi = \mathcal{F}^{-1}[\psi(\omega)] = \delta(t - t_S) - t\left[\left(\frac{\beta}{\alpha}\right)^2\frac{\mathcal{H}(t - t_P)}{t_P^2} - \frac{\mathcal{H}(t - t_S)}{t_S^2}\right] \tag{4.14}$$

[2] Aki, K. and Richards, P. G., 1980, *Quantitative seismology: theory and methods*, W. H. Freeman, 1980, Vol. I, p. 73, eq. 4.23.

$$X = \mathcal{F}^{-1}\left[\chi(\omega)\right] = \left(\frac{\beta}{\alpha}\right)^2 \delta(t - t_{\mathrm{P}}) - \delta(t - t_{\mathrm{S}})$$

$$+ 3t\left[\left(\frac{\beta}{\alpha}\right)^2 \frac{\mathcal{H}(t - t_{\mathrm{P}})}{t_{\mathrm{P}}^2} - \frac{\mathcal{H}(t - t_{\mathrm{S}})}{t_{\mathrm{S}}^2}\right] \tag{4.15}$$

$$\frac{\partial \Psi}{\partial R} = \frac{1}{\alpha}\frac{\partial \psi}{\partial t_{\mathrm{P}}} + \frac{1}{\beta}\frac{\partial \psi}{\partial t_{\mathrm{S}}} \tag{4.16}$$

$$\frac{\partial X}{\partial R} = \frac{1}{\alpha}\frac{\partial X}{\partial t_{\mathrm{P}}} + \frac{1}{\beta}\frac{\partial X}{\partial t_{\mathrm{S}}} \tag{4.17}$$

Arbitrary forcing function $\delta(x)\,\delta(y)\,\delta(z)\,P(t)$

$$Q(t, t_{\mathrm{P}}, t_{\mathrm{S}}) = \frac{\beta^2}{R^2}\int_{t_{\mathrm{P}}}^{t_{\mathrm{S}}} \tau\, P(t - \tau)\, d\tau \tag{4.18}$$

$$\Psi = P(t - t_{\mathrm{S}}) - Q(t, t_{\mathrm{P}}, t_{\mathrm{S}}) \tag{4.19}$$

$$X = \left(\frac{\beta}{\alpha}\right)^2 P(t - t_{\mathrm{P}}) - P(t - t_{\mathrm{S}}) + 3\, Q(t, t_{\mathrm{P}}, t_{\mathrm{S}}) \tag{4.20}$$

$$\frac{\partial P(t - t_{\mathrm{P}})}{\partial R} = -\frac{1}{\alpha}\frac{\partial P(t - t_{\mathrm{P}})}{\partial t} = -\frac{\dot{P}(t - t_{\mathrm{P}})}{\alpha}, \tag{4.21}$$

$$\frac{\partial P(t - t_{\mathrm{S}})}{\partial R} = -\frac{1}{\beta}\frac{\partial P(t - t_{\mathrm{S}})}{\partial t} = -\frac{\dot{P}(t - t_{\mathrm{S}})}{\beta} \tag{4.22}$$

$$\frac{\partial Q(t, t_{\mathrm{P}}, t_{\mathrm{S}})}{\partial R} = \frac{\beta^2}{R^2}\left[\frac{t_{\mathrm{S}}}{\beta}P(t - t_{\mathrm{S}}) - \frac{t_{\mathrm{P}}}{\alpha}P(t - t_{\mathrm{P}}) - \frac{2}{R}\int_{t_{\mathrm{P}}}^{t_{\mathrm{S}}} \tau\, P(t - \tau)\, d\tau\right]$$

$$= \frac{1}{R}(Q - X) \tag{4.23}$$

$$\frac{\partial \Psi}{\partial R} = \frac{1}{R}(X - Q) - \frac{1}{\beta}\frac{\partial P(t - t_{\mathrm{S}})}{\partial t} \tag{4.24}$$

$$\frac{\partial X}{\partial R} = -\left(\frac{\beta}{\alpha}\right)^2 \frac{1}{\alpha}\frac{\partial P(t - t_{\mathrm{P}})}{\partial t} + \frac{1}{\beta}\frac{\partial P(t - t_{\mathrm{S}})}{\partial t} + 3\frac{Q - X}{R}$$

$$= 2\frac{Q - X}{R} - \frac{\partial \Psi}{\partial R} - \left(\frac{\beta}{\alpha}\right)^2 \frac{1}{\alpha}\frac{\partial P(t - t_{\mathrm{P}})}{\partial t} \tag{4.25}$$

4.3 Tension cracks

A tension point dipole is applied at the origin; see Fig. 2.2. Response functions are listed only in the frequency domain; for time domain solutions, simply replace ψ, χ by the Ψ, X

given earlier in Section 4.2:

$$\mathbf{G}_{xx} = -g_{ix,x}\hat{\mathbf{e}}_i = -\frac{1}{4\pi\mu R}\left\{\left[\frac{\partial\psi}{\partial R} - \frac{\psi}{R} + \frac{\chi}{R}\right]\delta_{ix}\gamma_x + \left[\frac{\partial\chi}{\partial R} - \frac{3\chi}{R}\right]\gamma_i\gamma_x^2 + \frac{\chi}{R}\gamma_i\right\}\hat{\mathbf{e}}_i$$

$$(4.26)$$

and similar expressions for yy, zz, replacing indices as appropriate. The functions in brackets needed in these expressions are listed at the end of this section. In spherical coordinates, the displacement components are:

Crack xx

$$G_{Rxx} = -\frac{1}{4\pi\mu R}\left\{\left[\frac{\partial\psi}{\partial R} + \frac{\partial\chi}{\partial R} - \frac{\psi}{R} - \frac{2\chi}{R}\right]\sin^2\phi\,\cos^2\theta + \frac{\chi}{R}\right\}$$

$$(4.27a)$$

$$G_{\phi xx} = -\frac{1}{4\pi\mu R}\left[\frac{\partial\psi}{\partial R} - \frac{\psi}{R} + \frac{\chi}{R}\right]\tfrac{1}{2}\sin 2\phi\,\cos^2\theta$$

$$(4.27b)$$

$$G_{\theta xx} = -\frac{1}{4\pi\mu R}\left[\frac{\partial\psi}{\partial R} - \frac{\psi}{R} + \frac{\chi}{R}\right]\left(-\tfrac{1}{2}\sin\phi\,\sin 2\theta\right)$$

$$(4.27c)$$

Crack yy

$$G_{Ryy} = -\frac{1}{4\pi\mu R}\left\{\left[\frac{\partial\psi}{\partial R} - \frac{\psi}{R} + \frac{\partial\chi}{\partial R} - \frac{2\chi}{R}\right]\sin^2\phi\,\sin^2\theta + \frac{\chi}{R}\right\}$$

$$(4.28a)$$

$$G_{\phi yy} = -\frac{1}{4\pi\mu R}\left[\frac{\partial\psi}{\partial R} - \frac{\psi}{R} + \frac{\chi}{R}\right]\tfrac{1}{2}\sin 2\phi\,\sin^2\theta$$

$$(4.28b)$$

$$G_{\theta yy} = -\frac{1}{4\pi\mu R}\left[\frac{\partial\psi}{\partial R} - \frac{\psi}{R} + \frac{\chi}{R}\right]\tfrac{1}{2}\sin\phi\,\sin 2\theta$$

$$(4.28c)$$

Crack zz

$$G_{Rzz} = -\frac{1}{4\pi\mu R}\left\{\left[\frac{\partial\psi}{\partial R} - \frac{\psi}{R} + \frac{\partial\chi}{\partial R} - \frac{2\chi}{R}\right]\cos^2\phi + \frac{\chi}{R}\right\}$$

$$(4.29a)$$

$$G_{\phi zz} = -\frac{1}{4\pi\mu R}\left[\frac{\partial\psi}{\partial R} - \frac{\psi}{R} + \frac{\chi}{R}\right]\left(-\tfrac{1}{2}\sin 2\phi\right)$$

$$(4.29b)$$

$$G_{\theta zz} = 0$$

$$(4.29c)$$

In the preceding expressions, the factors in brackets are as follows:

Frequency domain

$$\frac{\partial\psi}{\partial R} - \frac{\psi}{R} + \frac{\chi}{R} = \frac{1}{R}\left\{\left(\frac{\beta}{\alpha}\right)^2 e^{-i\Omega_P}\left[2 - \frac{6i}{\Omega_P} - \frac{6}{\Omega_P^2}\right] - e^{-i\Omega_S}\left[3 + i\Omega_S - \frac{6i}{\Omega_S} - \frac{6}{\Omega_S^2}\right]\right\}$$

$$(4.30)$$

$$\frac{\partial \chi}{\partial R} - \frac{3\chi}{R} = \frac{1}{R} \left\{ e^{-i\Omega_S} \left[6 + i\,\Omega_S - \frac{15\,i}{\Omega_S} - \frac{15}{\Omega_S^2} \right] - \left(\frac{\beta}{\alpha} \right)^2 e^{-i\Omega_P} \left[6 + i\,\Omega_P - \frac{15\,i}{\Omega_P} - \frac{15}{\Omega_P^2} \right] \right\}$$

(4.31)

$$\frac{\partial \psi}{\partial R} - \frac{\psi}{R} + \frac{\partial \chi}{\partial R} - \frac{2\chi}{R} = \frac{1}{R} \left\{ e^{-i\Omega_S} \left[3 - \frac{9\,i}{\Omega_S} - \frac{9}{\Omega_S^2} \right] - \left(\frac{\beta}{\alpha} \right)^2 e^{-i\Omega_P} \left[4 + i\,\Omega_P - \frac{9\,i}{\Omega_P} - \frac{9}{\Omega_P^2} \right] \right\}$$

(4.32)

Time domain

Source function $= \delta(x)\,\delta(y)\,\delta(z)\,M(t)$ (4.33)

$$Q(t, t_P, t_S) = \frac{\beta^2}{R^2} \int_{t_P}^{t_S} \tau\,M(t - \tau)\,d\tau$$

(4.34)

$$\frac{\partial \Psi}{\partial R} - \frac{\Psi}{R} + \frac{X}{R} = 2 \left(\frac{\beta}{\alpha} \right)^2 \frac{M(t - t_P)}{R} - \left[3\frac{M(t - t_S)}{R} + \frac{1}{\beta}\frac{\partial M(t - t_S)}{\partial t} \right] + 6\frac{Q}{R}$$

(4.35)

$$\frac{\partial X}{\partial R} - \frac{3X}{R} = \left[6\frac{M(t - t_S)}{R} + \frac{1}{\beta}\frac{\partial M(t - t_S)}{\partial t} \right] - \left(\frac{\beta}{\alpha} \right)^2 \left[6\frac{M(t - t_P)}{R} + \frac{1}{\alpha}\frac{\partial M(t - t_P)}{\partial t} \right] - 15\frac{Q}{R}$$

(4.36)

$$\frac{\partial \Psi}{\partial R} - \frac{\Psi}{R} + \frac{\partial X}{\partial R} - \frac{2X}{R} = 3\frac{M(t - t_S)}{R} - \left(\frac{\beta}{\alpha} \right)^2 \left[4\frac{M(t - t_P)}{R} + \frac{1}{\alpha}\frac{\partial M(t - t_P)}{\partial t} \right] - 9\frac{Q}{R}$$

(4.37)

4.4 Double couples (seismic moments)

A seismic moment (point source) of unit strength is applied at the origin. Response functions are listed only in the frequency domain; for a time domain solution, simply replace ψ, χ by their Fourier transforms Ψ, X. See previous Section 4.3 for the expressions in brackets needed in either case. With reference to Fig. 2.4, we have

Cartesian coordinates

$$\mathbf{M}_{jk} = M_{ijk}\hat{\mathbf{e}}_i = -\left(g_{ij,k} + g_{ik,j} \right) \hat{\mathbf{e}}_i$$

$$= -\frac{1}{4\pi\mu R} \left\{ \left[\frac{\partial \psi}{\partial R} - \frac{\psi}{R} + \frac{\chi}{R} \right] (\delta_{ij}\gamma_k + \delta_{ik}\gamma_j) + \left[\frac{\partial \chi}{\partial R} - \frac{3\chi}{R} \right] 2\gamma_i\gamma_j\gamma_k + \frac{2\chi}{R}\delta_{jk}\gamma_i \right\} \hat{\mathbf{e}}_i$$

(4.38)

Observe that when $j = k$, $\mathbf{M}_{kk} = 2\mathbf{G}_{kk}$ we recover twice the response functions for tension cracks. For seismic moments, the last term vanishes.

Spherical coordinates

Note: Instead of the double indices jk identifying the planes in which the double couples lie, we use in the following a single index identifying the direction of the normal to these planes. Also, observe that the M_{jk} are *displacements* due to a seismic moment, and not moments.

Double couple in plane yz (normal x)

$$M_{Rx} = -\frac{1}{4\pi\mu R}\left[\frac{\partial\psi}{\partial R} - \frac{\psi}{R} + \frac{\partial\chi}{\partial R} - \frac{2\chi}{R}\right]\sin 2\phi\,\sin\theta \tag{4.39a}$$

$$M_{\phi x} = -\frac{1}{4\pi\mu R}\left[\frac{\partial\psi}{\partial R} - \frac{\psi}{R} + \frac{\chi}{R}\right]\cos 2\phi\,\sin\theta \tag{4.39b}$$

$$M_{\theta x} = -\frac{1}{4\pi\mu R}\left[\frac{\partial\psi}{\partial R} - \frac{\psi}{R} + \frac{\chi}{R}\right]\cos\phi\,\cos\theta \tag{4.39c}$$

Double couple in plane xz (normal y)

$$M_{Ry} = -\frac{1}{4\pi\mu R}\left[\frac{\partial\psi}{\partial R} - \frac{\psi}{R} + \frac{\partial\chi}{\partial R} - \frac{2\chi}{R}\right]\sin 2\phi\,\cos\theta \tag{4.40a}$$

$$M_{\phi y} = -\frac{1}{4\pi\mu R}\left[\frac{\partial\psi}{\partial R} - \frac{\psi}{R} + \frac{\chi}{R}\right]\cos 2\phi\,\cos\theta \tag{4.40b}$$

$$M_{\theta y} = -\frac{1}{4\pi\mu R}\left[\frac{\partial\psi}{\partial R} - \frac{\psi}{R} + \frac{\chi}{R}\right](-\cos\phi\,\sin\theta) \tag{4.40c}$$

Double couple in plane xy (normal z)

$$M_{Rz} = -\frac{1}{4\pi\mu R}\left[\frac{\partial\psi}{\partial R} - \frac{\psi}{R} + \frac{\partial\chi}{\partial R} - \frac{2\chi}{R}\right]\sin^2\phi\,\sin 2\theta \tag{4.41a}$$

$$M_{\phi z} = -\frac{1}{4\pi\mu R}\left[\frac{\partial\psi}{\partial R} - \frac{\psi}{R} + \frac{\chi}{R}\right]\frac{1}{2}\sin 2\phi\,\sin 2\theta \tag{4.41b}$$

$$M_{\theta z} = -\frac{1}{4\pi\mu R}\left[\frac{\partial\psi}{\partial R} - \frac{\psi}{R} + \frac{\chi}{R}\right]\sin\phi\,\cos 2\theta \tag{4.41c}$$

4.5 Torsional point source

A torsional point source of unit strength is applied at the origin. Response functions are listed only in the frequency domain; for a time domain solution, replace ψ, χ by their

Fourier transforms Ψ, X, or use the expressions listed below for an arbitrary source. With reference to Fig. 2.3 we have:

Cartesian coordinates

Note: Choose $jk = yz, zx, xy$ for a positive (right-handed) torsional moment, and observe that the reversed sequence $jk = zy, xz, yx$ leads to a negative moment, and $j = k$ gives zero.

$$\mathbf{T}_{jk} = T_{ijk}\hat{\mathbf{e}}_i = -\frac{1}{8\pi\mu R}\left[\frac{\partial\psi}{\partial R} - \frac{\psi}{R} - \frac{\chi}{R}\right](\delta_{ik}\gamma_j - \delta_{ij}\gamma_k)\,\hat{\mathbf{e}}_i \tag{4.42}$$

$$\frac{\partial\psi}{\partial R} - \frac{\psi}{R} - \frac{\chi}{R} = -\frac{1}{R}e^{-i\Omega_S}(1 + i\,\Omega_S) \tag{4.43}$$

$$\frac{\partial\Psi}{\partial R} - \frac{\Psi}{R} - \frac{X}{R} = -\left[\frac{M(t - t_S)}{R} + \frac{1}{\beta}\frac{\partial M(t - t_S)}{\partial t}\right] \tag{4.44}$$

Spherical coordinates

Note: Instead of the double indices jk identifying the planes in which the torsional moment lies, we use in the following a single index identifying the direction of the normal to these planes. Also, observe that the T_{jk} are the *displacements* due to unit torsional moments.

Torsional moment in plane yz (normal x)

$$T_{Rx} = 0 \tag{4.45a}$$

$$T_{\phi x} = -\frac{1}{8\pi\mu R}\left[\frac{\partial\psi}{\partial R} - \frac{\psi}{R} - \frac{\chi}{R}\right](-\sin\theta) \tag{4.45b}$$

$$T_{\theta x} = -\frac{1}{8\pi\mu R}\left[\frac{\partial\psi}{\partial R} - \frac{\psi}{R} - \frac{\chi}{R}\right](-\cos\phi\cos\theta) \tag{4.45c}$$

Torsional moment in plane xz (normal y)

$$T_{Ry} = 0 \tag{4.46a}$$

$$T_{\phi y} = -\frac{1}{8\pi\mu R}\left[\frac{\partial\psi}{\partial R} - \frac{\psi}{R} - \frac{\chi}{R}\right]\cos\theta \tag{4.46b}$$

$$T_{\theta y} = -\frac{1}{8\pi\mu R}\left[\frac{\partial\psi}{\partial R} - \frac{\psi}{R} - \frac{\chi}{R}\right](-\cos\phi\sin\theta) \tag{4.46c}$$

Torsional moment in plane xy (normal z)

$$T_{Rz} = T_{\phi z} = 0 \tag{4.47a}$$

$$T_{\theta z} = -\frac{1}{8\pi\mu R}\left[\frac{\partial\psi}{\partial R} - \frac{\psi}{R} - \frac{\chi}{R}\right]\sin\phi \tag{4.47b}$$

See also the next section.

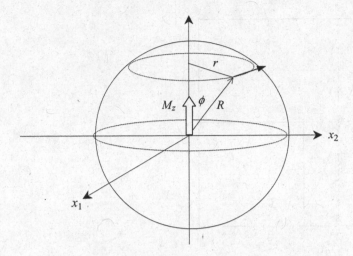

Figure 4.3: Torque about vertical axis.

4.6 Torsional point source with vertical axis

Elicits SH waves that propagate spherically. Particle motions are tangential to horizontal circles. Response given in spherical coordinates.

Preliminary definitions

$$\Omega_S = \frac{\omega R}{\beta}, \qquad t_S = \frac{R}{\beta}, \qquad \tau = \frac{t\beta}{R} \tag{4.48}$$

$$\sin\phi = \frac{r}{R}, \qquad \cos\phi = \frac{z}{R} \tag{4.49}$$

Frequency domain

$$u_\theta = M_z(\omega) \frac{\sin\phi}{8\pi\mu R^2} (1 + i\,\Omega_S)\, e^{-i\Omega_S} \tag{4.50}$$

Time domain, arbitrary causal variation of source with time

$$u_\theta = \frac{\sin\phi}{8\pi\mu} \left[\frac{M_z(t-t_S)}{R^2} + \frac{1}{\beta R} \frac{\partial M_z(t-t_S)}{\partial t} \right] \tag{4.51}$$

or in dimensionless time

$$u_\theta = \frac{\sin\phi}{8\pi\mu R^2} \left[M_z(\tau-1) + \frac{\partial}{\partial\tau} M_z(\tau-1) \right] \tag{4.52}$$

Do notice, however, that in the dimensionless form, τ alone is not sufficient to fully define the response, because the source depends in addition on some other parameters. Thus, the dimensionless response $u_\theta(\tau)\, R^2$ is not invariant with respect to R. For example, for

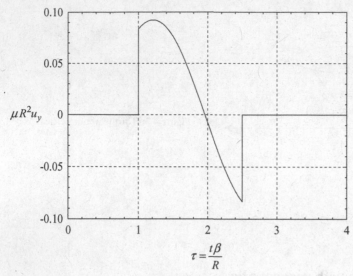

$$\tau = \frac{t\beta}{R}$$

Figure 4.4: Response due to torsional pulse $M_z = \sin \pi t/t_d, 0 \le t \le t_d = 1.5R/\beta$.

a source $M_z = \sin \pi (t/t_d) = \sin \pi (\tau/\tau_d)$, with t_d being the duration, the derivative term contains a factor $\pi/\tau_d = \pi R/(\beta t_d)$, which changes the relative mix of the two terms. The response for such a source would be

$$\mu R^2 u_\theta (\tau, \tau_d) = \frac{\sin \phi}{8\pi} \left[\sin(\tau - 1) + \frac{\pi}{\tau_d} \cos(\tau - 1) \right] \left[\mathcal{H}(\tau - 1) - \mathcal{H}(\tau + \tau_d - 1) \right]$$

$$(4.53)$$

The two Heaviside terms arise from the discontinuities in the source. Figure 4.4 shows the response due to a single pulse $M_z = \sin \pi (\tau/\tau_d)$ when $\tau_d = 1.5$ ($\tau_s = 1$ is the arrival time). Observe that this response is *not* a delayed, scaled replica of the source.

More generally, for an arbitrary torsional source $M(t)$ of finite duration t_d, the formulas above imply that the wave field is entirely contained within two concentric spheres of constant separation βt_d, and that within this region it consists of two distinct wavelets that decay at *different rates* with increasing distance to the source. Since the total wave energy flux must be constant, this means that energy is gradually being transferred from one wavelet to the other, although it is not a priori obvious how this occurs. The explanation is that the individual wavelets cannot exist independently of one another, but both are needed to satisfy the wave equation in spherical coordinates, that is, there are not two wavelets, but only one. One implication is that the shape of the torsional wave pulse evolves as it moves into the far field, and looks different there from what it does in the near field. This can have practical consequences, for example, in the interpretation of cross-hole tests used to measure shear-wave velocities by means of torsional sources.

4.7 Point blast source

Such a source elicits solely P waves that propagate spherically. Particle motions are in the radial direction.

A singularly large pressure transient acts within an infinitesimally small, spherical cavity. The source has an implied space–time variation $p(R, t) = b(t)\, \delta(R)/4\pi R^2$, which, when integrated over the volume of the cavity, reduces to the source strength $b(t)$ used here. The response is given in spherical coordinates.

Frequency domain

$$u_R = b(\omega) \frac{3}{16\pi\mu R^2} (1 + i\,\Omega_P)\, e^{-i\,\Omega_P}, \qquad \Omega_P = \frac{\omega R}{\alpha} \tag{4.54}$$

Time domain, arbitrary causal variation of source with time:

$$u_R = \frac{3}{16\pi\mu R} \left[\frac{b(t - t_P)}{R} + \frac{1}{\alpha} \frac{\partial b(t - t_P)}{\partial t} \right], \qquad t_P = \frac{R}{\alpha} \tag{4.55}$$

Observe that, except for the wave speed, the lack of variation with polar angle, and the scaling factor, the response is very similar to that of a torsional point source. Remarkably, the scaling factor does not depend on the constrained modulus $\lambda + 2\mu$ (i.e., on Poisson's ratio) but solely on the shear modulus μ. By contrast, if the solution to this problem were found by superposition of the three tensile cracks in an infinite space, the same solution would be obtained, but with a factor $1/4(\lambda + 2\mu)$ instead of $3/16\mu$. This is because of the effect of the infinitesimally small particle that fills the cavity, which arises when the solution is found using the dipoles for tensile cracks. Contrary to intuition, the particle's effect does not vanish as its size shrinks to zero.[3]

4.8 Spherical cavity subjected to arbitrary pressure

A harmonic or transient pressure acts within a spherical cavity of radius R_0. This pressure elicits P waves that propagate spherically. Particle motions are in the radial direction. In contrast to the difficult-to-solve 2-D case of a cylindrical cavity, the 3-D case is amenable to closed-form solution.

a) Motion of cavity wall

Frequency domain

Frequency response function

$$g_0(\omega) = p(\omega) \frac{R_0}{4\mu} \left[\frac{1 + i\,\Omega_0}{1 + i\,\Omega_0 - (\Omega_0\,\alpha/2\beta)^2} \right], \qquad \Omega_0 = \frac{\omega R_0}{\alpha} \tag{4.56}$$

Notice that the factor in square brackets is identical to the transfer function due to support motion of a single degree of freedom oscillator with natural frequency ω_n, fraction

[3] Kausel, E., 1998, Blast loads versus point loads: the missing factor, *Journal of Engineering Mechanics*, Vol. 124, No. 2 (Feb.), pp. 243–244.

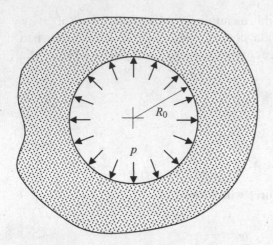

Figure 4.5: Spherical cavity subjected to pressure.

of viscous critical damping ξ, and damped frequency ω_d:

$$\frac{1 + i\,\Omega_0}{1 + i\,\Omega_0 - \frac{1}{4}\left(\frac{\Omega_0\,\alpha}{\beta}\right)^2} = \frac{1 + 2\xi i\dfrac{\omega}{\omega_n}}{1 + 2\xi i\dfrac{\omega}{\omega_n} - \left(\dfrac{\omega}{\omega_n}\right)^2} \tag{4.57}$$

This implies an equivalent oscillator with the following parameters:

$$\omega_n = \frac{2\beta}{R_0}, \qquad \xi = \frac{\beta}{\alpha},$$

$$\omega_d = \omega_n\sqrt{1 - \xi^2} = \frac{2\beta}{R_0}\sqrt{1 - \left(\frac{\beta}{\alpha}\right)^2} = \frac{2\beta}{R_0\sqrt{2(1 - \nu)}} \tag{4.58}$$

Observe that the higher the Poisson's ratio, the smaller the critical damping and thus the larger the response will be. Also, it is remarkable that the resonant frequency should depend only on the shear modulus, not the dilatational (constrained) modulus, so it is independent of Poisson's ratio. The above analogy leads us immediately to the impulse response function $u_0(t)$ given below.

Time domain

Impulse response function

$$u_0(t) = \frac{R_0}{4\mu}\frac{\omega_n^2}{\omega_d}\,e^{-\xi\omega_n t}\cos\left(\omega_d t - \phi\right)\mathcal{H}(t)$$

$$= \frac{1}{2\rho\beta}\sqrt{2(1 - \nu)}\,e^{-\xi\omega_n t}\cos\left(\omega_d t - \phi\right)\mathcal{H}(t) \tag{4.59}$$

$$\sin\phi = \frac{\nu}{1 - \nu} = \frac{\lambda}{\lambda + 2\mu}, \qquad \cos\phi = \frac{\sqrt{1 - 2\nu}}{1 - \nu} \tag{4.60}$$

Response to arbitrary pressure

$$u_R(t, R_0) = u_0 * p = \int_0^t u_0(\tau)\,p(t - \tau)\,d\tau \tag{4.61}$$

b) Response beyond the cavity

Frequency domain

Frequency response function

$$g_R(\omega) = g_0(\omega) \frac{h_1^{(2)}(\Omega_P)}{h_1^{(2)}(\Omega_0)}$$

$$= g_0(\omega) \left(\frac{R_0}{R}\right)^2 \frac{1 + i\Omega_P}{1 + i\Omega_0} e^{-i(R/R_0 - 1)\Omega_0}, \qquad \Omega_0 = \frac{\omega R_0}{\alpha}, \qquad \Omega_P = \frac{\omega R}{\alpha}$$

$$= p(\omega) R_0^3 \frac{1}{4\mu R^2} \left[\frac{1 + i\Omega_P}{1 + i\Omega_0 - (\Omega_0 \alpha/2\beta)^2} \right] e^{-i(R/R_0 - 1)\Omega_0} \tag{4.62}$$

with $h_1^{(2)}$ the second spherical Hankel function of first order, and $g_0(\omega)$ the frequency response function at the cavity wall, as before.

For a harmonic pressure of the form $p(\omega)V_0 = 1$, in which $V_0 = \frac{4}{3}\pi R_0^3$ is the volume of the cavity (i.e., with the same strength as a point blast load), the frequency response function is

$$g_R(\omega) = \frac{3}{16\pi \mu R^2} \left[\frac{1 + i\Omega_P}{1 + i\Omega_0 - (\Omega_0 \alpha/2\beta)^2} \right] e^{-i(\Omega_P - \Omega_0)} \tag{4.63}$$

Setting in this formula $\Omega_0 = 0$, we recover the point blast source. A plot of this response function is given in Fig. 4.6a.

Time domain

We obtain this response by considering a fictitious blast point source in an infinite medium beginning to act at an earlier time $t_0 = R_0/\alpha$, and matching the displacement that it causes at $R = R_0$ with the displacement produced by the impulsive pressure on the cavity wall, that is,

$$\frac{3}{16\pi \mu R_0} \left[\frac{b(t)}{R_0} + \frac{1}{\alpha} \frac{\partial b(t)}{\partial t} \right] = \frac{1}{2\rho\beta} \sqrt{2(1 - \nu)} e^{-\xi \omega_n t} \cos\left[\omega_d t - \phi\right] \mathcal{H}(t) \tag{4.64}$$

The solution of this differential equation is

$$b(t) = \frac{16\pi \mu}{3} \frac{R_0^2}{2\rho\beta} \sqrt{2(1 - \nu)} e^{-\xi \omega_n t} \sin(\omega_d t) \, \mathcal{H}(t) \tag{4.65}$$

The impulse response function at points beyond the cavity wall is then

$$u_R = \frac{3}{16\pi \mu} \frac{1}{R} \left[\frac{b(t - t')}{R} + \frac{1}{\alpha} \frac{\partial b(t - t')}{\partial t} \right] \tag{4.66}$$

with

$$t' = \frac{R - R_0}{\alpha} \tag{4.67}$$

from which we obtain the response given below due to a unit impulsive pressure.

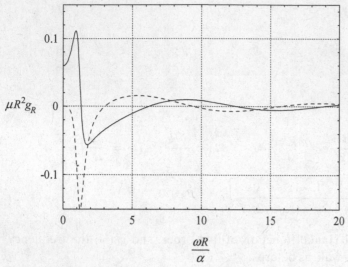

Figure 4.6a: Spherical cavity of radius R_0 in a full space with $\nu = 0.45$ subjected to a harmonic pressure $p\frac{4}{3}\pi R_0^3 = 1$. Response at $R/R_0 = 2$. Solid line, real part; dashed line, imaginary part.

Impulse response function

$$u_R(t) = \frac{\sqrt{2(1-\nu)}}{2\rho\beta}\, e^{-\xi\omega_n(t-t')}\frac{R_0}{R}\left\{\cos\left[\omega_d(t-t')-\phi\right] - \left(1 - \frac{R_0}{R}\right)\right.$$

$$\left. \times \sin\left[\omega_d(t-t')\right]\right\}\mathcal{H}(t-t') \tag{4.68}$$

Observe that $t' = 0$ for $R = R_0$, in which case we recover the impulse response function at the cavity wall. Notice that the phase lag does not enter into the sine term. By an appropriate

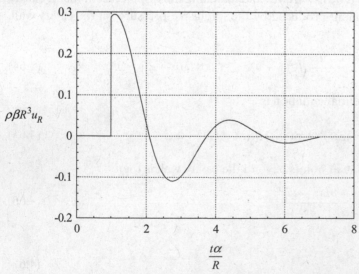

Figure 4.6b: Spherical cavity of radius R_0 in a full space with $\nu = 0.45$ subjected to an impulsive pressure $p\frac{4}{3}\pi R_0^3 = \delta(t)$. Response at $R/R_0 = 2$.

convolution, we can also obtain the response due to a pressure with an arbitrary variation in time.

The magnitude of the discontinuity at the wave front is obtained by setting $t = t'$, which gives

$$u_R(t') = \frac{1}{\rho\alpha}\frac{R_0}{R} \tag{4.69}$$

which is similar to that of the cylindrical cavity (Section 3.7), except that it decays faster with increasing R.

In the special case of an impulsive pressure of unit strength $pV = \delta(t)$, i.e., $p = \delta(t)/\frac{4}{3}\pi R_0^3$, the impulse response function is

$$u_R(t) = \frac{3\sqrt{2(1-\nu)}}{8\pi\rho\beta R_0^2 R} e^{-\xi\omega_n(t-t')}\left\{\cos\left[\omega_d(t-t') - \phi\right] - \left(1 - \frac{R_0}{R}\right)\right.$$

$$\left. \times \sin\left[\omega_d(t-t')\right]\right\}\mathcal{H}(t-t') \tag{4.70}$$

A plot of this response function is given in Fig. 4.6b.

4.9 Spatially harmonic line source (2½-D problem)[4]

Consider an infinite, homogeneous three-dimensional space subjected at the origin of coordinates to a harmonic, spatially sinusoidal line load of the form $b_j(x) = \delta(x)\delta(y)\exp\mathrm{i}(\omega t - k_z z)$ acting in the jth spatial direction ($j = 1, 2, 3$, or x,y,z), as illustrated in Fig. 4.7. This is often referred to as the 2.5-D problem.[5] By an appropriate inverse Fourier transformation, the response to loads with arbitrary spatial variation in z can be obtained, including the response to moving loads.[6]

Preliminary definitions

$A = \dfrac{1}{4\,\mathrm{i}\rho\omega^2}$	Amplitude	(4.71)
$r = \sqrt{x^2 + y^2} \equiv \sqrt{x_1^2 + x_2^2}$	Source–receiver distance	(4.72)
$\gamma_i = \dfrac{\partial r}{\partial x_i} = \dfrac{x_i}{r}, \quad i = 1,2$	Direction cosines in the x,y plane	(4.73)
$\delta_{ii} = \delta_{11} + \delta_{22} = 2, \quad \gamma_i\gamma_i = \gamma_1^2 + \gamma_2^2 = 1$	Implied summations	(4.74)
$\gamma_{i,j} = \dfrac{1}{r}(\delta_{ij} - \gamma_i\gamma_j)$	First derivatives of γ_i, i, $j = 1,2$	(4.75)

[4] Tadeu, A. and E. Kausel, 2000, Green's functions for two-and-a-half dimensional elastodynamic problems, *Journal of Engineering Mechanics*, Vol. 126, No. 10 (Oct.), pp. 1093–1097.

[5] Papageorgiou, A. S. and Pei, D., 1998, A discrete wavenumber boundary element method for study of the 3-D response of 2-D scatterers, *Earthquake Engineering and Structural Dynamics*, Vol. 27, pp. 619–638.

[6] Pedersen, H. A., Sánchez-Sesma, F. J., and Campillo, M., 1994, Three-dimensional scattering by two-dimensional topographies, *Bulletin of the Seismological Society of America*, Vol. 84, pp. 1169–1183.

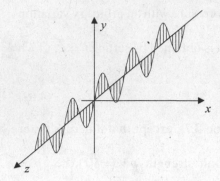

Figure 4.7: Spatially harmonic line source in a full 3-D space.

$$\gamma_{i,jk} = \frac{1}{r^2}\left(3\gamma_i\,\gamma_j\,\gamma_k - \gamma_i\,\delta_{jk} - \gamma_j\,\delta_{ki} - \gamma_k\,\delta_{ij}\right) \qquad \text{Second derivatives} \qquad (4.76)$$

$$k_z \qquad\qquad\qquad\qquad\qquad\qquad\qquad \text{Wavenumber in } z \text{ direction} \qquad (4.77)$$

$$k_{\mathrm{P}} = \frac{\omega}{\alpha} \qquad\qquad\qquad\qquad\qquad \text{Wavenumber for P waves} \qquad (4.78)$$

$$k_{\mathrm{S}} = \frac{\omega}{\beta} \qquad\qquad\qquad\qquad\qquad \text{Wavenumber for S waves} \qquad (4.79)$$

$$k_\alpha = \sqrt{k_{\mathrm{P}}^2 - k_z^2}, \qquad k_\beta = \sqrt{k_{\mathrm{S}}^2 - k_z^2} \qquad \text{More wavenumbers} \qquad (4.80)$$

$$H_{n\alpha} = H_n^{(2)}(k_\alpha r), \qquad H_{n\beta} = H_n^{(2)}(k_\beta r) \qquad \text{Shorthand for Hankel functions} \qquad (4.81)$$

$$H_{0\beta,l} = -k_\beta\,\gamma_l\,H_{1\beta} \qquad\qquad\qquad \begin{array}{l}\text{Derivative of Hankel function} \\ \text{in the plane } (l = 1,2)\end{array} \qquad (4.82)$$

$$H_{0\beta,z} = -\mathrm{i}\,k_z\,H_{0\beta} \qquad\qquad\qquad \begin{array}{l}\text{Derivative of Hankel function} \\ \text{in the } z \text{ direction}\end{array} \qquad (4.83)$$

$$B_n = k_\beta^n\,H_{n\beta} - k_\alpha^n\,H_{n\alpha} \qquad\qquad \begin{array}{l}B_n \text{ functions (see end of section} \\ \text{for definitions and properties)}\end{array} \qquad (4.84)$$

$$\hat{\nabla}^2 = \frac{\partial^2}{\partial x^2} + \frac{\partial^2}{\partial y^2} - k_z^2 \qquad\qquad \text{Laplacian} \qquad (4.85)$$

Green's functions

Cartesian coordinates

The displacement in direction i due to a unit load applied in direction j, which has an implied factor $\exp\mathrm{i}(\omega t - k_z z)$, can be written compactly as $g_{ij} = A(\delta_{ij}\,k_s^2\,H_{0\beta} + B_{0,ij})$, or in full

$$g_{xx} = A\left[k_s^2 H_{0\beta} - \frac{1}{r}B_1 + \gamma_x^2\,B_2\right] \qquad (4.86a)$$

$$g_{yy} = A\left[k_s^2 H_{0\beta} - \frac{1}{r}B_1 + \gamma_y^2\,B_2\right] \qquad (4.86b)$$

$$g_{zz} = A\left[k_s^2 H_{0\beta} - k_z^2 B_0\right] \tag{4.86c}$$

$$g_{xy} = g_{yx} = \gamma_x \gamma_y A B_2 \tag{4.86d}$$

$$g_{xz} = g_{zx} = i k_z \gamma_x A B_1 \tag{4.86e}$$

$$g_{yz} = g_{zy} = i k_z \gamma_y A B_1 \tag{4.86f}$$

Cylindrical coordinates

Let r be the source-receiver distance in the x, y plane, with θ being the angle between r and x, so that $\cos\theta = \gamma_x$, $\sin\theta = \gamma_y$, $\gamma_z = 0$. Also, define the parameters u, v, w, U, W as

$$u = A\left(k_S^2 H_{0\beta} - \frac{1}{r} B_1 + B_2\right), \tag{4.87a}$$

$$v = A\left(k_S^2 H_{0\beta} - \frac{1}{r} B_1\right) \tag{4.87b}$$

$$w = Ai\, k_z B_1, \tag{4.87c}$$

$$U = Ai\, k_z B_1 \tag{4.87d}$$

$$W = A\left(k_S^2 H_{0\beta} - k_z^2 B_0\right) \tag{4.87e}$$

The radial, tangential, and axial components of the Green's functions are then

$$g_{rx} = u\,(\cos\theta), \qquad g_{ry} = u\,(\sin\theta), \qquad g_{rz} = U \tag{4.88a}$$

$$g_{\theta x} = v\,(-\sin\theta), \qquad g_{\theta y} = v\,(\cos\theta), \qquad g_{\theta z} = 0 \tag{4.88b}$$

$$g_{zx} = w\,(\cos\theta), \qquad g_{zy} = w\,(\sin\theta), \qquad g_{zz} = W \tag{4.88c}$$

These can be used in turn to obtain the Green's functions for 2.5-D dipoles, torsional moments, and seismic moments, using for this purpose the expressions given in Section 2.6. For example, the Green's function for a 2.5-D blast load in a homogeneous full space is

$$B_{2.5\text{-D}} = -\left\{\left(\frac{\partial u}{\partial r} + \frac{u - v}{r} + \frac{\partial U}{\partial z}\right)\hat{\mathbf{r}} + \left(\frac{\partial w}{\partial r} + \frac{w}{r} + \frac{\partial W}{\partial z}\right)\hat{\mathbf{k}}\right\} \tag{4.89}$$

except that we have omitted here the factor $3(\lambda + 2\mu)/4\mu$ associated with the presence of an infinitesimal cavity within which the blast load acts, because there is no such thing as a 2.5-D cavity. Thus, when the above expression is specialized to the case of a line load ($k_z = 0 \Rightarrow k_\beta = k_S$), we recover the Green function $B_{2\text{-D}}$ for a line blast source of Section 3.5, but multiplied by $\mu/(\lambda + 2\mu)$. Similarly, a Fourier transform could be used to recover the point blast source $B_{3\text{-D}}$ of section 4.7, but multiplied by $4\mu/3(\lambda + 2\mu)$. Thus, the above expression is for a full solid, including the infinitesimal particle. Observe also that we have changed the axial derivatives with respect to the source into axial derivatives with respect to the receiver, i.e., $\partial/\partial z' = -\partial/\partial z$, since the space is homogeneous. All other Green's functions for dipoles are as in section 2.6, so they need not be repeated.

Strain components ($l = x, y, z = direction\ of\ load$)

$$\varepsilon^l_{(ii)} = g_{il,(i)} = k_S^2\ A\delta_{il}\ H_{0\beta,(i)} + A\,B_{0,(ii)l}, \qquad i = x, y, z \tag{4.90a}$$

$$\varepsilon^l_{ij} = g_{il,j} + g_{jl,i} = k_S^2\ A(\delta_{il}\ H_{0\beta,j} + \delta_{jl}\ H_{0\beta,i}) + 2A\,B_{0,ijl}, \qquad i \neq j \tag{4.90b}$$

$$\varepsilon^l_{\mathrm{Vol}} = g_{xl,x} + g_{yl,y} + g_{zl,z} = A\frac{\partial}{\partial x_l}\left[k_S^2 H_{0\beta} + \hat{\nabla}^2 B_0\right] \tag{4.90c}$$

a) Strains for loads in the plane, $l = x, y$

$$\varepsilon^l_{xx} = \gamma_l\ A\left(\left(\frac{2}{r}B_2 - k_S^2\,k_\beta\,H_{1\beta}\right)\delta_{xl} + \frac{1}{r}B_2 - \gamma_x^2\,B_3\right) \tag{4.91a}$$

$$\varepsilon^l_{yy} = \gamma_l\ A\left(\left(\frac{2}{r}B_2 - k_S^2\,k_\beta\,H_{1\beta}\right)\delta_{yl} + \frac{1}{r}B_2 - \gamma_y^2\,B_3\right) \tag{4.91b}$$

$$\varepsilon^l_{zz} = \gamma_l\,k_z^2\,A\,B_1 \tag{4.91c}$$

$$\varepsilon^l_{xy} = 2A\left(\left(\frac{1}{r}B_2 - \frac{1}{2}k_S^2\,k_\beta\,H_{1\beta}\right)(\delta_{xl}\,\gamma_y + \delta_{yl}\,\gamma_x) - \gamma_x\,\gamma_y\,\gamma_l\,B_3\right) \tag{4.91d}$$

$$\varepsilon^l_{xz} = 2\mathrm{i}\,k_z\,A\left(\left(\frac{1}{r}B_1 - \frac{1}{2}k_S^2\,H_{0\beta}\right)\delta_{xl} - \gamma_x\,\gamma_l\,B_2\right) \tag{4.91e}$$

$$\varepsilon^l_{yz} = 2\mathrm{i}\,k_z\,A\left(\left(\frac{1}{r}B_1 - \frac{1}{2}k_S^2\,H_{0\beta}\right)\delta_{yl} - \gamma_y\,\gamma_l\,B_2\right) \tag{4.91f}$$

$$\varepsilon^l_{\mathrm{Vol}} = \gamma_l\,A\left(-k_S^2\,k_\beta\,H_{1\beta} + k_z^2\,B_1 + \frac{4}{r}B_2 - B_3\right) \tag{4.91g}$$

b) Strain for axial loads, $l = z$

$$\varepsilon^z_{xx} = \mathrm{i}\,k_z\,A\left(\frac{1}{r}B_1 - \gamma_x^2\,B_2\right) \tag{4.92a}$$

$$\varepsilon^z_{yy} = \mathrm{i}\,k_z\,A\left(\frac{1}{r}B_1 - \gamma_y^2\,B_2\right) \tag{4.92b}$$

$$\varepsilon^z_{zz} = \mathrm{i}\,k_z\,A(-k_S^2\,H_{0\beta} + k_z^2\,B_0) \tag{4.92c}$$

$$\varepsilon^z_{xy} = -2\mathrm{i}\,k_z\,\gamma_x\,\gamma_y\,A\,B_2 \tag{4.92d}$$

$$\varepsilon^z_{xz} = 2\gamma_x\,A\left(-\frac{1}{2}k_S^2\,k_\beta\,H_{1\beta} + k_z^2\,B_1\right) \tag{4.92e}$$

$$\varepsilon^z_{yz} = 2\gamma_y\,A\left(-\frac{1}{2}k_S^2\,k_\beta\,H_{1\beta} + k_z^2\,B_1\right) \tag{4.92f}$$

$$\varepsilon^z_{\mathrm{Vol}} = \mathrm{i}\,k_z\,A\left(-k_S^2\,H_{0\beta} + k_z^2\,B_0 + \frac{2}{r}B_1 - B_2\right) \tag{4.92g}$$

The B_n functions

Definitions

$$B_0 = H_0^{(2)}(k_\beta r) - H_0^{(2)}(k_\alpha r) \tag{4.93a}$$

$$B_n = k_\beta^n H_n^{(2)}(k_\beta r) - k_\alpha^n H_n^{(2)}(k_\alpha r) \qquad \text{with } k_\alpha^n \equiv (k_\alpha)^n, \, k_\beta^n \equiv (k_\beta)^n \tag{4.93b}$$

Properties

$$B_n' = \frac{dB_n}{dr} = \frac{n}{r} B_n - B_{n+1} \qquad \textit{Recursion equation} \tag{4.94}$$

$$B_{n,i} = \gamma_i B_n' = \gamma_i \left(\frac{n}{r} B_n - B_{n+1} \right) \tag{4.95}$$

$$B_{0,i} = -\gamma_i B_1, \qquad B_{0,z} = -i k_z B_0 \tag{4.96}$$

$$B_{0,ij} = -\frac{1}{r} B_1 \delta_{ij} + \gamma_i \gamma_j B_2, \qquad B_{0,iz} = i k_z \gamma_i B_1 \tag{4.97}$$

$$B_{0,ii} = -\frac{2}{r} B_1 + B_2, \qquad B_{0,zz} = -k_z^2 B_0 \tag{4.98}$$

$$\hat{\nabla}^2 B_0 = -k_z^2 B_0 - \frac{2}{r} B_1 + B_2 \tag{4.99}$$

$$\frac{\partial}{\partial x_i} \left(\frac{1}{r} B_1 \right) = -\gamma_i \frac{1}{r} B_2 \tag{4.100}$$

$$B_{0,ijk} = (\gamma_i \delta_{jk} + \gamma_j \delta_{ki} + \gamma_k \delta_{ij}) \frac{1}{r} B_2 - \gamma_i \gamma_j \gamma_k B_3 \tag{4.101}$$

$$B_{0,ijj} = \gamma_i \left(\frac{4}{r} B_2 - B_3 \right) \tag{4.102}$$

$$\frac{\partial}{\partial x_i} \hat{\nabla}^2 B_0 = \gamma_i \left(k_z^2 B_1 + \frac{4}{r} B_2 - B_3 \right) \tag{4.103}$$

$$\frac{\partial}{\partial z} \hat{\nabla}^2 B_0 = i k_z \left(k_z^2 B_0 + \frac{2}{r} B_1 - B_2 \right) \tag{4.104}$$

5 Two-dimensional problems in homogeneous half-spaces

5.1 Half-plane, SH line source and receiver anywhere

A unit impulsive anti-plane line load is applied at some arbitrary point (x_0, z_0) in the interior of an elastic half-space. The displacement at any other arbitrary point (x, z) is obtained from the full space solution by the method of images. The origin of coordinates is taken at the free surface.

Preliminary definitions

$$r_1 = \sqrt{(x - x_0)^2 + (z - z_0)^2}, \qquad \Omega_1 = \frac{\omega r_1}{\beta}, \qquad t_1 = \frac{r_1}{\beta} \tag{5.1}$$

$$r_2 = \sqrt{(x - x_0)^2 + (z + z_0)^2}, \qquad \Omega_2 = \frac{\omega r_2}{\beta}, \qquad t_2 = \frac{r_2}{\beta} \tag{5.2}$$

Frequency domain

$$g_{yy}(r_1, r_2, \omega) = -\frac{i}{4\mu} \left[H_0^{(2)}(\Omega_1) + H_0^{(2)}(\Omega_2) \right] \tag{5.3}$$

Time domain

$$u_{yy}(x, z, t) = \frac{1}{2\pi\mu} \left[\frac{\mathcal{H}(t - t_1)}{\sqrt{t^2 - t_1^2}} + \frac{\mathcal{H}(t - t_2)}{\sqrt{t^2 - t_2^2}} \right] \tag{5.4}$$

Mixed wavenumber–time domain[1]

A half-plane is subjected to a spatially harmonic anti-plane SH source at the surface with horizontal wavenumber k. The source is impulsive in time. The exact solution for the

---

[1] Park, J. and Kausel, E., 2004, Impulse response of elastic half-space in the wavenumber–time domain, *Journal of Engineering Mechanics*, Vol. 130, No. 10, pp. 1211–1222.

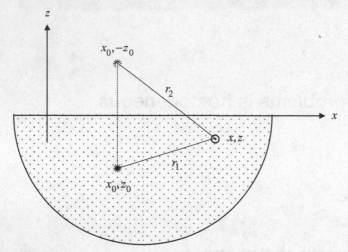

Figure 5.1: SH source in a half-space at depth x_0, z_0; receiver at x, z. Method of images.

impulse response function at depth z is

$$u_{yy}(k, z, t) = \frac{p_y(k)}{\rho\beta} J_0\left(k\beta\sqrt{t^2 - t_z^2}\right) \mathcal{H}(t - t_z), \qquad t_z = z/\beta \tag{5.5}$$

in which J_0 is the Bessel function of the first kind, and

$$p_y(k) = \int_{-\infty}^{+\infty} p_y(x)\, e^{ikx} dx \tag{5.6}$$

is the spatial Fourier transform of the anti-plane load. Hence,

$$u_y(x, z, t) = \frac{1}{2\pi} \int_{-\infty}^{+\infty} u_{yy}(k, z, t)\, e^{-ikx} dk \tag{5.7}$$

5.2 SH line source in an orthotropic half-plane

Scale the vertical coordinate as done in Section 3.3 for the full space. After reducing the problem to the isotropic form, apply the formulas above in Section 5.1, observing that scaled vertical coordinates must be used to define r_1, r_2, t_1, t_2. The final equations are (for further details, see Section 3.3)

$$\tilde{r}_1 = \sqrt{(x - x_0)^2 + \frac{\mu_x}{\mu_z}(z - z_0)^2}, \qquad \tilde{r}_2 = \sqrt{(x - x_0)^2 + \frac{\mu_x}{\mu_z}(z + z_0)^2} \tag{5.8}$$

$$\tilde{\Omega}_j = \frac{\omega\tilde{r}_j}{\beta_x}, \qquad \tilde{t}_j = \frac{\tilde{r}_j}{\beta_x}, \qquad j = 1, 2 \tag{5.9}$$

$$g_{yy}(\tilde{r}_1, \tilde{r}_2, \omega) = -\frac{i}{4\mu_x}\left[H_0^{(2)}(\tilde{\Omega}_1) + H_0^{(2)}(\tilde{\Omega}_2)\right] \tag{5.10}$$

$$u_{yy}(x, z, t) = \frac{1}{2\pi\mu_x}\left[\frac{\mathcal{H}(t - \tilde{t}_1)}{\sqrt{t^2 - \tilde{t}_1^2}} + \frac{\mathcal{H}(t - \tilde{t}_2)}{\sqrt{t^2 - \tilde{t}_2^2}}\right] \tag{5.11}$$

5.3 Half-plane, SV-P source and receiver at surface (Lamb's problem)[2]

An impulsive in-plane line source P is applied on the surface of a *lower* half-space, and displacements are observed there. For an *upper* half-space $z > 0$, reverse the sign of the coupling terms. The source has dimensions $[F][T]/[L] = [\text{impulse}]/[\text{length in } y \text{ direction}]$.

$$\tau = \frac{t\beta}{|x|}, \quad \tau_R = \frac{\beta}{C_R}, \quad a = \frac{\beta}{\alpha} = \frac{1 - 2\nu}{2(1 - \nu)}, \quad \delta = \text{Dirac delta} \tag{5.12}$$

$$u_{xx} = \frac{P\beta}{\pi\mu|x|}\begin{cases} 0 & \tau < a \\[2mm] \dfrac{4\tau^2(1 - \tau^2)\sqrt{\tau^2 - a^2}}{(2\tau^2 - 1)^4 + 16\tau^4(\tau^2 - a^2)(1 - \tau^2)} & a \le \tau \le 1 \\[4mm] \dfrac{-\sqrt{\tau^2 - 1}}{(2\tau^2 - 1)^2 - 4\tau^2\sqrt{\tau^2 - a^2}\sqrt{\tau^2 - 1}} & \tau > 1 \end{cases} \tag{5.13a}$$

$$u_{zz} = \frac{P\beta}{\pi\mu|x|}\begin{cases} 0 & \tau < a \\[2mm] \dfrac{-(2\tau^2 - 1)^2\sqrt{\tau^2 - a^2}}{(2\tau^2 - 1)^4 + 16\tau^4(\tau^2 - a^2)(1 - \tau^2)} & a \le \tau \le 1 \\[4mm] \dfrac{-\sqrt{\tau^2 - a^2}}{(2\tau^2 - 1)^2 - 4\tau^2\sqrt{\tau^2 - a^2}\sqrt{\tau^2 - 1}} & \tau > 1 \end{cases} \tag{5.13b}$$

$$u_{xz} = \frac{P\beta}{\pi\mu x}\begin{cases} \dfrac{2\tau(2\tau^2 - 1)\sqrt{\tau^2 - a^2}\sqrt{1 - \tau^2}}{(2\tau^2 - 1)^4 + 16\tau^4(\tau^2 - a^2)(1 - \tau^2)} & a \le \tau \le 1 \\[4mm] \dfrac{\pi(2\tau_R^2 - 1)^3}{4\left(1 - 4\tau_R^2 + 8\tau_R^6(1 - a^2)\right)}\delta(\tau - \tau_R) & \text{else} \end{cases} \tag{5.13c}$$

$$u_{zx} = -u_{xz} \tag{5.13d}$$

$$u_{yy} = \frac{P\beta}{\pi\mu|x|}\frac{1}{\sqrt{\tau^2 - 1}}\mathcal{H}(\tau - 1) \tag{5.13e}$$

Note: Observe that x in u_{xz} has no absolute sign. Also,

$$(\beta/x)\delta(\tau - \tau_R) = \text{sgn}\, x\ \delta(t - t_R).$$

[2] Eringen, A. C. and Suhubi, S. S., 1975, *Elastodynamics*, Academic Press, Vol. II, p. 617. *Note:* eq. 7.16.7 in that book has an error that affects the coefficient of the Dirac delta term in u_{xz}. This has been corrected here.

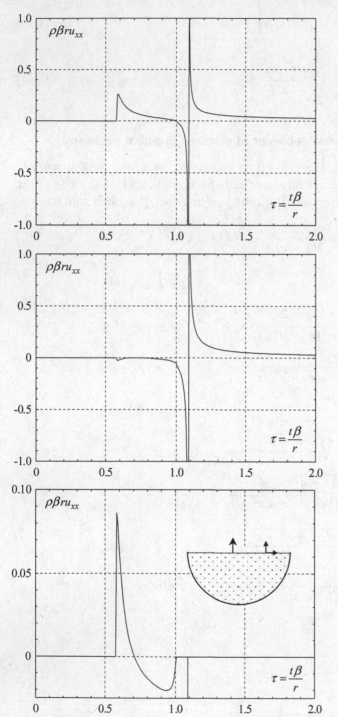

Figure 5.2: Lamb's problem in 2-D, $\nu = 0.25$.

5.4 Half-plane, SV-P source on surface, receiver at interior, or vice versa[3]

An impulsive line load is applied on the surface of an elastic half-space, and displacements are observed at an interior point, or vice versa. Solution fails if $z = 0$, in which case the formulas of Section 5.3 must be used.

Preliminary definitions

$$\tau = \frac{\beta t}{r}, \quad \sin\theta_z = \frac{|x|}{r}, \quad \cos\theta_z = \frac{|z|}{r}, \quad r = \sqrt{x^2 + z^2}, \quad a = \frac{\beta}{\alpha} = \sqrt{\frac{1 - 2v}{2(1 - v)}} \tag{5.14}$$

$$R(q^2) = (1 + 2q^2)^2 - 4q^2\sqrt{q^2 + a^2}\sqrt{q^2 + 1} = \text{Rayleigh function} \tag{5.15}$$

where q denotes a dimensionless parameter. The Cagniard–De Hoop paths for P and S waves are defined by

$$\tau + i\,q_\alpha \sin\theta_z - \cos\theta_z\sqrt{a^2 + q_\alpha^2} = 0, \qquad \tau + i\,q_\beta \sin\theta_z - \cos\theta_z\sqrt{1 + q_\beta^2} = 0 \tag{5.16}$$

with the constraint that the dimensionless time τ must remain real. From here, it follows that

$$q_\alpha = i\,\tau\sin\theta_z + \cos\theta_z\sqrt{\tau^2 - a^2}, \qquad\qquad q_\beta = i\,\tau\sin\theta_z + \cos\theta_z\sqrt{\tau^2 - 1} \tag{5.17}$$

$$\sqrt{q_\alpha^2 + a^2} = \tau\cos\theta_z + i\,\tau\sin\theta_z\sqrt{\tau^2 - a^2}, \qquad \sqrt{q_\beta^2 + 1} = \tau\cos\theta_z + i\,\tau\sin\theta_z\sqrt{\tau^2 - 1}$$

$$\tag{5.18}$$

$$\frac{\partial q_\alpha}{\partial\tau} = i\sin\theta_z + \frac{\tau\cos\theta_z}{\sqrt{\tau^2 - a^2}} = \frac{\sqrt{q_\alpha^2 + a^2}}{\sqrt{\tau^2 - a^2}}, \qquad \frac{\partial q_\beta}{\partial\tau} = i\sin\theta_z + \frac{\tau\cos\theta_z}{\sqrt{\tau^2 - 1}} = \frac{\sqrt{q_\beta^2 + 1}}{\sqrt{\tau^2 - 1}}$$

$$\tag{5.19}$$

These parameters must satisfy

$$\text{Re}\,q_a \geq 0, \qquad\qquad \text{Re}\,q_\beta \geq 0 \tag{5.20a}$$

$$\text{Re}\sqrt{q_\alpha^2 + a^2} \geq 0, \qquad \text{Re}\sqrt{q_\beta^2 + a^2} \geq 0 \tag{5.20b}$$

$$\text{Re}\sqrt{q_\alpha^2 + 1} \geq 0, \qquad \text{Re}\sqrt{q_\beta^2 + 1} \geq 0 \tag{5.20c}$$

$$\text{Im}\sqrt{\tau^2 - a^2} \leq 0, \qquad \text{Im}\sqrt{\tau^2 - 1} \leq 0 \tag{5.20d}$$

that is, the terms involving square roots in time must have a negative imaginary part when τ is small. This is most easily accomplished in the computer by defining $\tau_\alpha' = \text{conj}\sqrt{\tau^2 - a^2}$ and $\tau_\beta' = \text{conj}\sqrt{\tau^2 - 1}$. Finally, the critical dimensionless time is

$$\tau_c = \begin{cases} \cos(\theta_z - \theta_c) & \theta_z \leq \theta_c \\ 1 & \theta_z > \theta_c \end{cases} \quad \text{with} \quad \theta_c = \arcsin a = \text{critical angle} \tag{5.21}$$

---

[3] Pilant, W. L. (1979), *Elastic waves in the earth*, Elsevier, Chapter 11.

The ensuing formulas for displacements at $z \neq 0$ apply to:

- A lower half-space $z < 0$, with the load applied on the *surface* at $x = 0, z = 0$ and the displacements observed in its interior at $z < 0$.
- An upper half-space with the load applied on the *surface* at $x = 0, z = 0$ and the displacements observed in its interior at $z > 0$.

The formulas also may be used for either a lower or an upper half-space in which the load is applied in its *interior* at $x = 0, z \neq 0$, and the displacements are observed on the surface $z = 0$. In this case, replace $u_{xz} \rightarrow -u_{zx}$ and $u_{zx} \rightarrow -u_{xz}$, that is, exchange the coupling terms and reverse their signs (i.e., apply the reciprocity and symmetry principles, and reverse the signs to take account of the location of a receiver to the right of the point of application of the buried load). Observe that in general, $u_{xz} \neq -u_{zx}$, i.e., the coupling terms are not antisymmetric.

$$u_{xx} = \frac{\beta}{\pi \mu r} \text{Re} \left\{ -\frac{2q_\alpha^2 \sqrt{1 + q_\alpha^2}}{R\left(q_\alpha^2\right)} \frac{\partial q_\alpha}{\partial \tau} \mathcal{H}(\tau - a) + \frac{\left(1 + 2q_\beta^2\right)\sqrt{1 + q_\beta^2}}{R\left(q_\beta^2\right)} \frac{\partial q_\beta}{\partial \tau} \mathcal{H}(\tau - \tau_c) \right\} \quad (5.22a)$$

$$u_{zx} = \frac{\beta}{\pi \mu r} \text{Im} \left\{ \frac{2q_\alpha \sqrt{1 + q_\alpha^2}\sqrt{a^2 + q_\alpha^2}}{R\left(q_\alpha^2\right)} \frac{\partial q_\alpha}{\partial \tau} \mathcal{H}(\tau - a) - \frac{q_\beta \left(1 + 2q_\beta^2\right)}{R\left(q_\beta^2\right)} \frac{\partial q_\beta}{\partial \tau} \mathcal{H}(\tau - \tau_c) \right\} \text{sgn}(x)\text{sgn}(z) \quad (5.22b)$$

$$u_{xz} = \frac{\beta}{\pi \mu r} \text{Im} \left\{ \frac{q_\alpha \left(1 + 2q_\alpha^2\right)}{R\left(q_\alpha^2\right)} \frac{\partial q_\alpha}{\partial \tau} \mathcal{H}(\tau - a) - \frac{2q_\beta \sqrt{a^2 + q_\beta^2}\sqrt{1 + q_\beta^2}}{R\left(q_\beta^2\right)} \frac{\partial q_\beta}{\partial \tau} \mathcal{H}(\tau - \tau_c) \right\} \text{sgn}(x)\text{sgn}(z) \quad (5.22c)$$

$$u_{zz} = \frac{\beta}{\pi \mu r} \text{Re} \left\{ \frac{\left(1 + 2q_\alpha^2\right)\sqrt{a^2 + q_\alpha^2}}{R\left(q_\alpha^2\right)} \frac{\partial q_\alpha}{\partial \tau} \mathcal{H}(\tau - a) - \frac{2q_\beta^2 \sqrt{a^2 + q_\beta^2}}{R\left(q_\beta^2\right)} \frac{\partial q_\beta}{\partial \tau} \mathcal{H}(\tau - \tau_c) \right\} \quad (5.22d)$$

Motions on epicentral line ($x = 0$)

$$\tau = \frac{t\beta}{|z|} \quad (5.23)$$

$$D_\alpha(\tau) = \left(\tau^2 - a^2 + \tfrac{1}{2}\right)^2 - \tau(\tau^2 - a^2)\sqrt{\tau^2 - a^2 + 1} \quad (5.24)$$

$$D_\beta(\tau) = \left(\tau^2 - \tfrac{1}{2}\right)^2 - \tau(\tau^2 - 1)\sqrt{\tau^2 - 1 + a^2} \quad (5.25)$$

$$u_{xx}(0, z, t) = \frac{\beta}{2\pi \mu |z|} \left\{ \frac{\tau^2 \left(\tau^2 - \tfrac{1}{2}\right)}{D_\beta \sqrt{\tau^2 - 1}} \mathcal{H}(\tau - 1) - \frac{\tau\sqrt{\tau^2 - a^2}\sqrt{\tau^2 - a^2 + 1}}{D_\alpha} \mathcal{H}(\tau - a) \right\} \quad (5.26a)$$

$$u_{zz}(0, z, t) = \frac{\beta}{2\pi \mu |z|} \left\{ \frac{\tau^2 \left(\tau^2 - a^2 + \tfrac{1}{2}\right)}{D_\alpha \sqrt{\tau^2 - a^2}} \mathcal{H}(\tau - a) - \frac{\tau\sqrt{\tau^2 - 1}\sqrt{\tau^2 - 1 + a^2}}{D_\beta(\tau)} \mathcal{H}(\tau - 1) \right\} \quad (5.26b)$$

$$u_{zx} = u_{xz} = 0 \quad (5.26c)$$

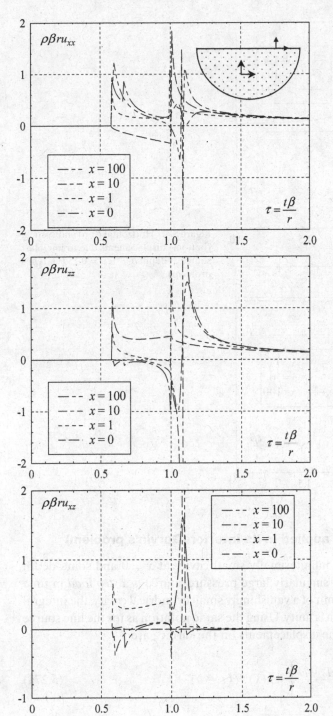

Figure 5.3: Displacements at surface due to impulsive line loads at depth $h = 1$, $v = 0.25$ (Lamb's problem).

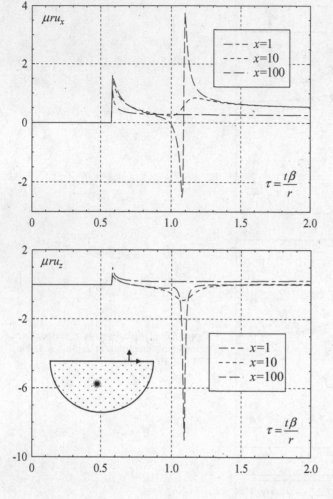

Figure 5.4: Horizontal (top) and vertical (bottom) displacements due to line blast source at depth $h = 1$, $\nu = 0.25$ (Garvin's problem).

5.5 Half-plane, line blast load applied in its interior (Garvin's problem)[4]

A line blast source acts within an infinitesimally small cavity at $x = 0$ and some depth $h = |z|$ below the free surface. The singularly large pressure source is a *step load* in time and of magnitude such that in the limit of a vanishingly small cylindrical cavity, the integral of the pressure over the cross section is unity. Using the same notation as for the line source in the Section 5.4, it is found that the displacements on the surface are

$$u_x(x, 0, t) = \frac{1}{\pi \mu r} \text{Im} \left\{ \frac{2 q_\alpha \sqrt{1 + q_\alpha^2}}{R(q_\alpha^2)} \frac{\partial q_\alpha}{\partial \tau} \right\} \text{sgn}(x)\, \mathcal{H}(\tau - a) \tag{5.27a}$$

$$u_z(x, 0, t) = \frac{-1}{\pi \mu r} \text{Re} \left\{ \frac{(1 + 2 q_\alpha^2)}{R(q_\alpha^2)} \frac{\partial q_\alpha}{\partial \tau} \right\} \text{sgn}(z)\, \mathcal{H}(\tau - a) \tag{5.27b}$$

---

[4] Garvin, W. W. (1956), Exact transient solution of the buried line source problem, *Proceedings of the Royal Society of London, Series A, Mathematical and Physical Sciences*, Vol. 234, No. 1199 (Mar.), pp. 528–541.

This formula is valid for both an upper ($z > 0$) and a lower ($z < 0$) half-space, with z being the coordinate of the source relative to the surface. For a lower half-space and at large times, these displacements converge asymptotically to the static solution

$$u_x(x, 0, \infty) = \frac{2(1 - \nu)}{\pi \mu r} \sin \theta_z \qquad (5.28a)$$

$$u_y(x, 0, \infty) = \frac{2(1 - \nu)}{\pi \mu r} \cos \theta_z \qquad (5.28b)$$

6 Three-dimensional problems in homogeneous half-spaces

6.1 3-D half-space, suddenly applied vertical point load on its surface (Pekeris-Mooney's problem)[1]

A vertical point source that varies as a step function in time is applied onto the surface. Displacements are also observed on the surface at a range r.

Definitions

$$r = \sqrt{x^2 + y^2} \tag{6.1}$$

$$a^2 = \left(\frac{\beta}{\alpha}\right)^2 = \frac{1 - 2v}{2(1 - v)}, \quad \tau = \frac{t\beta}{r}, \tag{6.2}$$

$$\mathcal{H}(t - t_0) = \begin{cases} 1 & t > t_0 \\ \frac{1}{2} & t = t_0 = \textit{Heaviside step function} \\ 0 & t < t_0 \end{cases} \tag{6.3}$$

$$K(k) = \int_0^{\pi/2} \frac{d\theta}{\sqrt{1 - k^2 \sin^2 \theta}}, \tag{6.4}$$

$$\Pi(n, k) = \int_0^{\pi/2} \frac{d\theta}{(1 + n \sin^2 \theta)\sqrt{1 - k^2 \sin^2 \theta}} \tag{6.5}$$

K and Π are complete elliptical integrals of the first and third kind, respectively.
Consider the Rayleigh function

$$R(\xi^2) = (1 - 2\xi^2)^2 + 4\sqrt{\xi^2 - 1}\sqrt{\xi^2 - a^2} = 0 \tag{6.6}$$

in which $\xi = \beta/c$ is a dimensionless wave slowness, with c a wave velocity. Multiplying by $(1 - 2\xi^2)^2 - 4\sqrt{\xi^2 - 1}\sqrt{\xi^2 - a^2}$ and dividing by ξ^2, one obtains the bicubic (i.e., cubic in ξ^2) equation

$$1 - 8\xi^2 + 8\xi^4(3 - 2a^2) - 16\xi^6(1 - a^2) = 0 \tag{6.7}$$

---

[1] Eringen, A. C. and Suhubi, S. S., 1975, *Elastodynamics*, Academic Press, Vol. II, p. 748. Solution up to $v = 0.2631$.

which has three roots $[\xi_1^2, \xi_2^2, \xi_3^2]$, the first two of which are non-physical solutions of the rationalized Rayleigh function, while $\xi_3 \equiv \beta/C_R$ is the actual true root. When $\nu < \nu_0 = 0.2631$, all three roots are real and satisfy $0 < \xi_1^2 < \xi_2^2 < a^2 < 1 < \xi_3^2$. The transition value ν_0 is the root of the discriminant $D(\nu) = 32\nu^3 - 16\nu^2 + 21\nu - 5 = 0$ in the interval $[0 \leq \nu \leq 0.5]$. It defines the point beyond which the false roots turn complex. When $\nu = \nu_0$, the false roots are repeated, i.e., $\xi_1 = \xi_2$.

a) $\nu < 0.2631$

$$u_{zz} = \frac{P(1-\nu)}{2\pi\mu r}\begin{cases} 0 & \tau < a \\[2ex] \dfrac{1}{2}\left[1 - \displaystyle\sum_{i=1}^{3}\frac{A_i}{\sqrt{|\tau^2 - \xi_i^2|}}\right] & a < \tau < 1 \\[3ex] 1 - \dfrac{A_3}{\sqrt{\xi_3^2 - \tau^2}}\,\mathcal{H}(\xi_3 - \tau) & \tau > 1 \end{cases} \tag{6.8a}$$

$$u_{rz} = \frac{P\tau}{8\pi^2\mu r}\begin{cases} 0 & \tau < a \\[2ex] \dfrac{1}{(1-a^2)^{3/2}}\left[2K(k) - \displaystyle\sum_{i=1}^{3}B_i\,\Pi(k^2 n_i, k)\right] & a < \tau < 1 \\[3ex] \dfrac{k^{-1}}{(1-a^2)^{3/2}}\left[2K(k^{-1}) - \displaystyle\sum_{i=1}^{3}B_i\,\Pi(n_i, k^{-1})\right] + \dfrac{2\pi C}{\sqrt{\tau^2 - \xi_3^2}}\,\mathcal{H}(\tau - \xi_3) & \tau > 1 \end{cases}$$

$$\tag{6.8b}$$

in which

$$A_i = \frac{\left(1 - 2\xi_i^2\right)^2\sqrt{|a^2 - \xi_i^2|}}{4\left(\xi_i^2 - \xi_j^2\right)\left(\xi_i^2 - \xi_k^2\right)}, \qquad B_i = \frac{\left(1 - 2\xi_i^2\right)\left(1 - \xi_i^2\right)}{\left(\xi_i^2 - \xi_j^2\right)\left(\xi_i^2 - \xi_k^2\right)} \qquad \xi_i \neq \xi_j \neq \xi_k \tag{6.9}$$

$$k^2 = \frac{\tau^2 - a^2}{1 - a^2}, \qquad n_i = \frac{1 - a^2}{a^2 - \xi_i^2}, \qquad C = \frac{\left(2\xi_3^2 - 1\right)^3}{1 - 4\xi_3^2 + 8(1 - a^2)\xi_3^6} \tag{6.10}$$

Observe that $n_1 > 0$, $n_2 > 0$, but $n_3 < 0$.

b) $\nu > 0.2631$ *(ξ_1, ξ_2 are complex conjugates)*[2]

u_{rz} not available $\hspace{10cm}$ (6.11a)

$$u_{zz} = \frac{P(1-\nu)}{16\pi\mu r}\begin{cases} 0 & \tau < a \\[2ex] 8\,\mathrm{Re}\left[\dfrac{\left(1 - 2\xi_1^2\right)^2\left(a^2 - \xi_1^2\right)}{\left(\xi_1^2 - \xi_2^2\right)\left(\xi_1^2 - \xi_3^2\right)}\dfrac{1}{\left(Q_1 - Q_1^{-1}\right)}\right] + \dfrac{A_3}{\sqrt{\xi_3^2 - \tau^2}} - 4 & a < \tau < 1 \\[3ex] \dfrac{2A_3}{\sqrt{\xi_3^2 - \tau^2}}\,\mathcal{H}(\xi_3 - \tau) - 8 & \tau > 1 \end{cases}$$

$$\tag{6.11b}$$

---

[2] Mooney, H. M., 1974, Some numerical solutions for Lamb's problem, *Bulletin of the Seismological Society of America*, Vol. 64, No. 2, pp. 473–491. Mooney presents the solution for arbitrary values of Poisson's ratio, but does so only for the vertical component.

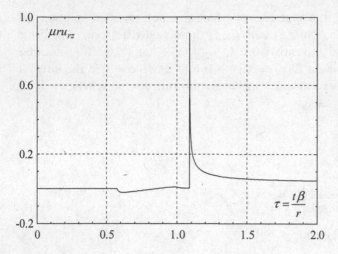

Figure 6.1a: Radial displacement at surface due to suddenly applied vertical point load, $\nu = 0.25$ (Lamb–Pekeris–Mooney problem).

in which A_3 is as before, and

$$Q_1(\tau) = 1 + 2z + \sqrt{z^2 + z}, \qquad z = \frac{a^2 - \xi_1^2}{\tau^2 - a^2} \tag{6.12}$$

provided that $|Q_1| < 1$. If $|Q_1| > 1$, replace $Q_1 \to 1/Q_1$. Observe that ξ_1^2 is complex and that $Q_2(\tau) = 1 + 2z - \sqrt{z^2 + z}$ satisfies $Q_1 Q_2 = 1$, that is, $Q_1^{-1} = Q_2$. Hence, the test $|Q_1| < 1$ implicitly assigns the proper sign to the difference $Q_1 - Q_1^{-1}$. However, it is not enough to simply always add the square root term, because the proper sign of the square root depends also on the complex quadrant that z lies in, which may change with time τ. Thus, it behooves one to check the absolute value of Q_1, and replace it with its reciprocal as needed.

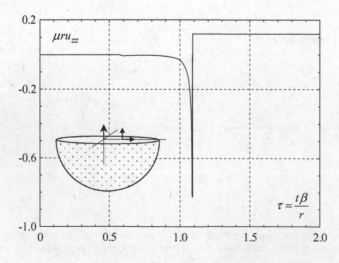

Figure 6.1b: Vertical displacement at surface due to suddenly applied vertical point load, $\nu = 0.25$ (Lamb–Pekeris–Mooney problem).

6.2 3-D half-space, suddenly applied horizontal point load on its surface (Chao's problem)[3]

A horizontal point source that varies as a step function in time is applied onto the surface. Displacements are also observed on the surface at a range r.

$v = 0.25$ *only* (solutions for other values of Poisson's ratio unknown).

$$a = \tfrac{1}{3}\sqrt{3}, \qquad \xi_1^2 = \tfrac{1}{4}, \qquad \xi_2^2 = \tfrac{1}{4}(3 - \sqrt{3}), \qquad \xi_3^2 = \tfrac{1}{4}(3 + \sqrt{3}) \tag{6.13}$$

$$C_1 = \tfrac{3}{4}\sqrt{3}, \qquad C_2 = \tfrac{1}{8}\sqrt{6\sqrt{3} + 10}, \qquad C_3 = \tfrac{1}{8}\sqrt{6\sqrt{3} - 10}, \qquad r = \sqrt{x^2 + y^2} \tag{6.14}$$

$$u_{rx} = \frac{P}{2\pi\mu r} \begin{cases} 0 & \tau < a \\[2mm] \tau^2 \left[\dfrac{C_1}{\sqrt{\tau^2 - \xi_1^2}} - \dfrac{C_2}{\sqrt{\tau^2 - \xi_2^2}} - \dfrac{C_3}{\sqrt{\xi_3^2 - \tau^2}} \right] & a < \tau < 1 \\[4mm] 1 - \dfrac{2\tau^2 C_3}{\sqrt{\xi_3^2 - \tau^2}}\left[1 - \mathcal{H}(\tau - \xi_3)\right] & \tau > 1 \end{cases} \tag{6.15a}$$

$$u_{\theta x} = \frac{-3P}{8\pi\mu r} \begin{cases} 0 & \tau < a \\[2mm] \left[\tfrac{1}{2} - \tfrac{4}{3}\left(C_1\sqrt{\tau^2 - \xi_1^2} - C_2\sqrt{\tau^2 - \xi_2^2} + C_3\sqrt{\xi_3^2 - \tau^2}\right)\right] & a < \tau < 1 \\[2mm] 1 - \tfrac{8}{3}C_3\sqrt{\xi_3^2 - \tau^2}\left[1 - II(\tau - \xi_3)\right] & \tau > 1 \end{cases}$$

$$\tag{6.15b}$$

$$u_{zx} = -u_{xz} : \text{see vertical load} \tag{6.15c}$$

The actual Cartesian displacements produced by a horizontal load in direction x are then

$$u_x = u_{rx}\cos^2\theta - u_{\theta x}\sin^2\theta, \qquad u_y = (u_{rx} + u_{\theta x})\sin\theta \cos\theta, \qquad u_z = u_{zx}\sin\theta \tag{6.16}$$

in which θ is the azimuth.

Observation: Experiments demonstrate that the coefficients C_i are of the form $C_1 = \tfrac{3}{64}\sqrt{6}\,D_1$, $-C_2 = \tfrac{2}{64}\sqrt{6}\,D_2$, $-C_3 = \tfrac{2}{64}\sqrt{6}\,D_3$, $D_i = \sqrt{|1 - 2\xi_i^2|/[4(\xi_i^2 - \xi_j^2)(\xi_i^2 - \xi_k^2)]}$. If the coefficients in front of the D_i were all equal to the same constant, it would mean that this constant would depend only on Poisson's ratio. Unfortunately, this is not the case, so these expressions do not help in generalizing the expressions above for arbitrary Poisson's ratio. In addition, the sign of C_3 in $u_{\theta x}$ is opposite to that in u_{rx}.

[3] Chao, C. C., 1960, Dynamical response of an elastic half-space to tangential surface loadings, *Journal of Applied Mechanics*, Vol. 27 (Sept.) pp. 559–567.

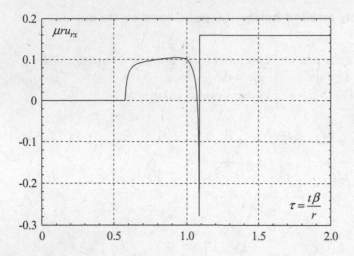

Figure 6.2a: Radial displacement at surface due to suddenly applied horizontal point load, $\nu = 0.25$ (Chao's problem). Varies as $\cos\theta$.

Displacements on the epicentral axis ($r = 0$, $z \neq 0$)

At all times, $u_{xx}(0, z, t) = u_{rx}(0, z, t)$, $u_{\theta x}(0, z, t) = -u_{rx}(0, z, t)$, $u_z(0, z, t) = 0$. Define the auxiliary functions

$$f(\tau) = \frac{\tau\left(\tau^2 - \frac{1}{3}\right)\sqrt{\tau^2 + \frac{2}{3}}}{\left(2\tau^2 + \frac{1}{3}\right)^2 - 4\tau\left(\tau^2 - \frac{1}{3}\right)\sqrt{\tau^2 + \frac{2}{3}}} \tag{6.17}$$

$$g(\tau) = \frac{\tau^2(\tau^2 - 1)\left(2\tau\sqrt{\tau^2 - \frac{2}{3}} - (2\tau^2 - 1)\right)}{(2\tau^2 - 1)^2 - 4\tau(\tau^2 - 1)\sqrt{\tau^2 - \frac{2}{3}}} \tag{6.18}$$

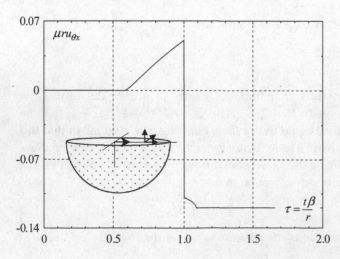

Figure 6.2b: Tangential displacement at surface due to suddenly applied horizontal point load, $\nu = 0.25$ (Chao's problem). Varies as $-\sin\theta$.

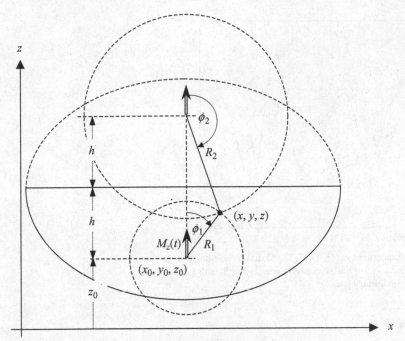

Figure 6.3: Torsional point source in elastic half-space.

in which $\tau = t\beta/|z|$. The horizontal displacement on the epicentral axis is then

$$u_{xx}(0, z, t) = \frac{P}{2\pi\mu|z|} \begin{cases} 0 & \tau < a \\ -f(\tau) & a < \tau < 1 \\ \frac{1}{2}(\tau^2 + 1) - f(\tau) + g(\tau) & 1 < \tau \end{cases} \qquad (6.19)$$

As written, each function in the last expression grows as τ^2, but the sum tends to the constant value 5/8. Cancellation errors can be avoided by using asymptotic expansions for each term.

6.3 3-D half-space, buried torsional point source with vertical axis

Consider a torsional point source with vertical axis (i.e., perpendicular to the free surface) applied at an arbitrary point (x_0, y_0, z_0). The response is observed at any other arbitrary point (x, y, z). The distance from the point of application of the torque to the free surface is $h \geq 0$. If the origin is taken at the free surface, then $z_0 = -h$. This problem is similar to that of a torsional point source in a full space. Its solution is obtained by application of the method of virtual images, as shown in Fig. 6.3.

Note: If the axis of the torque is not vertical, then reflections at the surface may partially convert to waves other than SH waves, in which case the solution provided below breaks down. On the other hand, a transversely anisotropic (or cross-isotropic) half-space can be considered by recourse to the scaling method presented earlier for anti-plane loads in a 2-D full space.

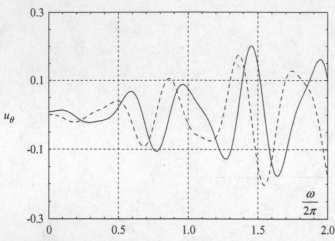

Figure 6.4a: Tangential displacement) at $(r, z) = (2, -1)$ due to unit harmonic torsional source at $z_0 = -2$ in half-space with $\mu = 1$, $\beta = 1$. Solid line, real part; dashed line, imaginary part.

Preliminary definitions

$$R_1 = \sqrt{(x - x_0)^2 + (y - y_0)^2 + (z - z_0)^2} \tag{6.20}$$

$$R_2 = \sqrt{(x - x_0)^2 + (y - y_0)^2 + (z - z_0 - 2h)^2} \tag{6.21}$$

$$\sin \phi_1 = \sqrt{1 - \left(\frac{z - z_0}{R_1}\right)^2}, \qquad \sin \phi_2 = \frac{R_1}{R_2} \sin \phi_1 \tag{6.22}$$

$$\Omega_j = \frac{\omega R_j}{\beta}, \quad t_j = \frac{R_j}{\beta}, \qquad j = 1, 2 \tag{6.23}$$

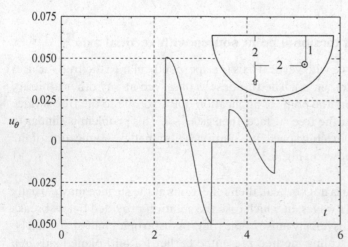

Figure 6.4b: Tangential displacement) at $(r, z) = (2, -1)$ due to single torsional sine pulse of duration $t_d = 1$ at $z_0 = -2$ in half-space with $\mu = 1$, $\beta = 1$.

Frequency domain

$$u_\theta = M_z(\omega) \frac{1}{8\pi\mu} \left[\frac{\sin\phi_1}{R_1^2} (1 + i\,\Omega_1)\, e^{-i\Omega_1} + \frac{\sin\phi_2}{R_2^2} (1 + i\,\Omega_2)\, e^{-i\Omega_2} \right] \tag{6.24}$$

Time domain, arbitrary causal variation of source with time

$$u_\theta = \frac{1}{8\pi\mu} \left\{ \frac{\sin\phi_1}{R_1} \left[\frac{M_z(t - t_1)}{R_1} + \frac{1}{\beta} \frac{\partial M_z(t - t_1)}{\partial t} \right] \right.$$
$$\left. + \frac{\sin\phi_2}{R_2} \left[\frac{M_z(t - t_2)}{R_2} + \frac{1}{\beta} \frac{\partial M_z(t - t_2)}{\partial t} \right] \right\} \tag{6.25}$$

These solutions are valid for both an upper and a lower half-space. Figure 6.4 illustrates these response functions for a source at depth $h = 2$, and a receiver at depth $d = 0.4h$ and range $r = 0.8h$. The origin is at the surface of the half-space with $\mu = 1, \beta = 1$, and the torsional source is a single sine pulse of the form $M(t) = \sin\pi\frac{t}{t_d}, 0 \leq t \leq t_d = 1$. Observe that the response consists of a direct pulse and a reflected pulse, and is dominated by the wavelet in the derivative of the moment. The wavelet in M is apparent only close to the source.

7 Two-dimensional problems in homogeneous plates and strata

7.1 Plate subjected to SH line source

Consider a homogeneous plate of infinite length and arbitrary thickness h subjected to a harmonic $\exp(i\omega t)$ or impulsive $\delta(t)$ line source of the form $b(x, z) = \delta(x - x_0)\,\delta(z - z_0)$, where x_0, z_0 is the point of application of the load, with $0 \leq z_0 \leq h$, see Figure 7.1. The origin of coordinates is at the bottom surface. Both external surfaces are traction-free.

7.1.1 Solution using the method of images

$$r_{1j} = \sqrt{(x - x_0)^2 + (z - z_0 - 2jh)^2}, \qquad r_{2j} = \sqrt{(x - x_0)^2 + (z + z_0 - 2jh)^2} \tag{7.1}$$

$$\Omega_{1j} = \frac{\omega r_{1j}}{\beta}, \qquad t_{1j} = \frac{r_{1j}}{\beta}, \qquad \Omega_{2j} = \frac{\omega r_{2j}}{\beta}, \qquad t_{2j} = \frac{r_{2j}}{\beta} \tag{7.2}$$

$$g_{yy}(\omega, x, z, x_0, z_0) = -\frac{i}{4\mu} \sum_{j=-\infty}^{\infty} \left[H_0^{(2)}(\Omega_{1j}) + H_0^{(2)}(\Omega_{2j}) \right] \tag{7.3}$$

$$u_{yy}(t, x, z, x_0, z_0) = \frac{1}{2\pi\mu} \sum_{j=-\infty}^{\infty} \left[\frac{\mathcal{H}(t - t_{1j})}{\sqrt{t^2 - t_{1j}^2}} + \frac{\mathcal{H}(t - t_{2j})}{\sqrt{t^2 - t_{2j}^2}} \right] \tag{7.4}$$

Observe that when $z_0 = 0$, then $r_{1j} = r_{2j}$, in which case each pair in the summations consists of identical terms.

7.1.2 Normal mode solution

$$\xi = \frac{x - x_0}{h}, \qquad \zeta = \frac{z}{h}, \qquad \zeta_0 = \frac{z_0}{h}, \qquad \Omega_S = \frac{\omega h}{\beta} \tag{7.5}$$

$$\kappa_j = \sqrt{\Omega_S^2 - \pi^2 j^2}, \qquad \text{Im}(\kappa_j) \leq 0 \qquad \text{Dimensionless modal wavenumber} \tag{7.6}$$

$$\phi_j = \cos j\pi\zeta, \qquad j = 0, 1, 2, \ldots \qquad \text{Normal mode} \tag{7.7}$$

$$g_{yy}(\omega, x, z, x_0, z_0) = -\frac{i}{\mu} \left\{ \frac{e^{-i\Omega_S|\xi|}}{2\Omega_S} + \sum_{j=1}^{\infty} \frac{\cos(\pi j\zeta)\cos(\pi j\zeta_0)\,e^{-i\kappa_j|\xi|}}{\kappa_j} \right\} \tag{7.8}$$

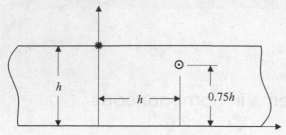

Figure 7.1: Plate subjected to SH line source.

Observe that the high modes have negative imaginary characteristic wavenumber (They are evanescent modes), so these decay exponentially in x. Hence, the summation can safely be truncated above $j \gg \Omega_S/\pi$, except near the source. For small x, the index of the first neglected mode may be established by setting a threshold of exponential decay, say, $\exp(-i\kappa_j\xi) < \varepsilon$, where ε is some small number, and ξ is the shortest range (source–receiver distance). The mode for $j = 0$, which carries weight $1/2$, is referred to as the *rigid body mode*. Also, $\kappa_j = 0$ at each of the cutoff frequencies of the plate, so the last mode to change from imaginary to real contributes the most.

7.2 Stratum subjected to SH line source

Consider a homogeneous stratum of infinite length and arbitrary thickness h subjected to a harmonic $\exp(i\omega t)$ or impulsive $\delta(t)$ line source of the form $b(x, z) = \delta(x - x_0)\,\delta(z - z_0)$, where x_0, z_0 is the point of application of the load, with $0 \leq z_0 \leq h$, see Figure 7.3. The origin of coordinates is at the rigid bottom interface, where displacements are zero. The upper surface is traction-free.

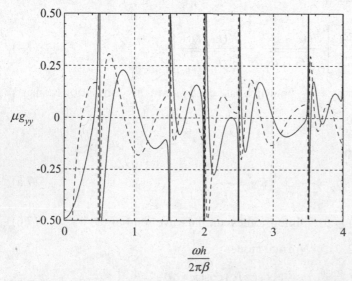

Figure 7.2a: Homogeneous plate, Green's function at $(x, z) = (h, 0.75h)$ due to SH line load at elevation $z_0 = h$ (i.e., upper surface).

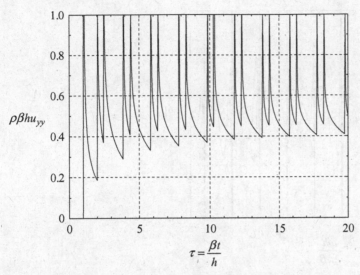

$$\tau = \frac{\beta t}{h}$$

Figure 7.2b: Figure Homogeneous plate, impulse response function at $(x, z) = (h, 0.75h)$ due to SH line load at elevation $z_0 = h$.

7.2.1 Solution using the method of images

$$r_{1j} = \sqrt{(x - x_0)^2 + (z - z_0 - 2jh)^2}, \qquad r_{2j} = \sqrt{(x - x_0)^2 + (z + z_0 - 2jh)^2} \qquad (7.9)$$

$$\Omega_{1j} = \frac{\omega r_{1j}}{\beta}, \quad t_{1j} = \frac{r_{1j}}{\beta}, \qquad \Omega_{2j} = \frac{\omega r_{2j}}{\beta}, \quad t_{2j} = \frac{r_{2j}}{\beta} \qquad (7.10)$$

$$g_{yy}(\omega, x, z, x_0, z_0) = -\frac{i}{4\mu} \sum_{j=-\infty}^{\infty} (-1)^j \left[H_0^{(2)}(\Omega_{1j}) - H_0^{(2)}(\Omega_{2j}) \right] \qquad (7.11)$$

$$u_{yy}(t, x, z, x_0, z_0) = \frac{1}{2\pi \mu} \sum_{j=-\infty}^{\infty} (-1)^j \left[\frac{\mathcal{H}(t - t_{1j})}{\sqrt{t^2 - t_{1j}^2}} - \frac{\mathcal{H}(t - t_{2j})}{\sqrt{t^2 - t_{2j}^2}} \right] \qquad (7.12)$$

Observe that the response is null when $z_0 = 0$, because then $r_{1j} = r_{2j}$, so each pair in the summations cancels out.

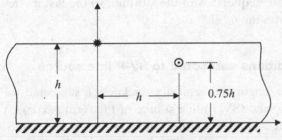

Figure 7.3: Stratum subjected to SH line source.

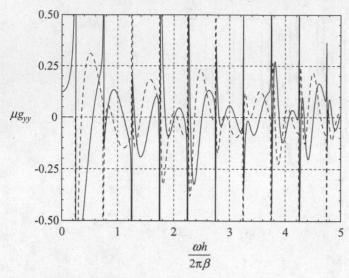

μg_{yy}

$$\frac{\omega h}{2\pi\beta}$$

Figure 7.4a: Homogeneous stratum, Green's function at $(x, z) = (h, 0.75h)$ due to SH line load at elevation $z_0 = h$ (i.e., upper surface). Solid line, real part; dashed line, imaginary part.

7.2.2 Normal mode solution

$$\xi = \frac{x - x_0}{h}, \zeta = \frac{z}{h}, \qquad \zeta_0 = \frac{z_0}{h}, \qquad \Omega_S = \frac{\omega h}{\beta} \tag{7.13}$$

$$l = \tfrac{1}{2}(2j - 1) = j - \tfrac{1}{2}, \quad j = 1, 2, 3 \ldots \qquad \text{Modal indices} \tag{7.14}$$

$$\kappa_j = \sqrt{\Omega_S^2 - \pi^2 l^2}, \qquad \text{Im}(\kappa_j) \le 0 \qquad \text{Modal wavenumber} \tag{7.15}$$

$$\phi_j = (-1)^j \sin l\pi\xi \qquad \qquad \text{Normal mode} \tag{7.16}$$

$$g_{yy}(\omega, x, z, x_0, z_0) = -\frac{i}{\mu} \sum_{j=1}^{\infty} \frac{\sin(\pi l\zeta) \sin(\pi l\zeta_0) \, e^{-i\kappa_j|\xi|}}{\kappa_j} \tag{7.17}$$

Observe again that the high modes are evanescent and thus decay rapidly in x. Hence, the summation can safely be truncated above $l \gg \Omega_S/\pi$. For small x, the index of the first neglected mode may be established by setting a threshold of exponential decay, say, $\exp(-i\kappa_j\xi) < \varepsilon$, where ε is some small number, and ξ is the shortest range (source–receiver distance). Also, $\kappa_j = 0$ at each of the cutoff frequencies of the stratum, so the last mode to change from imaginary to real contributes the most.

7.3 Plate with mixed boundary conditions subjected to SV-P line source

Consider a homogeneous plate of infinite length and arbitrary thickness h subjected to a harmonic $\exp(i\omega t)$ or impulsive $\delta(t)$ in-plane (SV-P) line source of the form $b(x, z) = \delta(x - x_0)\,\delta(z - z_0)$, where x_0, z_0 is the point of application of the load, with $0 \le z_0 \le h$, see Figure 7.5. The origin of coordinates is at the bottom surface. The plate has *mixed boundary conditions*, that is, either the normal or the tangential displacements are restrained, but

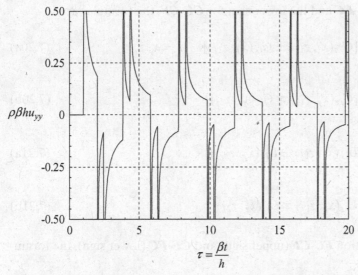

Figure 7.4b: Homogeneous stratum, impulse response function at $(x, z) = (h, 0.75h)$ due to SH line load at elevation $z_0 = h$.

not both at the same time on the same surface. Denoting a *free* condition as F, and a *constrained* condition as C, and listing first the tangential (horizontal) and then the normal (vertical) boundary condition, then a plate with mixed boundary conditions can have any of the following four combinations: **FC-FC, CF-CF, FC-CF, CF-FC**, with the bottom pair of conditions being listed first. *Note*: Other boundary combinations, such as **FF-FF** (*Mindlin plate*) or **CC-FF** (stratum), cannot be solved by the methods used to obtain the formulas that follow.

7.3.1 Solution using the method of images

Let $G_{xx}(x, z)$, $G_{xz}(x, z)$, $G_{zx}(x, z)$, $G_{zz}(x, z)$ and $U_{xx}(x, z)$, $U_{xz}(x, z)$, $U_{zx}(x, z)$, $U_{zz}(x, z)$ be, respectively, the frequency response functions and the impulse response functions for an in-plane line load in a homogeneous full space; see Section 3.4. Also, define the receiver coordinates relative to the location of the image sources as

$$x_j = x - x_0, \qquad z_{1j} = z - z_0 - 2jh, \qquad z_{2j} = z + z_0 - 2jh \qquad (7.18)$$

$$r_{nj} = \sqrt{x_j^2 + z_{nj}^2}, \qquad \gamma_{nx} = x_j/r_{nj}, \qquad \gamma_{nz} = z_j/r_{nj}, \qquad n = 1, 2 \qquad (7.19)$$

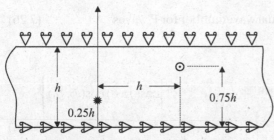

Figure 7.5: Plate with mixed boundary conditions.

Using the upper sign for **FC-FC** and the lower sign for **CF-CF**, the formulas are

$$g_{ix}(\omega, x, z, x_0, z_0) = \sum_{j=-\infty}^{\infty} [G_{ix}(x_j, z_{1j}) \pm G_{ix}(x_j, z_{2j})], \quad i = x, z \tag{7.20a}$$

$$g_{iz}(\omega, x, z, x_0, z_0) = \sum_{j=-\infty}^{\infty} [G_{iz}(x_j, z_{1j}) \mp G_{iz}(x_j, z_{2j})] \tag{7.20b}$$

$$u_{ix}(t, x, z, x_0, z_0) = \sum_{j=-\infty}^{\infty} [U_{ix}(x_j, z_{1j}) \pm U_{ix}(x_j, z_{2j})] \tag{7.21a}$$

$$u_{iz}(t, x, z, x_0, z_0) = \sum_{j=-\infty}^{\infty} [U_{iz}(x_j, z_{1j}) \mp U_{iz}(x_j, z_{2j})] \tag{7.21b}$$

For a mixed boundary condition **FC-CF** (upper sign) and **CF-FC** (lower sign), the formulas are

$$g_{ix}(\omega, x, z, x_0, z_0) = \sum_{j=-\infty}^{\infty} (-1)^j [G_{ix}(x_j, z_{1j}) \pm G_{ix}(x_j, z_{2j})], \quad i = x, z \tag{7.22a}$$

$$g_{iz}(\omega, x, z, x_0, z_0) = \sum_{j=-\infty}^{\infty} (-1)^j [G_{iz}(x_j, z_{1j}) \mp G_{iz}(x_j, z_{2j})] \tag{7.22b}$$

$$u_{ix}(t, x, z, x_0, z_0) = \sum_{j=-\infty}^{\infty} (-1)^j [U_{ix}(x_j, z_{1j}) \pm U_{ix}(x_j, z_{2j})] \tag{7.23a}$$

$$u_{iz}(t, x, z, x_0, z_0) = \sum_{j=-\infty}^{\infty} (-1)^j [U_{iz}(x, z_{1j}) \mp U_{iz}(x, z_{2j})] \tag{7.23b}$$

7.3.2 Normal mode solution

Preliminary definitions

$$\xi = \frac{x - x_0}{h}, \qquad \zeta = \frac{z}{h}, \qquad \zeta_0 = \frac{z_0}{h}, \qquad \Omega_P = \frac{\omega h}{\alpha}, \qquad \Omega_S = \frac{\omega h}{\beta} \tag{7.24}$$

Definitions for cases a), b)

$$\kappa_{Sj} = \sqrt{\Omega_S^2 - \pi^2 j^2}, \quad \mathrm{Im}(\kappa_{Sj}) \leq 0 \quad \text{Modal wavenumber for S waves} \tag{7.25}$$

$$\kappa_{Pj} = \sqrt{\Omega_P^2 - \pi^2 j^2}, \quad \mathrm{Im}(\kappa_{Pj}) \leq 0 \quad \text{Modal wavenumber for P waves} \tag{7.26}$$

a) FC-FC

$$u_{xx} = \frac{1}{i\mu \, \Omega_S^2} \left\{ \frac{1}{2}\Omega_P \, e^{-i\Omega_P|\xi|} + \sum_{j=1}^{\infty} \left[\kappa_{Pj} \, e^{-i\kappa_{Pj}|\xi|} + \frac{j^2\pi^2}{\kappa_{Sj}} e^{-i\kappa_{Sj}|\xi|} \right] \cos(j\pi\zeta) \cos(j\pi\zeta_0) \right\}$$

$$\tag{7.27a}$$

$$u_{zx} = -\frac{\mathrm{sgn}(\xi)}{\mu \, \Omega_S^2} \sum_{j=1}^{\infty} j\pi \left(e^{-\mathrm{i}\kappa_{Pj}|\xi|} - e^{-\mathrm{i}\kappa_{Sj}|\xi|} \right) \sin(j\pi\zeta) \cos(j\pi\zeta_0) \tag{7.27b}$$

$$u_{xz} = +\frac{\mathrm{sgn}(\xi)}{\mu \, \Omega_S^2} \sum_{j=1}^{\infty} j\pi \left(e^{-\mathrm{i}\kappa_{Pj}|\xi|} - e^{-\mathrm{i}\kappa_{Sj}|\xi|} \right) \cos(j\pi\zeta) \sin(j\pi\zeta_0) \tag{7.27c}$$

$$u_{zz} = \frac{1}{\mathrm{i}\mu \, \Omega_S^2} \sum_{j=1}^{\infty} \left[\frac{j^2\pi^2}{\kappa_{Pj}} e^{-\mathrm{i}\kappa_{Pj}|\xi|} + \kappa_{Sj} \, e^{-\mathrm{i}\kappa_{Sj}|\xi|} \right] \sin(\pi j\zeta) \sin(\pi j\zeta_0) \tag{7.27d}$$

b) CF-CF

$$u_{xx} = \frac{1}{\mathrm{i}\mu \, \Omega_S^2} \sum_{j=1}^{\infty} \left[\kappa_{Pj} \, e^{-\mathrm{i}\kappa_{Pj}|\xi|} + \frac{j^2\pi^2}{\kappa_{Sj}} e^{-\mathrm{i}\kappa_{Sj}|\xi|} \right] \sin(\pi j\zeta) \sin(\pi j\zeta_0) \tag{7.28a}$$

$$u_{zx} = +\frac{\mathrm{sgn}(\xi)}{\mu \, \Omega_S^2} \sum_{j=1}^{\infty} j\pi \left(e^{-\mathrm{i}\kappa_{Pj}|\xi|} - e^{-\mathrm{i}\kappa_{Sj}|\xi|} \right) \cos(j\pi\zeta) \sin(j\pi\zeta_0) \tag{7.28b}$$

$$u_{xz} = -\frac{\mathrm{sgn}(\xi)}{\mu \, \Omega_S^2} \sum_{j=1}^{\infty} j\pi \left(e^{-\mathrm{i}\kappa_{Pj}|\xi|} - e^{-\mathrm{i}\kappa_{Sj}|\xi|} \right) \sin(j\pi\zeta) \cos(j\pi\zeta_0) \tag{7.28c}$$

$$u_{zz} = \frac{1}{\mathrm{i}\mu \, \Omega_S^2} \left\{ \tfrac{1}{2}\Omega_S \, e^{-\mathrm{i}\Omega_S|\xi|} + \sum_{j=1}^{\infty} \left[\frac{j^2\pi^2}{\kappa_{Pj}} e^{-\mathrm{i}\kappa_{Pj}|\xi|} + \kappa_{Sj} \, e^{-\mathrm{i}\kappa_{Sj}|\xi|} \right] \cos(j\pi\zeta) \cos(j\pi\zeta_0) \right\} \tag{7.28d}$$

Definitions for cases c) and d)

$$l = \tfrac{1}{2}(2j-1) = \tfrac{1}{2}, \tfrac{3}{2}, \tfrac{5}{2}, \cdots \tag{7.29}$$

$$\kappa_{Sl} = \sqrt{\Omega_S^2 - \pi^2 l^2}, \quad \mathrm{Im}\,(\kappa_{Sl}) \le 0 \quad \text{Modal wavenumber for S waves} \tag{7.30}$$

$$\kappa_{Pl} = \sqrt{\Omega_P^2 - \pi^2 l^2}, \quad \mathrm{Im}\,(\kappa_{Pl}) \le 0 \quad \text{Modal wavenumber for P waves} \tag{7.31}$$

c) FC-CF

$$u_{xx} = \frac{1}{\mathrm{i}\mu \, \Omega_S^2} \sum_{j=1}^{\infty} \left[\kappa_{Pl} \, e^{-\mathrm{i}\kappa_{Pl}|\xi|} + \frac{l^2\pi^2}{\kappa_{Sl}} e^{-\mathrm{i}\kappa_{Sl}|\xi|} \right] \cos(\pi l\zeta) \cos(\pi l\zeta_0) \tag{7.32a}$$

$$u_{zx} = -\frac{\mathrm{sgn}(\xi)}{\mu \, \Omega_S^2} \sum_{j=1}^{\infty} l\pi \left(e^{-\mathrm{i}\kappa_{Pl}|\xi|} - e^{-\mathrm{i}\kappa_{Sl}|\xi|} \right) \sin(l\pi\zeta) \cos(l\pi\zeta_0) \tag{7.32b}$$

$$u_{xz} = +\frac{\mathrm{sgn}(\xi)}{\mu \, \Omega_S^2} \sum_{j=1}^{\infty} l\pi \left(e^{-\mathrm{i}\kappa_{Pl}|\xi|} - e^{-\mathrm{i}\kappa_{Sl}|\xi|} \right) \cos(l\pi\zeta) \sin(l\pi\zeta_0) \tag{7.32c}$$

$$u_{zz} = \frac{1}{\mathrm{i}\mu \, \Omega_S^2} \sum_{j=1}^{\infty} \left[\frac{l^2\pi^2}{\kappa_{Pl}} e^{-\mathrm{i}\kappa_{Pl}|\xi|} + \kappa_{Sl} \, e^{-\mathrm{i}\kappa_{Sl}|\xi|} \right] \sin(\pi l\zeta) \sin(\pi l\zeta_0) \tag{7.32d}$$

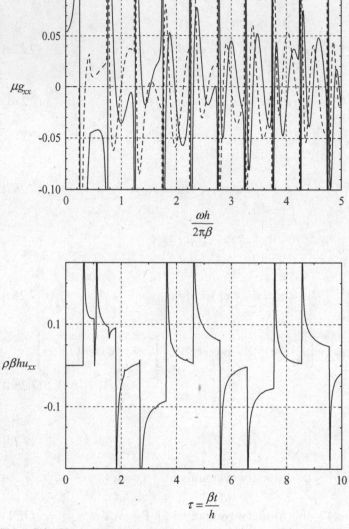

Figure 7.6a: Homogeneous plate with mixed boundary conditions **FC-CF**, $\nu = 0.25$. Horizontal response at $(x, z) = (h, 0.75h)$ due to horizontal SVP line load at elevation $z_0 = 0.25h$. Top: frequency domain. Solid line, real part; dashed line, imaginary part. Bottom: time domain.

d) CF-FC

$$u_{xx} = \frac{1}{i\mu\,\Omega_S^2} \sum_{j=1}^{\infty} \left[\kappa_{Pl}\, e^{-i\kappa_{Pl}|\xi|} + \frac{l^2\pi^2}{\kappa_{Sl}} e^{-i\kappa_{Sl}|\xi|} \right] \sin(\pi l\zeta)\sin(\pi l\zeta_0) \qquad (7.33a)$$

$$u_{zx} = +\frac{\text{sgn}(\xi)}{\mu\,\Omega_S^2} \sum_{j=1}^{\infty} l\pi \left(e^{-i\kappa_{Pl}|\xi|} - e^{-i\kappa_{Sl}|\xi|} \right) \cos(l\pi\zeta)\sin(l\pi\zeta_0) \qquad (7.33b)$$

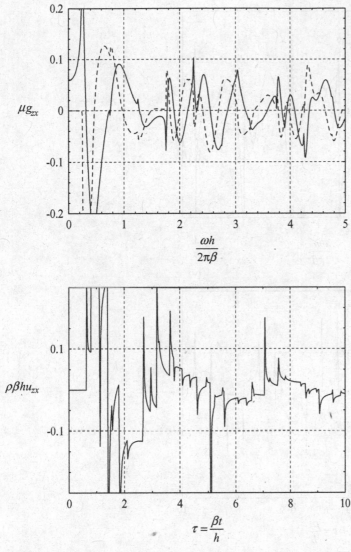

Figure 7.6b: Homogeneous plate with mixed boundary conditions *FC-CF*, $\nu = 0.25$. Vertical response at $(x,z) = (h, 0.75h)$ due to horizontal SVP line load at elevation $z_0 = 0.25h$. Top: frequency domain. Solid line, real part; dashed line, imaginary part. Bottom: time domain.

$$u_{xz} = -\frac{\text{sgn}(\xi)}{\mu\,\Omega_S^2} \sum_{j=1}^{\infty} l\pi \left(e^{-i\kappa_{Pl}|\xi|} - e^{-i\kappa_{Sl}|\xi|} \right) \sin(l\pi\zeta)\cos(l\pi\zeta_0) \qquad (7.33c)$$

$$u_{zz} = \frac{1}{i\mu\,\Omega_S^2} \sum_{j=1}^{\infty} \left[\frac{l^2\pi^2}{\kappa_{Pl}} e^{-i\kappa_{Pl}|\xi|} + \kappa_{Sl}\, e^{-i\kappa_{Sl}|\xi|} \right] \cos(\pi l\zeta)\cos(\pi l\zeta_o) \qquad (7.33d)$$

Note: The modal solution is *much* faster and more accurate than the frequency domain solution with the method of images. Figures 7.6a, 7.6b, 7.6c shows the results for a plate

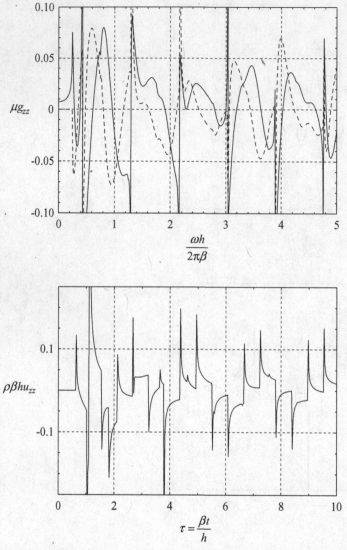

Figure 7.6c: Homogeneous plate with mixed boundary conditions **FC-CF**, $\nu = 0.25$. Vertical response at $(x,z) = (h, 0.75h)$ due to vertical SVP line load at elevation $z_0 = 0.25h$. Top: frequency domain. Solid line, real part; dashed line, imaginary part. Bottom: time domain.

with mixed boundary conditions **FC** (bottom) and **CF** (top), for Poison's ratio $\nu = 0.25$. The time domain solution uses solely the mirror images whose arrivals at the receiver fall within the observation window; the normal mode solution uses 20 modes. The frequency domain solution with the method of images and 50 images gives similar results to the modal method, but with some low amplitude ripples that decrease as the number of images is increased.

Read me first

Section V consists of three chapters, of which the first two present the theoretical framework and equations needed to obtain the practical numerical methods and formulations included in the third. Thus, Chapters 8 and 9 are for the most part meant as a reference to Chapter 10, and less so for their own utility. Nonetheless, we decided to include this material herein in full length because no readily available reference exists containing the detailed derivation of the powerful stiffness matrix method for all the three major coordinate systems. Thus, the reader may wish to either browse lightly Chapters 8 and 9 – especially the example problems in Chapter 9 – or skip these altogether at first and return later only as may be needed to clarify matters.

Chapter 8 begins with a summary of the solutions to the scalar and vector Helmholtz equations in three-dimensional space, and then proceeds to give full derivations to these equations and to the wave equation in all three coordinate systems. Unlike most books on the theory of elasticity, we use matrix algebra throughout, and manage to express the final results as products of matrices, each of which depends on one coordinate only (whether or not the systems are layered). This greatly simplifies applications to layered media.

Chapter 9 makes a compact introduction to the integral transform methods commonly used to analyze stratified media, such as finite and infinite plates, rods, or spheres. Examples are also given to illustrate the application of these concepts.

Finally, Chapter 10 contains the detailed equations needed to implement the *stiffness matrix method* for layered media in Cartesian, cylindrical, or spherical coordinates (also called by some the spectral element method). This is a powerful numerical tool that allows solving dynamically loaded laminated systems consisting of arbitrarily thick layers, or even infinite media. Examples are included that show what purposes it serves and how the method is used. A concise description is presented in the introduction and in Section 10.1.

8 Solution to the Helmholtz and wave equations

The solution to the wave equations in Cartesian coordinates can be found in most books on the theory of elasticity, but the corresponding solutions for both the Helmholtz and wave equations in cylindrical and (especially) spherical coordinates are much less common in the literature. Thus, to establish the fundamental methods, equations, and notation needed later on, we present in the ensuing the detailed solutions in the *frequency–wavenumber domain* for the following three equations:

Scalar Helmholtz equation	$\nabla^2 \Phi + k_P^2 \Phi = 0$	(8.1)
Vector Helmholtz equation	$\nabla \cdot \nabla \Psi + k_S^2 \Psi = \mathbf{0}$	(8.2)
Wave equation, isotropic medium	$(\lambda + 2\mu)\nabla\nabla \cdot \mathbf{u} - \mu\nabla \times \nabla \times \mathbf{u} = \rho\ddot{\mathbf{u}}$	(8.3)

We solve these equations by *separation of variables*, which is restricted to separable coordinate systems and to appropriate constitutive properties, such as isotropy and transverse isotropy. We begin with a summary of the results in all three coordinate systems, and provide thereafter the detailed solutions. All of these carry an implied exponential factor $e^{i\omega t}$, and the parameters a_j, b_j, c_j are arbitrary integration constants.

8.1 Summary of results

Cartesian coordinates

$$\Phi(x, y, z) = \left(a_1 e^{ik_x x} + a_2 e^{-ik_x x}\right)\left(a_3 e^{ik_y y} + a_4 e^{-ik_y y}\right)\left(a_5 e^{ik_z z} + a_6 e^{-ik_z z}\right) \tag{8.4}$$

$$\psi(x, y, z) = \left(b_1 e^{ik_x x} + b_2 e^{-ik_x x}\right)\left(b_3 e^{ik_y y} + b_4 e^{-ik_y y}\right)\left(b_5 e^{ik_z z} + b_6 e^{-ik_z z}\right) \tag{8.5}$$

$$\chi(x, y, z) = \left(c_1 e^{ik_x x} + c_2 e^{-ik_x x}\right)\left(c_3 e^{ik_y y} + c_4 e^{-ik_y y}\right)\left(c_5 e^{ik_z z} + c_6 e^{-ik_z z}\right) \tag{8.6}$$

$$\Psi = \psi\hat{\mathbf{k}} + \frac{1}{k_S^2}\nabla\left(\frac{\partial\psi}{\partial z}\right) + \frac{1}{k_S}\nabla \times \left(\chi\hat{\mathbf{k}}\right) \tag{8.7}$$

$$\mathbf{u} = \left[\frac{1}{k_P}\frac{\partial\Phi}{\partial x} + \frac{1}{k_S^2}\frac{\partial^2\chi}{\partial x\,\partial z} + \frac{1}{k_S}\frac{\partial\psi}{\partial y}\right]\hat{\mathbf{i}} + \left[\frac{1}{k_P}\frac{\partial\Phi}{\partial y} + \frac{1}{k_S^2}\frac{\partial^2\chi}{\partial y\,\partial z} - \frac{1}{k_S}\frac{\partial\psi}{\partial x}\right]\hat{\mathbf{j}}$$

$$+ \left[\frac{1}{k_P}\frac{\partial\Phi}{\partial z} + \frac{1}{k_S^2}\left(k_S^2\chi + \frac{\partial^2\chi}{\partial z^2}\right)\right]\hat{\mathbf{k}} \tag{8.8}$$

Plane strain, horizontally layered media

$$\bar{\mathbf{u}}(x, z, \omega) = [u_x \quad u_y \quad -iu_z]^T = \mathbf{u}\,e^{-ikx} \quad \text{(overbar for imaginary factor)} \tag{8.9}$$

$$\mathbf{u}(k, z, \omega) = \mathbf{R}_1\,\mathbf{E}_z^{-1}\mathbf{a}_1 + \mathbf{R}_2\,\mathbf{E}_z\mathbf{a}_2 \quad (\mathbf{a}_1, \mathbf{a}_2 : 3 \times 1 \text{ vectors of constants}) \tag{8.10}$$

$$\mathbf{E}_z = \mathbf{diag}\left\{e^{kpz} \quad e^{ksz} \quad e^{ksz}\right\} \tag{8.11}$$

$$p = \sqrt{1 - \left(\frac{\omega}{\alpha k}\right)^2} = \sqrt{1 - \left(\frac{c}{\alpha}\right)^2}, \qquad s = \sqrt{1 - \left(\frac{\omega}{\beta k}\right)^2} = \sqrt{1 - \left(\frac{c}{\beta}\right)^2} \tag{8.12}$$

$$k \equiv k_x = \text{horizontal wavenumber}, \qquad c = \text{phase velocity} \tag{8.13}$$

$$\mathbf{R}_1 = \left\{\begin{matrix} 1 & 0 & -s \\ 0 & 1 & 0 \\ -p & 0 & 1 \end{matrix}\right\}, \qquad \mathbf{R}_2 = \left\{\begin{matrix} 1 & 0 & s \\ 0 & 1 & 0 \\ p & 0 & 1 \end{matrix}\right\} \tag{8.14}$$

Cylindrical coordinates

$$\Phi(r, \theta, z) = (a_1 \cos n\theta + a_2 \sin n\theta)\left(a_3 J_n(k_\alpha r) + a_4 Y_n(k_\alpha r)\right)\left(a_5 e^{ik_z z} + a_6 e^{-ik_z z}\right) \tag{8.15}$$

$$\psi(r, \theta, z) = (b_1 \cos n\theta + b_2 \sin n\theta)\left(b_3 J_n(k_\alpha r) + b_4 Y_n(k_\alpha r)\right)\left(b_5 e^{ik_z z} + b_6 e^{-ik_z z}\right) \tag{8.16}$$

$$\chi(r, \theta, z) = (c_1 \cos n\theta + c_2 \sin n\theta)\left(c_3 J_n(k_\alpha r) + c_4 Y_n(k_\alpha r)\right)\left(c_5 e^{ik_z z} + c_6 e^{-ik_z z}\right) \tag{8.17}$$

$$\mathbf{\Psi} = \psi\hat{\mathbf{k}} + \frac{1}{k_S^2}\nabla\left(\frac{\partial \psi}{\partial z}\right) + \frac{1}{k_S}\nabla \times \left(\chi\hat{\mathbf{k}}\right) \tag{8.18}$$

$$\mathbf{u} = \left[\frac{1}{k_\alpha}\frac{\partial \Phi}{\partial r} + \frac{1}{k_\beta^2}\frac{\partial^2 \chi}{\partial r\,\partial z} + \frac{1}{k_\beta}\frac{1}{r}\frac{\partial \psi}{\partial \theta}\right]\hat{\mathbf{r}} + \left[\frac{1}{k_\alpha r}\frac{\partial \Phi}{\partial \theta} + \frac{1}{k_\beta^2}\frac{1}{r}\frac{\partial^2 \chi}{\partial \theta\,\partial z} - \frac{1}{k_\beta}\frac{\partial \psi}{\partial r}\right]\hat{\mathbf{t}}$$

$$+ \left[\frac{1}{k_\alpha}\frac{\partial \Phi}{\partial z} + \frac{1}{k_\beta^2}\left(k_S^2 \chi + \frac{\partial^2 \chi}{\partial z^2}\right)\right]\hat{\mathbf{k}} \tag{8.19}$$

$$\mathbf{T}_n = \mathbf{diag}\left[\begin{pmatrix} \cos n\theta \\ \sin n\theta \end{pmatrix} \begin{pmatrix} -\sin n\theta \\ \cos n\theta \end{pmatrix} \begin{pmatrix} \cos n\theta \\ \sin n\theta \end{pmatrix}\right] \quad \text{(either the top or bottom elements)} \tag{8.20}$$

Horizontally layered system
(Observe the similarities with Cartesian coordinates.)

$$\mathbf{u}(r, \theta, z) = [u_r \quad u_\theta \quad u_z]^T = \mathbf{T}_n\,\mathbf{C}_n\,\mathbf{u} \tag{8.21}$$

$$\mathbf{u}(k, z, \omega) = \mathbf{R}_1\mathbf{E}_z^{-1}\mathbf{a}_1 + \mathbf{R}_2\,\mathbf{E}_z\mathbf{a}_2 \quad (\mathbf{a}_1, \mathbf{a}_2 = 3 \times 1 \text{ vectors of constants}) \tag{8.22}$$

$$\mathbf{E}_z = \mathbf{diag}\left\{e^{kpz} \quad e^{ksz} \quad e^{ksz}\right\} \tag{8.23}$$

$$p = \sqrt{1 - \left(\frac{\omega}{\alpha k}\right)^2}, \qquad s = \sqrt{1 - \left(\frac{\omega}{\beta k}\right)^2}, \qquad k = \text{radial wavenumber} \tag{8.24}$$

$$
\mathbf{C}_n = \left\{ \begin{array}{ccc} J_n' & \frac{n}{kr} J_n & 0 \\ \frac{n}{kr} J_n & J_n' & 0 \\ 0 & 0 & J_n \end{array} \right\}, \qquad J_n' = \frac{dJ_n(kr)}{d(kr)} \tag{8.25}
$$

$$
\mathbf{R}_1 = \left\{ \begin{array}{ccc} 1 & 0 & -s \\ 0 & 1 & 0 \\ -p & 0 & 1 \end{array} \right\}, \qquad \mathbf{R}_2 = \left\{ \begin{array}{ccc} 1 & 0 & s \\ 0 & 1 & 0 \\ p & 0 & 1 \end{array} \right\} \tag{8.26}
$$

Cylindrically layered system

$$
\bar{\mathbf{u}}(r, \theta, z, \omega) = [u_r \quad u_\theta \quad -\mathrm{i}\, u_z]^T = \mathbf{T}_n \mathbf{u}\, e^{-\mathrm{i}k_z z} \quad \text{(overbar for imaginary factor)} \tag{8.27}
$$

$$
\mathbf{u}(r, k_z, \omega) = \mathbf{H}_n^{(1)} \mathbf{a}_1 + \mathbf{H}_n^{(2)} \mathbf{a}_2 \qquad (\mathbf{a}_1, \mathbf{a}_2 : 3 \times 1 \text{ vectors of constants}) \tag{8.28}
$$

$$
\mathbf{H}_n^{(1)} = \left\{ \begin{array}{ccc} \left(H_{\alpha n}^{(1)}\right)' & n\dfrac{H_{\beta n}^{(1)}}{k_\beta r} & \dfrac{k_z}{k_\beta}\left(H_{\beta n}^{(1)}\right)' \\[2mm] n\dfrac{H_{\alpha n}^{(1)}}{k_\alpha r} & \left(H_{\beta n}^{(1)}\right)' & n\dfrac{k_z}{k_\beta}\dfrac{H_{\beta n}^{(1)}}{k_\beta r} \\[2mm] -\dfrac{k_z}{k_\alpha} H_{\alpha n}^{(1)} & 0 & H_{\beta n}^{(1)} \end{array} \right\}, \tag{8.29a}
$$

$$
\mathbf{H}_n^{(2)} = \left\{ \begin{array}{ccc} \left(H_{\alpha n}^{(2)}\right)' & n\dfrac{H_{\beta n}^{(2)}}{k_\beta r} & \dfrac{k_z}{k_\beta}\left(H_{\beta n}^{(2)}\right)' \\[2mm] n\dfrac{H_{\alpha n}^{(2)}}{k_\alpha r} & \left(H_{\beta n}^{(2)}\right)' & n\dfrac{k_z}{k_\beta}\dfrac{H_{\beta n}^{(2)}}{k_\beta r} \\[2mm] -\dfrac{k_z}{k_\alpha} H_{\alpha n}^{(2)} & 0 & H_{\beta n}^{(2)} \end{array} \right\} \tag{8.29b}
$$

$$
k_\alpha = \sqrt{k_P^2 - k_z^2}, \qquad k_\beta = \sqrt{k_S^2 - k_z^2}, \qquad (k_z = \text{axial wavenumber}) \tag{8.30}
$$

$$
H_{\alpha n}^{(1)} = H_n^{(1)}(k_a r), \qquad H_{\alpha n}^{(2)} = H_n^{(2)}(k_a r), \qquad \left(H_{\alpha n}^{(1)}\right)' = \frac{dH_{\alpha n}^{(1)}}{d(k_\alpha r)}, \qquad \left(H_{\alpha n}^{(2)}\right)' = \frac{dH_{\alpha n}^{(2)}}{d(k_\alpha r)} \tag{8.31}
$$

$$
H_{\beta n}^{(1)} = H_n^{(1)}(k_\beta r), \qquad H_{\beta n}^{(2)} = H_n^{(2)}(k_\beta r), \qquad \left(H_{\beta n}^{(1)}\right)' = \frac{dH_{\beta n}^{(1)}}{d(k_\beta r)}, \qquad \left(H_{\beta n}^{(2)}\right)' = \frac{dH_{\beta n}^{(2)}}{d(k_\beta r)} \tag{8.32}
$$

Note: In the matrices $\mathbf{H}_n$ above, we could also have used the standard Bessel and Neumann functions J_n, Y_n in place of the Hankel functions. However, when the argument of the Bessel functions is complex, and especially when it is an imaginary number, J_n and Y_n are virtually proportional, and are thus not truly independent solutions to the Bessel equation.

Spherical coordinates

$$\Phi(R, \phi, \theta) = (a_1 \cos n\theta + a_2 \sin n\theta)\,(a_3\, j_m(kR) + a_4\, y_m(kR))\,(a_5\, P_m^n + a_6\, Q_m^n) \tag{8.33}$$

$$\psi(R, \phi, \theta) = (b_1 \cos n\theta + b_2 \sin n\theta)\,(b_3\, j_m(kR) + b_4\, y_m(kR))\,(b_5\, P_m^n + b_6\, Q_m^n) \tag{8.34}$$

$$\chi(R, \phi, \theta) = (c_1 \cos n\theta + c_2 \sin n\theta)\,(c_3\, j_m(kR) + c_4\, y_m(kR))\,(c_5\, P_m^n + c_6\, Q_m^n) \tag{8.35}$$

$$\mathbf{\Psi} = R\psi\hat{\mathbf{r}} + \frac{1}{k^2}\nabla\left(\frac{\partial R\psi}{\partial R}\right) + \frac{1}{k}\nabla \times (R\chi\hat{\mathbf{r}}) \tag{8.36}$$

$$\mathbf{u}(R, \phi, \theta, \omega) = \mathbf{T}_n \mathbf{L}_m^n (\mathbf{H}_m^{(1)}\mathbf{a}_1 + \mathbf{H}_m^{(2)}\mathbf{a}_2) \quad (\mathbf{a}_1, \mathbf{a}_2 : 3 \times 1 \text{ vectors of constants}) \tag{8.37}$$

$$\mathbf{T}_n = \mathrm{diag}\left\{ \begin{pmatrix} \cos n\theta \\ \sin n\theta \end{pmatrix} \begin{pmatrix} \cos n\theta \\ \sin n\theta \end{pmatrix} \begin{pmatrix} -\sin n\theta \\ \cos n\theta \end{pmatrix} \right\} \quad (\text{either the top or bottom elements}) \tag{8.38}$$

$$\mathbf{L}_m^n = \left\{ \begin{array}{ccc} P_m^n & 0 & 0 \\[2mm] 0 & \dfrac{dP_m^n}{d\phi} & \dfrac{n\,P_m^n}{\sin\phi} \\[3mm] 0 & \dfrac{n\,P_m^n}{\sin\phi} & \dfrac{dP_m^n}{d\phi} \end{array} \right\}, \tag{8.39}$$

$$\mathbf{H}_m = \left\{ \begin{array}{ccc} \dfrac{dh_{Pm}}{dz_P} & m(m+1)\dfrac{h_{Sm}}{z_S} & 0 \\[3mm] \dfrac{h_{Pm}}{z_P} & \dfrac{1}{z_S}\dfrac{d(z_S h_{Sm})}{dz_S} & 0 \\[3mm] 0 & 0 & h_{Sm} \end{array} \right\}, \tag{8.40}$$

$$z_P = k_P R, \quad z_S = k_S R \tag{8.41}$$

$$h_{Pm} = h_m^{(1)}(z_P), h_m^{(2)}(z_P), j_m(z_P) \text{ or } y_m(z_P), \quad h_{Sm} = h_m^{(1)}(z_S), h_m^{(2)}(z_S), j_m(z_S) \text{ or } y_m(z_S) \tag{8.42}$$

$$\mathbf{H}_m^{(1)} = \mathbf{H}_m(h_m^{(1)}(z_P), h_m^{(1)}(z_S)), \qquad \mathbf{H}_m^{(2)} = \mathbf{H}_m(h_m^{(2)}(z_P), h_m^{(2)}(z_S)) \tag{8.43}$$

8.2 Scalar Helmholtz equation in Cartesian coordinates

Consider the Helmholtz equation involving a scalar function Φ:

$$\nabla^2\Phi + k_P^2\,\Phi = 0 \tag{8.44}$$

with parameter $k_P = \omega/\alpha$. In Cartesian coordinates, this is expressed as

$$\frac{\partial^2\Phi}{\partial x^2} + \frac{\partial^2\Phi}{\partial y^2} + \frac{\partial^2\Phi}{\partial z^2} + k_P^2\,\Phi = 0 \tag{8.45}$$

where $\Phi(x, y, z)$ is a scalar function of the coordinates. Making an ansatz

$$\Phi(x, y, z) = e^{ik_x x} e^{ik_y y} e^{ik_z z} \tag{8.46}$$

and substituting into the Helmholtz equation, we find that it is satisfied if

$$k_x^2 + k_y^2 + k_z^2 = k_P^2 \tag{8.47}$$

The general solution to the scalar Helmholtz equation is then

$$\Phi(x, y, z, \omega) = \left(a_1 e^{ik_x x} + a_2 e^{-ik_x x}\right)\left(a_3 e^{ik_y y} + a_4 e^{-ik_y y}\right)\left(a_5 e^{ik_z z} + a_6 e^{-ik_z z}\right) \tag{8.48}$$

in which the a_i are arbitrary constants. In combination with the implied factor $e^{i\omega t}$, and for real wavenumbers k_x, k_y, k_z, this solution for Φ represents compressional (P) waves that propagate in the (locally homogeneous) medium with velocity α.

For two-dimensional problems in which the wave velocity α does not change in horizontal planes (i.e., for a laterally homogeneous medium), and considering only waves that propagate and/or decay in the positive x direction, the solution simplifies to

$$\Phi(x, z, \omega) = \left(a_1 e^{ik_z z} + a_2 e^{-ik_z z}\right) e^{-ik_x x}, \qquad k_z = \sqrt{k_P^2 - k_x^2} \tag{8.49}$$

8.3 Vector Helmholtz equation in Cartesian coordinates

Consider the vector Helmholtz equation

$$\nabla^2 \boldsymbol{\Psi} + k_S^2 \boldsymbol{\Psi} = \mathbf{0} \tag{8.50}$$

with parameter $k_s = \omega/\beta$. More precisely,

$$\nabla \cdot \nabla \boldsymbol{\Psi} + k_S^2 \boldsymbol{\Psi} = \mathbf{0} \tag{8.51}$$

which involves the vector $\boldsymbol{\Psi} = \psi_x \hat{\mathbf{i}} + \psi_y \hat{\mathbf{j}} + \psi_z \hat{\mathbf{k}}$. We shall assume $\boldsymbol{\Psi}$ to be *solenoidal*, which means that it satisfies the *gauge condition* $\nabla \cdot \boldsymbol{\Psi} = 0$. Hence, only two of the three components of $\boldsymbol{\Psi}$ are independent, so the solution can only contain two independent functions. However,

$$\nabla \cdot \nabla \boldsymbol{\Psi} = \nabla(\nabla \cdot \boldsymbol{\Psi}) - \nabla \times \nabla \times \boldsymbol{\Psi}$$
$$= -\nabla \times \nabla \times \boldsymbol{\Psi} \tag{8.52}$$

Hence

$$\nabla \times \nabla \times \boldsymbol{\Psi} = k_S^2 \boldsymbol{\Psi} \tag{8.53}$$

This equation admits solutions of the form

$$\begin{aligned}\boldsymbol{\Psi} &= \boldsymbol{\Psi}_1 + \boldsymbol{\Psi}_2 \\ &= \psi\,\hat{\mathbf{k}} + \frac{1}{k_S^2}\nabla\left(\frac{\partial \psi}{\partial z}\right) + \frac{1}{k_S}\nabla \times \left(\chi\hat{\mathbf{k}}\right)\end{aligned} \tag{8.54}$$

in which ψ, χ are solutions to the scalar Helmholtz equations

$$\boxed{\nabla^2 \psi + k_S^2 \psi = 0} \quad \text{and} \quad \boxed{\nabla^2 \chi + k_S^2 \chi = 0} \tag{8.55}$$

To prove that this is indeed the solution, we make use of some well-known vector operation identities, namely $\nabla \times \nabla f = 0$, $\nabla \cdot \nabla \times \mathbf{f} = 0$, which are true for any scalar and vector functions f and $\mathbf{f}$, respectively. Consider the first term

$$\nabla \times \boldsymbol{\Psi}_1 = \nabla \times \left(\psi \hat{\mathbf{k}}\right) + \frac{1}{k_S^2} \nabla \times \nabla \left(\frac{\partial \psi}{\partial z}\right) = \nabla \times \left(\psi \hat{\mathbf{k}}\right)$$

$$= \frac{\partial \psi}{\partial y} \hat{\mathbf{i}} - \frac{\partial \psi}{\partial x} \hat{\mathbf{j}} \tag{8.56}$$

$$\nabla \times \nabla \times \boldsymbol{\Psi}_1 = \nabla \times \left[\frac{\partial \psi}{\partial y} \hat{\mathbf{i}} - \frac{\partial \psi}{\partial x} \hat{\mathbf{j}}\right]$$

$$= \nabla \left(\frac{\partial \psi}{\partial z}\right) - \hat{\mathbf{k}} \nabla^2 \psi \tag{8.57}$$

Since $\boldsymbol{\Psi}_1$ must satisfy $\nabla \times \nabla \times \boldsymbol{\Psi}_1 = k_S^2 \boldsymbol{\Psi}_1$, then

$$\nabla \left(\frac{\partial \psi}{\partial z}\right) - \hat{\mathbf{k}} \nabla^2 \psi = k_S^2 \left[\psi \hat{\mathbf{k}} + \frac{1}{k_S^2} \nabla \left(\frac{\partial \psi}{\partial z}\right)\right] \tag{8.58}$$

which results in the condition $\nabla^2 \psi + k_S^2 \psi = 0$. Similarly, the second term gives

$$k_S \nabla \times \boldsymbol{\Psi}_2 = \nabla \left(\frac{\partial \chi}{\partial z}\right) - \hat{\mathbf{k}} \nabla^2 \chi \tag{8.59}$$

so

$$k_S \nabla \times \nabla \times \boldsymbol{\Psi}_2 = \nabla \times \nabla \left(\frac{\partial \chi}{\partial z}\right) - \nabla \times \left(\hat{\mathbf{k}} \nabla^2 \chi\right) = k_S^3 \boldsymbol{\Psi}_2$$

$$= -\nabla \times \left(\hat{\mathbf{k}} \nabla^2 \chi\right) = k_S^2 \nabla \times \left(\hat{\mathbf{k}} \chi\right) \tag{8.60}$$

$$\nabla \times \left[\hat{\mathbf{k}} \left(\nabla^2 \chi + k_S^2 \chi\right)\right] = \mathbf{0} \tag{8.61}$$

which implies $\nabla^2 \chi + k_S^2 \chi = 0$. Thus, the solution provided does indeed satisfy the vector Helmholtz equation. It remains to verify the gauge condition:

$$\nabla \cdot \boldsymbol{\Psi} = \nabla \cdot \left(\psi \hat{\mathbf{k}}\right) + \frac{1}{k_S^2} \nabla \cdot \nabla \left(\frac{\partial \psi}{\partial z}\right) + \frac{1}{k_S} \nabla \cdot \nabla \times \left(\chi \hat{\mathbf{k}}\right)$$

$$= \nabla \cdot \left(\psi \hat{\mathbf{k}}\right) + \frac{1}{k_S^2} \nabla \cdot \left[\nabla \times \nabla \times \left(\psi \hat{\mathbf{k}}\right) + \hat{\mathbf{k}} \nabla^2 \psi\right]$$

$$= \nabla \cdot \left(\psi \hat{\mathbf{k}}\right) - \nabla \cdot \left(\psi \hat{\mathbf{k}}\right) = \mathbf{0} \tag{8.62}$$

Hence, the solution to the vector Helmholtz equation can now be formed with the solutions to the two scalar Helmholtz equations for ψ and χ, using for this purpose the expressions provided in the previous section:

$$\boxed{\begin{aligned} \psi(x, y, z, \omega) &= \left(b_1 e^{ik_x x} + b_2 e^{-ik_x x}\right) \left(b_3 e^{ik_y y} + b_4 e^{-ik_y y}\right) \left(b_5 e^{ik_z z} + b_6 e^{-ik_z z}\right) \\ \chi(x, y, z, \omega) &= \left(c_1 e^{ik_x x} + c_2 e^{-ik_x x}\right) \left(c_3 e^{ik_y y} + c_4 e^{-ik_y y}\right) \left(c_5 e^{ik_z z} + c_6 e^{-ik_z z}\right) \end{aligned}} \tag{8.63}$$

in which the b_i, c_i are arbitrary constants, and

$$\boxed{k_x^2 + k_y^2 + k_z^2 = k_S^2}$$

(8.64)

In combination with the implied factor $e^{i\omega t}$, and for real wavenumbers k_x, k_y, k_z, these two solutions represent shear (S) waves that propagate in the medium with velocity β.

In the 2-D case and considering only waves propagating and/or decaying in the positive x direction, the solution simplifies to

$$\left.\begin{array}{l} \psi(x, z, \omega) = \left(b_1 e^{ik_z z} + b_2 e^{-ik_z z}\right) e^{-ik_x x} \\ \chi(x, z, \omega) = \left(c_1 e^{ik_z z} + c_2 e^{-ik_z z}\right) e^{-ik_x x} \end{array}\right\} \qquad k_z = \sqrt{k_S^2 - k_x^2}$$

(8.65)

8.4 Elastic wave equation in Cartesian coordinates

Consider the elastic wave equation for an isotropic medium,

$$(\lambda + 2\mu) \nabla\nabla \cdot \mathbf{u} - \mu \nabla \times \nabla \times \mathbf{u} = \rho \ddot{\mathbf{u}}$$

(8.66)

Defining

$$\varepsilon = \nabla \cdot \mathbf{u} = \text{volumetric strain}$$

(8.67)

$$\mathbf{\Omega} = \tfrac{1}{2}\nabla \times \mathbf{u} = \text{rotation vector, which satisfies } \nabla \cdot \mathbf{\Omega} = 0$$

(8.68)

we can write the elastic wave equation as

$$(\lambda + 2\mu) \nabla\varepsilon - 2\mu\nabla \times \mathbf{\Omega} = \rho \ddot{\mathbf{u}}$$

(8.69)

Applying in turn a gradient and then a curl to the above equation (but not both), we obtain after straightforward algebra

$$(\lambda + 2\mu) \nabla^2 \varepsilon = \rho \ddot{\varepsilon} \quad \text{and} \quad \mu\nabla^2\mathbf{\Omega} = \rho\ddot{\mathbf{\Omega}}$$

(8.70)

These constitute a scalar and a vector wave equation. The latter consists of three scalar wave equations, of which one depends on the other two (because of the condition $\nabla \cdot \mathbf{\Omega} = 0$). Hence, for harmonic motion, these reduce to the two Helmholtz equations

$$\nabla^2 \varepsilon + k_P^2 \varepsilon = 0, \quad k_P = \omega/\alpha, \quad \alpha = \sqrt{(\lambda + 2\mu)/\rho} \quad \Rightarrow \quad \text{P waves}$$

(8.71)

$$\nabla^2\mathbf{\Omega} + k_S^2\mathbf{\Omega} = \mathbf{0}, \quad k_S = \omega/\beta, \quad \beta = \sqrt{\mu/\rho} \quad \Rightarrow \quad \text{S waves}$$

(8.72)

From the previous sections, the solutions to these two equations are

$$\varepsilon = \Phi, \quad \nabla^2\Phi + k_P^2 \Phi = 0, \qquad \nabla^2\psi + k_S^2 \psi = 0, \quad \nabla^2\chi + k_S^2 \chi = 0$$

(8.73)

$$2\mathbf{\Omega} \equiv \mathbf{\Psi} = \psi\hat{\mathbf{k}} + \frac{1}{k_S^2}\nabla\left(\frac{\partial\psi}{\partial z}\right) + \frac{1}{k_S}\nabla \times \left(\chi\hat{\mathbf{k}}\right)$$

(8.74)

Also, for harmonic motion, the wave equation can be written as

$$(\lambda + 2\mu) \nabla\varepsilon - 2\mu\nabla \times \mathbf{\Omega} = -\rho\omega^2 \mathbf{u}$$

(8.75)

that is,

$$\mathbf{u} = \frac{\nabla \times \boldsymbol{\Psi}}{k_S^2} - \frac{\nabla \Phi}{k_P^2}$$

$$= \frac{1}{k_S^2} \nabla \times \left[\psi \hat{\mathbf{k}} + \frac{1}{k_S^2} \nabla \left(\frac{\partial \psi}{\partial z} \right) + \frac{1}{k_S} \nabla \times \left(\chi \hat{\mathbf{k}} \right) \right] - \frac{1}{k_P^2} \nabla \Phi \tag{8.76}$$

Carrying out the above differentiation operations, we obtain

$$\mathbf{u} = \frac{1}{k_S^2} \left\{ \left[\frac{1}{k_S} \frac{\partial^2 \chi}{\partial x \, \partial z} + \frac{\partial \psi}{\partial y} \right] \hat{\mathbf{i}} + \left[\frac{1}{k_S} \frac{\partial^2 \chi}{\partial y \, \partial z} - \frac{\partial \psi}{\partial x} \right] \hat{\mathbf{j}} + \left[k_S \chi + \frac{1}{k_S} \frac{\partial^2 \chi}{\partial z^2} \right] \hat{\mathbf{k}} \right\}$$

$$- \frac{1}{k_P^2} \left\{ \frac{\partial \Phi}{\partial x} \hat{\mathbf{i}} + \frac{\partial \Phi}{\partial y} \hat{\mathbf{j}} + \frac{\partial \Phi}{\partial z} \hat{\mathbf{k}} \right\} \tag{8.77}$$

Because the solutions for Φ, χ, and ψ contain arbitrary constants of integration, we can rescale these by any other constants. Hence, using rescaling factors k_P for Φ and k_S for ψ and χ, we can write the displacement vector as

$$\mathbf{u} = \left[\frac{1}{k_P} \frac{\partial \Phi}{\partial x} + \frac{1}{k_S^2} \frac{\partial^2 \chi}{\partial x \, \partial z} + \frac{1}{k_S} \frac{\partial \psi}{\partial y} \right] \hat{\mathbf{i}}$$

$$+ \left[\frac{1}{k_P} \frac{\partial \Phi}{\partial y} + \frac{1}{k_S^2} \frac{\partial^2 \chi}{\partial y \, \partial z} - \frac{1}{k_S} \frac{\partial \psi}{\partial x} \right] \hat{\mathbf{j}}$$

$$+ \left[\frac{1}{k_P} \frac{\partial \Phi}{\partial z} + \frac{1}{k_S^2} \left(k_S^2 \chi + \frac{\partial^2 \chi}{\partial z^2} \right) \right] \hat{\mathbf{k}} \tag{8.78}$$

in which the solutions for the scalar potentials Φ, χ, and ψ from the earlier sections must be used. In compact notation, these solutions can be written as

$$\Phi = e^{\pm ik_x x} e^{\pm ik_y y} e^{\pm ik_{z\alpha} z} A_\Phi \tag{8.79}$$

$$\chi = e^{\pm ik_x x} e^{\pm ik_y y} e^{\pm ik_{z\beta} z} A_\chi \tag{8.80}$$

$$\psi = e^{\pm ik_x x} e^{\pm ik_y y} e^{\pm ik_{z\beta} z} A_\psi \tag{8.81}$$

in which the A_Φ, A_χ, A_ψ are arbitrary integration constants, and the vertical wavenumbers are

$$k_{z\alpha} = \sqrt{k_P^2 - k_x^2 - k_y^2} \quad \text{and} \quad k_{z\beta} = \sqrt{k_S^2 - k_x^2 - k_y^2} \tag{8.82}$$

8.4.1 Horizontally stratified media, plane strain

We now restrict our analysis to plane-strain waves in a horizontally stratified medium defined in the two-dimensional x, z space, in which case the derivatives with respect to y must be discarded. Thus, the displacement vector reduces to

$$\mathbf{u} = \left[\frac{1}{k_P} \frac{\partial \Phi}{\partial x} + \frac{1}{k_S^2} \frac{\partial^2 \chi}{\partial x \, \partial z} \right] \hat{\mathbf{i}} - \frac{1}{k_S} \frac{\partial \psi}{\partial x} \hat{\mathbf{j}} + \left[\frac{1}{k_P} \frac{\partial \Phi}{\partial z} + \frac{1}{k_S^2} \left(k_S^2 \chi + \frac{\partial^2 \chi}{\partial z^2} \right) \right] \hat{\mathbf{k}} \tag{8.83}$$

The terms in Φ, χ are vertically polarized shear and pressure waves (SV-P waves); the term in ψ corresponds to horizontally polarized shear waves (SH waves).

From this point on, we consider only waves that propagate or decay in the positive x direction, and drop for simplicity the sub-index for the horizontal wavenumber k_x, in which case the potentials can be written in the form

$$\Phi = \left(A_1 e^{kpz} + A_2 e^{-kpz}\right) e^{-ikx}, \quad k \equiv k_x \tag{8.84}$$

$$\chi = \left(B_1 e^{ksz} + B_2 e^{-ksz}\right) e^{-ikx} \tag{8.85}$$

$$\psi = \left(C_1 e^{ksz} + C_2 e^{-ksz}\right) e^{-ikx} \tag{8.86}$$

in which we have defined

$$kp = i\, k_{z\alpha} = \sqrt{k^2 - k_{\mathrm{P}}^2}, \qquad ks = i\, k_{z\beta} = \sqrt{k^2 - k_{\mathrm{S}}^2} \tag{8.87}$$

that is,

$$p = \sqrt{1 - \left(\frac{\omega}{k\alpha}\right)^2}, \qquad s = \sqrt{1 - \left(\frac{\omega}{k\beta}\right)^2} \tag{8.88}$$

Introducing these potentials into the equation for the displacement vector, we obtain after brief algebra the solution to the vector wave equation in the frequency–horizontal-wavenumber domain for Cartesian coordinates as

$$u_x = -i\left[\frac{k}{k_{\mathrm{P}}}\left(A_1 e^{kpz} + A_2 e^{-kpz}\right) + s\left(\frac{k}{k_{\mathrm{S}}}\right)^2 \left(B_1 e^{ksz} - B_2 e^{-ksz}\right)\right] e^{-ikx} \tag{8.89}$$

$$u_y = i\frac{k}{k_{\mathrm{S}}}\left(C_1 e^{ksz} + C_2 e^{-ksz}\right) e^{-ikx} \tag{8.90}$$

$$u_z = \left[p\frac{k}{k_{\mathrm{P}}}\left(A_1 e^{kpz} - A_2 e^{-kpz}\right) + \left(\frac{k}{k_{\mathrm{S}}}\right)^2 \left(B_1 e^{ksz} + B_2 e^{-ksz}\right)\right] e^{-ikx} \tag{8.91}$$

Rescaling once more the (arbitrary) integration constants, we can write this in matrix form as

$$\bar{\mathbf{u}}(x, z, \omega) = \left\{\begin{array}{c} u_x \\ u_y \\ -i\, u_z \end{array}\right\}$$

$$= \left\{\begin{array}{cccccc} 1 & 0 & -s & 1 & 0 & s \\ 0 & 1 & 0 & 0 & 1 & 0 \\ -p & 0 & 1 & p & 0 & 1 \end{array}\right\} \left\{\begin{array}{cccccc} e^{kpz} & & & & & \\ & e^{ksz} & & & & \\ & & e^{ksz} & & & \\ & & & e^{-kpz} & & \\ & & & & e^{-ksz} & \\ & & & & & e^{-ksz} \end{array}\right\} \left\{\begin{array}{c} a_1 \\ a_2 \\ a_3 \\ a_4 \\ a_5 \\ a_6 \end{array}\right\} e^{-ikx}$$

$$\tag{8.92}$$

You will observe that we have added an imaginary factor in the vertical component, and have identified this modification by adding an overbar to the displacement vector. We have done this is for reasons of convenience:

- To avoid imaginary factors in the matrix of coefficients.
- To obtain expressions that are virtually identical to those in cylindrical coordinates; this greatly simplifies the transition between one and the other coordinate system.
- To obtain *symmetric* stiffness matrices for layered media, as will be seen in Chapter 10.

We can now write the *displacement field* compactly as

$$\bar{\mathbf{u}}(x, z, \omega) = \mathbf{u}(k, z, \omega)\, e^{-ikx}$$

$$\mathbf{u}(k, z, \omega) = \left\{ \begin{array}{c} \tilde{u}_x \\ \tilde{u}_y \\ -i\,\tilde{u}_z \end{array} \right\} = \mathbf{R}_1 \mathbf{E}_z^{-1} \mathbf{a}_1 + \mathbf{R}_2 \mathbf{E}_z \mathbf{a}_2$$

$$= \text{displcement in frequency–wavenumber domain}$$

$$\mathbf{a}_1, \mathbf{a}_2 = \text{arbitrary } 3 \times 1 \text{ vectors of integration constants (wave amplitudes)}$$

$$\mathbf{E}_z^{\pm 1} = \text{diag}\left\{ e^{\pm kpz} \quad e^{\pm ksz} \quad e^{\pm ksz} \right\}$$

$$\mathbf{R}_1 = \left\{ \begin{array}{ccc} 1 & 0 & -s \\ 0 & 1 & 0 \\ -p & 0 & 1 \end{array} \right\}, \qquad \mathbf{R}_2 = \left\{ \begin{array}{ccc} 1 & 0 & s \\ 0 & 1 & 0 \\ p & 0 & 1 \end{array} \right\}$$

(8.93)

Stresses in horizontal surfaces

We provide also the solution for stresses in horizontal planes, for these are necessary to enforce equilibrium across layer interfaces. From Section 1.4.1, these stresses are

$$\mathbf{s}_z = \left[\sigma_{xz} \quad \sigma_{yz} \quad \sigma_z \right]^T = \mathbf{D}_{zx}\frac{\partial \mathbf{u}}{\partial x} + \mathbf{D}_{zz}\frac{\partial \mathbf{u}}{\partial z} \tag{8.94}$$

To evaluate the stresses elicited by the displacement field previously found, we use the material matrices $\mathbf{D}_{zr}$, etc., for an isotropic medium, carry out the required differentiations, introduce a convenience factor $-i$, and obtain

$$\bar{\mathbf{s}}(x, z, \omega) = \left[\sigma_{xz} \quad \sigma_{yz} \quad -i\sigma_z \right]^T = \mathbf{s}\, e^{-ikx}$$

$$\mathbf{s}(k, z, \omega) = \left\{ \begin{array}{c} \tilde{\sigma}_{xz} \\ \tilde{\sigma}_{yz} \\ -i\tilde{\sigma}_z \end{array} \right\} = 2k\mu \left(-\mathbf{Q}_1 \mathbf{E}_z^{-1} \mathbf{a}_1 + \mathbf{Q}_2 \mathbf{E}_z \mathbf{a}_2 \right)$$

$$\mathbf{Q}_1 = \left\{ \begin{array}{ccc} 2p & 0 & -(1+s^2) \\ 0 & s & 0 \\ -(1+s^2) & 0 & 2s \end{array} \right\}, \qquad \mathbf{Q}_2 = \left\{ \begin{array}{ccc} 2p & 0 & 1+s^2 \\ 0 & s & 0 \\ 1+s^2 & 0 & 2s \end{array} \right\}$$

(8.95)

The two boxes above provide the solution for horizontally stratified media in Cartesian coordinates, expressed in the frequency–horizontal-wavenumber domain.

8.5 Scalar Helmholtz equation in cylindrical coordinates

Consider the Helmholtz equation involving a scalar function Φ:

$$\nabla^2 \Phi + k_{\mathrm{P}}^2 \Phi = 0 \tag{8.96}$$

with parameter $k_{\mathrm{P}} = \omega/\alpha$. In cylindrical coordinates, this is expressed as

$$\frac{\partial^2 \Phi}{\partial r^2} + \frac{1}{r}\frac{\partial \Phi}{\partial r} + \frac{1}{r^2}\frac{\partial^2 \Phi}{\partial \theta^2} + \frac{\partial^2 \Phi}{\partial z^2} + k_{\mathrm{P}}^2 \Phi = 0 \tag{8.97}$$

where $\Phi(r, \theta, z)$ is a scalar function of the coordinates. To solve this problem, we use separation of variables, and begin by assuming a harmonic variation in the azimuth of the form

$$\Phi(R, \theta, z) = F(r)\,(c_1 \cos n\theta + c_2 \sin n\theta)\,G(z) \tag{8.98}$$

where c_1, c_2 are constants that may depend on the parameter n. To satisfy continuity in the azimuth at the transition from 2π to 0, n must be an integer, non-negative number. This implies

$$\frac{\partial^2 \Phi}{\partial \theta^2} = -n^2 \Phi \tag{8.99}$$

Substituting into the differential equation, dividing by FG, and taking the terms in G to the right-hand side, we obtain

$$\frac{1}{F}\left[\frac{\partial^2 F}{\partial r^2} + \frac{1}{r}\frac{\partial F}{\partial r} + \left(k_{\mathrm{P}}^2 - \frac{n^2}{r^2}\right)F\right] = -\frac{1}{G}\frac{\partial^2 G}{\partial z^2} \tag{8.100}$$

The left-hand side is a function only of r; the right-hand side is a function only of z. Hence, each term must equal a constant k_z^2, which we recognize later as the square of the axial (i.e., vertical) wavenumber. It follows that

$$\frac{\partial^2 F}{\partial r^2} + \frac{1}{r}\frac{\partial F}{\partial r} + \left[k_{\mathrm{P}}^2 - k_z^2 - \frac{n^2}{r^2}\right]F = 0 \quad \Rightarrow \quad \text{Bessel equation} \tag{8.101}$$

$$\frac{\partial^2 G}{\partial z^2} + k_z^2 G = 0 \quad \Rightarrow \quad \text{1-D Helmholtz equation} \tag{8.102}$$

Defining $k_\alpha = \sqrt{k_{\mathrm{P}}^2 - k_z^2}$, the solution to these two equations is

$$F(r) = c_3\,H_n^{(1)}(k_\alpha r) + c_4\,H_n^{(2)}(k_\alpha r) \qquad \text{Hankel functions} \tag{8.103}$$

or

$$F(r) = c_3\,J_n(k_\alpha r) + c_4\,Y_n(k_\alpha r) \quad \text{Bessel functions} \tag{8.104}$$

and

$$G(z) = c_5\,e^{\mathrm{i}k_z z} + c_6\,e^{-\mathrm{i}k_z z} \tag{8.105}$$

in which c_3, c_4, c_5, c_6 are constants that may depend on the parameters n, k_z, and k_P. It follows that in cylindrical coordinates, *particular* solutions to the Helmholtz equation exist that are the form

$$\Phi(r, \theta, z) = (c_1 \cos n\theta + c_2 \sin n\theta)\,(c_3\, J_n(k_\alpha r) + c_4\, Y_n(k_\alpha r))\,(c_5\, e^{ik_z z} + c_6\, e^{-ik_z z}) \tag{8.106}$$

in which $k_\alpha = \sqrt{k_P^2 - k_z^2}$, and k_z is the axial wavenumber.

8.6 Vector Helmholtz equation in cylindrical coordinates

Consider the vector Helmholtz equation

$$\nabla^2 \Psi + k_S^2\, \Psi = 0 \tag{8.107}$$

with parameter $k_S = \omega/\beta$. More precisely,

$$\nabla \cdot \nabla \Psi + k_S^2\, \Psi = 0 \tag{8.108}$$

which involves the vector $\Psi = \psi_r \hat{\mathbf{r}} + \psi_\theta \hat{\mathbf{t}} + \psi_z \hat{\mathbf{k}}$. We shall assume Φ to be *solenoidal*, which means that it satisfies the *gauge condition* $\nabla \cdot \Psi = 0$. Hence, only two of the three components of Ψ are independent, so the solution can only contain two independent functions. However,

$$\nabla \cdot \nabla \Psi = \nabla(\nabla \cdot \Psi) - \nabla \times \nabla \times \Psi$$
$$= -\nabla \times \nabla \times \Psi \tag{8.109}$$

Hence

$$\nabla \times \nabla \times \Psi = k_S^2\, \Psi \tag{8.110}$$

This equation admits solutions of the form

$$\Psi = \Psi_1 + \Psi_2$$
$$= \psi \hat{\mathbf{k}} + \frac{1}{k_S^2} \nabla\left(\frac{\partial \psi}{\partial z}\right) + \frac{1}{k_S} \nabla \times \left(\chi \hat{\mathbf{k}}\right) \tag{8.111}$$

in which ψ, χ are solutions to the scalar Helmholtz equations

$$\boxed{\nabla^2 \psi + k_S^2\, \psi = 0} \quad \text{and} \quad \boxed{\nabla^2 \chi + k_S^2\, \chi = 0} \tag{8.112}$$

To prove that this is indeed the solution, we make use of some well-known vector operation identities, namely $\nabla \times \nabla f = 0$, $\nabla \cdot \nabla \times \mathbf{f} = 0$, which are true for any scalar and vector functions f and $\mathbf{f}$, respectively. Consider the first term

$$\nabla \times \Psi_1 = \nabla \times \left(\psi \hat{\mathbf{k}}\right) + \frac{1}{k_S^2} \nabla \times \nabla\left(\frac{\partial \psi}{\partial z}\right) = \nabla \times \left(\psi \hat{\mathbf{k}}\right)$$

$$= \frac{1}{r}\frac{\partial \psi}{\partial \theta}\hat{\mathbf{r}} - \frac{\partial \psi}{\partial r}\hat{\mathbf{t}} \tag{8.113}$$

$$\nabla \times \nabla \times \Psi_1 = \nabla \times \left[\frac{1}{r}\frac{\partial \psi}{\partial \theta}\hat{\mathbf{r}} - \frac{\partial \psi}{\partial r}\hat{\mathbf{t}}\right]$$

$$= \nabla\left(\frac{\partial \psi}{\partial z}\right) - \hat{\mathbf{k}}\,\nabla^2 \psi \tag{8.114}$$

and since Ψ_1 must satisfy $\nabla \times \nabla \times \Psi_1 = k_S^2 \Psi_1$, then

$$\nabla\left(\frac{\partial \psi}{\partial z}\right) - \hat{\mathbf{k}} \nabla^2 \psi = k_S^2 \left[\psi \hat{\mathbf{k}} + \frac{1}{k_S^2} \nabla\left(\frac{\partial \psi}{\partial z}\right)\right] \tag{8.115}$$

which results in the condition $\nabla^2 \psi + k_S^2 \psi = 0$. Similarly, the second term gives

$$k_S \nabla \times \Psi_2 = \nabla\left(\frac{\partial \chi}{\partial z}\right) - \hat{\mathbf{k}} \nabla^2 \chi \tag{8.116}$$

so

$$k_S \nabla \times \nabla \times \Psi_2 = \nabla \times \nabla\left(\frac{\partial \chi}{\partial z}\right) - \nabla \times \left(\hat{\mathbf{k}} \nabla^2 \chi\right) = k_S^3 \Psi_2$$

$$= -\nabla \times \left(\hat{\mathbf{k}} \nabla^2 \chi\right) = k_S^2 \nabla \times \left(\hat{\mathbf{k}} \chi\right) \tag{8.117}$$

$$\nabla \times \left[\hat{\mathbf{k}} \left(\nabla^2 \chi + k_S^2 \chi\right)\right] = \mathbf{0} \tag{8.118}$$

which implies $\nabla^2 \chi + k_S^2 \chi = 0$. Thus, the solution provided does indeed satisfy the vector Helmholtz equation. It remains to verify the gauge condition:

$$\nabla \cdot \Psi = \nabla \cdot \left(\psi \hat{\mathbf{k}}\right) + \frac{1}{k_S^2} \nabla \cdot \nabla\left(\frac{\partial \psi}{\partial z}\right) + \frac{1}{k_S} \nabla \cdot \nabla \times \left(\chi \hat{\mathbf{k}}\right)$$

$$= \nabla \cdot \left(\psi \hat{\mathbf{k}}\right) + \frac{1}{k_S^2} \nabla \cdot \left[\nabla \times \nabla \times \left(\psi \hat{\mathbf{k}}\right) + \hat{\mathbf{k}} \nabla^2 \psi\right]$$

$$= \nabla \cdot \left(\psi \hat{\mathbf{k}}\right) - \nabla \cdot \left(\psi \hat{\mathbf{k}}\right) = \mathbf{0} \tag{8.119}$$

Hence, the solution to the vector Helmholtz equation can now be formed with the solutions to the two scalar Helmholtz equations for ψ and χ, using for this purpose the expressions provided in the previous section:

$$\psi(r, \theta, z) = (c_1 \cos n\theta + c_2 \sin n\theta)(c_3 J_n(k_\beta r) + c_4 Y_n(k_\beta r))(c_5 e^{ik_z z} + c_6 e^{-ik_z z}) \tag{8.120}$$

$$\chi(r, \theta, z) = (d_1 \cos n\theta + d_2 \sin n\theta)(d_3 J_n(k_\beta r) + d_4 Y_n(k_\beta r))(d_5 e^{ik_z z} + d_6 e^{-ik_z z}) \tag{8.121}$$

in which the c_i, d_i are arbitrary constants, and

$$k_\beta = \sqrt{k_S^2 - k_z^2} \tag{8.122}$$

8.7 Elastic wave equation in cylindrical coordinates

Consider the elastic wave equation for an isotropic medium

$$(\lambda + 2\mu) \nabla\nabla \cdot \mathbf{u} - \mu \nabla \times \nabla \times \mathbf{u} = \rho \ddot{\mathbf{u}} \tag{8.123}$$

Defining

$$\varepsilon = \nabla \cdot \mathbf{u} = \text{volumetric strain} \tag{8.124}$$

$$\Omega = \tfrac{1}{2} \nabla \times \mathbf{u} = \text{rotation vector, which satisfies } \nabla \cdot \Omega = 0 \tag{8.125}$$

we can write the elastic wave equation as

$$(\lambda + 2\mu) \nabla\varepsilon - 2\mu \nabla \times \Omega = \rho \ddot{\mathbf{u}} \tag{8.126}$$

Applying in turn a gradient and then a curl to the above equation (but not both), we obtain after straightforward algebra

$$(\lambda + 2\mu)\,\nabla^2\varepsilon = \rho\,\ddot{\varepsilon} \quad \text{and} \quad \mu\nabla^2\boldsymbol{\Omega} = \rho\ddot{\boldsymbol{\Omega}} \tag{8.127}$$

These constitute a scalar and a vector wave equation. The latter consists of three scalar wave equations, of which one depends on the other two (because of the condition $\nabla \cdot \boldsymbol{\Omega} = 0$). Hence, for harmonic motion, these reduce to the two Helmholtz equations

$$\nabla^2\varepsilon + k_P^2\,\varepsilon = 0, \quad k_P = \omega/\alpha, \quad \alpha = \sqrt{(\lambda + 2\mu)/\rho} \quad \Rightarrow \quad \text{P waves} \tag{8.128}$$

$$\nabla^2\boldsymbol{\Omega} + k_S^2\boldsymbol{\Omega} = \mathbf{0}, \quad k_S = \omega/\beta, \quad \beta = \sqrt{\mu/\rho} \qquad \Rightarrow \quad \text{S waves} \tag{8.129}$$

From the previous sections, the solutions to these two equations are

$$\varepsilon = \Phi, \quad \nabla^2\Phi + k_P^2\,\Phi = 0, \quad \nabla^2\psi + k_S^2\,\psi = 0, \quad \nabla^2\chi + k_S^2\,\chi = 0 \tag{8.130}$$

$$2\boldsymbol{\Omega} \equiv \boldsymbol{\Psi} = \psi\hat{\mathbf{k}} + \frac{1}{k_S^2}\nabla\left(\frac{\partial\psi}{\partial z}\right) + \frac{1}{k_S}\nabla\times\left(\chi\hat{\mathbf{k}}\right) \tag{8.131}$$

Also, for harmonic motion, the wave equation can be written as

$$(\lambda + 2\mu)\,\nabla\varepsilon - 2\mu\nabla\times\boldsymbol{\Omega} = -\rho\omega^2\,\mathbf{u} \tag{8.132}$$

that is,

$$\mathbf{u} = \frac{\nabla\times\boldsymbol{\Psi}}{k_S^2} - \frac{\nabla\Phi}{k_P^2}$$

$$= \frac{1}{k_S^2}\nabla\times\left[\psi\hat{\mathbf{k}} + \frac{1}{k_S^2}\nabla\left(\frac{\partial\psi}{\partial z}\right) + \frac{1}{k_S}\nabla\times\left(\chi\hat{\mathbf{k}}\right)\right] - \frac{1}{k_P^2}\nabla\Phi \tag{8.133}$$

and using the previous developments in the solution of the vector Helmholtz equation, we have

$$\mathbf{u} = \frac{1}{k_S^2}\left\{\left[\frac{1}{k_S}\frac{\partial^2\chi}{\partial r\,\partial z} + \frac{1}{r}\frac{\partial\psi}{\partial\theta}\right]\hat{\mathbf{r}} + \left[\frac{1}{k_S\,r}\frac{\partial^2\chi}{\partial\theta\,\partial z} - \frac{\partial\psi}{\partial r}\right]\hat{\mathbf{t}} + \left[k_S\chi + \frac{1}{k_S}\frac{\partial^2\chi}{\partial z^2}\right]\hat{\mathbf{k}}\right\} - \frac{1}{k_P^2}\nabla\Phi \tag{8.134}$$

Since the solutions for Φ, χ, and ψ contain arbitrary constants of integration, we can rescale these by any other constant. Hence, applying rescaling factors k_P^2/k_α, k_S^3/k_β^2, and k_S^2/k_β, we can write the displacement vector as

$$\mathbf{u} = \left[\frac{1}{k_\alpha}\frac{\partial\Phi}{\partial r} + \frac{1}{k_\beta^2}\frac{\partial^2\chi}{\partial r\,\partial z} + \frac{1}{k_\beta\,r}\frac{\partial\psi}{\partial\theta}\right]\hat{\mathbf{r}}$$

$$+ \left[\frac{1}{k_\alpha\,r}\frac{\partial\Phi}{\partial\theta} + \frac{1}{k_\beta^2\,r}\frac{\partial^2\chi}{\partial\theta\,\partial z} - \frac{1}{k_\beta}\frac{\partial\psi}{\partial r}\right]\hat{\mathbf{t}}$$

$$+ \left[\frac{1}{k_\alpha}\frac{\partial\Phi}{\partial z} + \frac{1}{k_\beta^2}\left(k_S^2\chi + \frac{\partial^2\chi}{\partial z^2}\right)\right]\hat{\mathbf{k}} \tag{8.135}$$

Observe that $k_S^2\chi = -\nabla^2\chi$, which vanishes at zero frequency.

Now, from the earlier section on the scalar Helmholtz equation, the solutions for the scalar potentials Φ, χ, and ψ for waves that propagate in either the positive or negative z direction can be written in compact form as

$$\Phi = \begin{pmatrix} \cos n\theta \\ \sin n\theta \end{pmatrix} H_{\alpha n}\, e^{\mp ik_{z\alpha}z}\, (A_{\Phi}) \tag{8.136}$$

$$\chi = \begin{pmatrix} \cos n\theta \\ \sin n\theta \end{pmatrix} H_{\beta n}\, e^{\mp ik_{z\beta}z}\, (A_{\chi}) \tag{8.137}$$

$$\psi = \begin{pmatrix} -\sin n\theta \\ \cos n\theta \end{pmatrix} H_{\beta n}\, e^{\mp ik_{z\beta}z}\, (-A_{\psi}) \tag{8.138}$$

in which A_{Φ}, A_{χ}, A_{ψ} are (conveniently scaled) integration constants; $H_{\alpha n}$, $H_{\beta n}$ is a short-hand for *any* of the Bessel functions of order n and argument $k_{\alpha}r$, $k_{\beta}r$; $k_{z\alpha}$, $k_{z\beta}$ are the vertical wavenumbers for P and S waves, respectively; and either the upper or lower trigonometric terms in parenthesis are used. The upper terms are called for in situations where the displacement field is known to be symmetric with respect to the vertical x, z plane (say, because of the spatial symmetry of the loads); the lower is for cases where this field is known to be antisymmetric with respect to this plane. Non-symmetric fields can be obtained by linear combinations of these two cases.

At this point, we must make a distinction between *horizontally* and *cylindrically* stratified media. In the first case, compatibility and equilibrium of adjoining layers require the radial wavenumber to be common to all layers; in the second, it is the vertical wavenumber that must be common. We consider each case in turn.

8.7.1 Horizontally stratified media

In this case, the radial wavenumber k must be common to all layers, which requires

$$k_{\alpha} = k_{\beta} \equiv k \tag{8.139}$$

It follows that

$$k = \sqrt{k_{P}^{2} - k_{z\alpha}^{2}} = \sqrt{k_{S}^{2} - k_{z\beta}^{2}} \tag{8.140}$$

that is

$$k_{z\alpha} = \sqrt{k_{P}^{2} - k^{2}} \quad \text{and} \quad k_{z\beta} = \sqrt{k_{S}^{2} - k^{2}} \tag{8.141}$$

where $k_{z\alpha}$, $k_{z\beta}$ are the vertical wavenumbers for P and S waves, which in this case differ from one another. For convenience, we choose to express these indirectly in term of parameters p, s as

$$kp = \sqrt{k^{2} - k_{P}^{2}} = i\, k_{z\alpha}, \quad \text{i.e.,} \quad p = \sqrt{1 - \left(\frac{\omega}{\alpha k}\right)^{2}} \tag{8.142}$$

and

$$ks = \sqrt{k^{2} - k_{S}^{2}} = i\, k_{z\beta}, \quad \text{i.e.,} \quad s = \sqrt{1 - \left(\frac{\omega}{\beta k}\right)^{2}} \tag{8.143}$$

Introducing the solutions for the scalar potentials Φ, χ, ψ into the solution for displacements and making use of the above parameters for the vertical wavenumbers, after some algebra we find expressions of the form

$$u_R = \begin{pmatrix} \cos n\theta \\ \sin n\theta \end{pmatrix} \left\{ H_n' \left[e^{\mp kpz} A_\Phi \mp s\, e^{\mp ksz} A_\chi \right] + \frac{n}{kr} H_n\, e^{\mp ksz} A_\psi \right\} \tag{8.144}$$

$$u_\theta = \begin{pmatrix} -\sin n\theta \\ \cos n\theta \end{pmatrix} \left\{ n\frac{H_n}{kr} \left[e^{\mp kpz} A_\Phi \mp s\, e^{\mp ksz} A_\chi \right] + H_n'\, e^{\mp ksz} A_\psi \right\} \tag{8.145}$$

$$u_z = \begin{pmatrix} \cos n\theta \\ \sin n\theta \end{pmatrix} \left\{ H_n \left[\mp p\, e^{\mp kpz} A_\Phi + e^{\mp ksz} A_\chi \right] \right\} \tag{8.146}$$

in which H_n is shorthand for *any* of the Bessel functions of order n and argument kr, and $H_n' = dH_n/d(kr)$. Combining the solutions for both negative and positive vertical wavenumbers and choosing $H_n \equiv J_n(kr)$, we obtain the solution for displacements in horizontally stratified media for cylindrical coordinates in the frequency–radial-wavenumber domain, expressed in compact matrix form as

---

$$\mathbf{u}(r, \theta, z) = \begin{bmatrix} u_r & u_\theta & u_z \end{bmatrix}^T = \mathbf{T}_n\, \mathbf{C}_n\, \mathbf{u}$$

$$\mathbf{u}(k, z, \omega) = \begin{Bmatrix} \tilde{u}_r \\ \tilde{u}_\theta \\ \tilde{u}_z \end{Bmatrix} = \mathbf{R}_1 \mathbf{E}_z^{-1} \mathbf{a}_1 + \mathbf{R}_2\, \mathbf{E}_z \mathbf{a}_2$$

$\mathbf{a}_1, \mathbf{a}_2 = $ arbitrary 3×1 vectors of integration constants (wave amplitudes)

$$\mathbf{T}_n = \mathrm{diag}\left[\begin{pmatrix} \cos n\theta \\ \sin n\theta \end{pmatrix} \quad \begin{pmatrix} -\sin n\theta \\ \cos n\theta \end{pmatrix} \quad \begin{pmatrix} \cos n\theta \\ \sin n\theta \end{pmatrix} \right] \quad \text{(upper or lower elements used)}$$

$$\mathbf{E}_z^{\pm 1} = \mathrm{diag}\left\{ e^{\pm kpz} \quad e^{\pm ksz} \quad e^{\pm ksz} \right\},$$

$$\mathbf{C}_n = \begin{Bmatrix} J_n' & \dfrac{n}{kr} J_n & 0 \\ \dfrac{n}{kr} J_n & J_n' & 0 \\ 0 & 0 & J_n \end{Bmatrix}, \quad J_n' = \frac{dJ_n(kr)}{d(kr)}$$

$$\mathbf{R}_1 = \begin{Bmatrix} 1 & 0 & -s \\ 0 & 1 & 0 \\ -p & 0 & 1 \end{Bmatrix}, \quad \mathbf{R}_2 = \begin{Bmatrix} 1 & 0 & s \\ 0 & 1 & 0 \\ p & 0 & 1 \end{Bmatrix}$$

$$\tag{8.147}$$

Stresses in horizontal surfaces

We provide also the solution for stresses in horizontal planes, for these are necessary to enforce equilibrium across layer interfaces. From Section 1.4.2, these stresses are

$$\mathbf{s}_z = \begin{bmatrix} \sigma_{rz} & \sigma_{\theta z} & \sigma_z \end{bmatrix}^T = \mathbf{D}_{zr}\frac{\partial \mathbf{u}}{\partial r} + \mathbf{D}_{z\theta}\frac{1}{r}\frac{\partial \mathbf{u}}{\partial \theta} + \mathbf{D}_{zz}\frac{\partial \mathbf{u}}{\partial z} + \mathbf{D}_{z1}\frac{\mathbf{u}}{r} \tag{8.148}$$

To evaluate the stresses elicited by the particular solution previously found, we use the material matrices $\mathbf{D}_{zr}$, etc., for an isotropic medium, carry out the required differentiations, and take into account the differential equation for the Bessel functions. The final result is

$$\mathbf{s}_z = k\mu\,\mathbf{T}_n\,\mathbf{C}_n\,\mathfrak{s}$$

$$\mathfrak{s}(k, z, \omega) = \left\{ \begin{array}{c} \tilde{\sigma}_{rz} \\ \tilde{\sigma}_{\theta z} \\ \tilde{\sigma}_z \end{array} \right\} = -\mathbf{Q}_1\mathbf{E}_z^{-1}\mathbf{a}_1 + \mathbf{Q}_2\,\mathbf{E}_z\mathbf{a}_2$$

$$\mathbf{Q}_1 = \left\{ \begin{array}{ccc} 2p & 0 & -(1+s^2) \\ 0 & s & 0 \\ -(1+s^2) & 0 & 2s \end{array} \right\}, \quad \mathbf{Q}_2 = \left\{ \begin{array}{ccc} 2p & 0 & 1+s^2 \\ 0 & s & 0 \\ 1+s^2 & 0 & 2s \end{array} \right\}$$

$$\text{(8.149)}$$

Observe that $\mathbf{u}$, $\mathfrak{s}$ do not depend explicitly on the azimuthal index n.

8.7.2 Radially stratified media

In this case, the vertical wavenumber k_z must be common to P and S waves. Hence, the radial wavenumbers for P and S waves will be distinct, i.e., $k_\alpha \neq k_\beta$. Introducing the solutions for the scalar potentials Φ, χ, and ψ for waves that propagate in the positive z direction (i.e., positive k_z) into the solution for displacements in the expression for $\mathbf{u}$, changing A_χ into i A_χ (a trivial change in an arbitrary constant), carrying out the differentiation operations, and defining

$$H'_{\alpha n} = \frac{dH_{\alpha n}}{d(k_\alpha r)} \qquad H'_{\beta n} = \frac{dH_{\beta n}}{d(k_\beta r)} \tag{8.150}$$

with $H_{\alpha n}$, $H_{\beta n}$ being *any* of the Bessel functions of order n and argument $k_\alpha r$, $k_\beta r$, we obtain

$$u_R = \left(\begin{array}{c} \cos n\theta \\ \sin n\theta \end{array} \right) \left\{ H'_{\alpha n}A_\Phi + n\frac{H_{\beta n}}{k_\beta r}\,A_\psi + \frac{k_z}{k_\beta}H'_{\beta n}\,A_\chi \right\} e^{-\mathrm{i}k_z z} \tag{8.151}$$

$$u_\theta = \left(\begin{array}{c} -\sin n\theta \\ \cos n\theta \end{array} \right) \left\{ n\frac{H_{\alpha n}}{k_\alpha r}\,A_\Phi + H'_{\beta n}A_\psi + n\frac{k_z}{k_\beta}\frac{H_{\beta n}}{k_\beta r}\,A_\chi \right\} e^{-\mathrm{i}k_z z} \tag{8.152}$$

$$u_z = \left(\begin{array}{c} \cos n\theta \\ \sin n\theta \end{array} \right) \left\{ -\mathrm{i}\frac{k_z}{k_\alpha}H_{\beta n}A_\Phi + \mathrm{i}\,H_{\beta n}A_\chi \right\} e^{-\mathrm{i}k_z z} \tag{8.153}$$

Combining all three into a matrix equation, we obtain

$$\begin{Bmatrix} u_R \\ u_\theta \\ u_z \end{Bmatrix} = \begin{Bmatrix} \begin{pmatrix} \cos n\theta \\ \sin n\theta \end{pmatrix} & 0 & 0 \\[2mm] 0 & \begin{pmatrix} -\sin n\theta \\ \cos n\theta \end{pmatrix} & 0 \\[2mm] 0 & 0 & \begin{pmatrix} \cos n\theta \\ \sin n\theta \end{pmatrix} \end{Bmatrix}$$

$$\times \begin{Bmatrix} H'_{\alpha n} & n\dfrac{H_{\beta n}}{k_\beta r} & \dfrac{k_z}{k_\beta} H'_{\beta n} \\[3mm] n\dfrac{H_{\alpha n}}{k_\alpha r} & H'_{\beta n} & n\dfrac{k_z}{k_\beta}\dfrac{H_{\beta n}}{k_\beta r} \\[3mm] -\mathrm{i}\dfrac{k_z}{k_\alpha} H_{\alpha n} & 0 & \mathrm{i}\, H_{\beta n} \end{Bmatrix} \begin{Bmatrix} A_\Phi \\ A_\psi \\ A_\chi \end{Bmatrix} e^{-\mathrm{i}k_z z} \qquad (8.154)$$

Multiplying the third component by $-\mathrm{i}$, considering either the upper or lower trigonometric element, and taking a linear combination of the first and second Hankel functions, we conclude that a solution to the elastic wave equation in cylindrical coordinates for a radially layered system is of the form (overbar is a reminder of imaginary factor; tilde refers to frequency–wavenumber domain)

$$\bar{\mathbf{u}}(r,\theta,z) = \begin{bmatrix} u_r & u_\theta & -\mathrm{i}\,u_z \end{bmatrix}^T = \mathbf{T}_n\, \mathbf{u}\, e^{-\mathrm{i}k_z z}$$

$$\mathbf{u}(r,n,k_z,\omega) = \begin{Bmatrix} \tilde{u}_r \\ \tilde{u}_\theta \\ -\mathrm{i}\,\tilde{u}_z \end{Bmatrix} = \mathbf{H}_n^{(1)}\,\mathbf{a}_1 + \mathbf{H}_n^{(2)}\,\mathbf{a}_2$$

$\mathbf{a}_1, \mathbf{a}_2 =$ arbitrary vectors of integration constants

$$\mathbf{H}_n = \begin{Bmatrix} (H_{\alpha n})' & n\dfrac{H_{\beta n}}{k_\beta r} & \dfrac{k_z}{k_\beta}(H_{\beta n})' \\[3mm] n\dfrac{H_{\alpha n}}{k_\alpha r} & (H_{\beta n})' & n\dfrac{k_z}{k_\beta}\dfrac{H_{\beta n}}{k_\beta r} \\[3mm] -\dfrac{k_z}{k_\alpha} H_{\alpha n} & 0 & H_{\beta n} \end{Bmatrix},$$

$$\mathbf{H}_n^{(1)} = \mathbf{H}_n\left(H_n^{(1)}(k_\alpha r),\, H_n^{(1)}(k_\beta r) \right), \quad \mathbf{H}_n^{(2)} = \mathbf{H}_n\left(H_n^{(2)}(k_\alpha r),\, H_n^{(2)}(k_\beta r) \right)$$

$$k_\alpha = \sqrt{k_P^2 - k_z^2}, \qquad k_\beta = \sqrt{k_S^2 - k_z^2}, \quad k_z = \text{axial wavenumber}$$

$$H_{\alpha n}^{(i)} = H_n^{(i)}(k_\alpha r), \qquad \left(H_{\alpha n}^{(i)} \right)' = \frac{d H_{\alpha n}^{(i)}}{d(k_\alpha r)}, \quad i = 1,2$$

$$H_{\beta n}^{(i)} = H_n^{(i)}(k_\beta r), \qquad \left(H_{\beta n}^{(i)} \right)' = \frac{d H_{\beta n}^{(i)}}{d(k_\beta r)}$$

$$\mathbf{T}_n = \mathbf{diag}\left[\begin{pmatrix} \cos n\theta \\ \sin n\theta \end{pmatrix} \begin{pmatrix} -\sin n\theta \\ \cos n\theta \end{pmatrix} \begin{pmatrix} \cos n\theta \\ \sin n\theta \end{pmatrix} \right] \quad \text{(upper or lower elements used)}$$

$$(8.155)$$

This is the solution for the displacements in a radially layered system in cylindrical coordinates, expressed in the frequency–vertical-wavenumber domain.

Stresses in cylindrical surfaces

From Section 1.4.2, the stresses in cylindrical surfaces are given by

$$\mathbf{s}_r = [\,\sigma_r \quad \sigma_{r\theta} \quad \sigma_{rz}\,]^T = \mathbf{D}_{rr}\frac{\partial \mathbf{u}}{\partial r} + \mathbf{D}_{r\theta}\frac{1}{r}\frac{\partial \mathbf{u}}{\partial \theta} + \mathbf{D}_{rz}\frac{\partial \mathbf{u}}{\partial z} + \mathbf{D}_{r1}\frac{\mathbf{u}}{r}$$

To evaluate the stresses elicited by the particular solution previously found, we must first remove the imaginary factor in the vertical component of the displacement vector, that is, $\mathbf{u} = \text{diag}\{1 \quad 1 \quad +\mathrm{i}\}\bar{\mathbf{u}}$. Thereafter, we use the material matrices for an isotropic medium, carry out the required differentiation, take into account the differential equation for the Bessel functions, and re-apply a factor $-\mathrm{i}$ to the third component, which in this case is the radial–axial shearing stress. The result is

$$\bar{\mathbf{s}}_r = [\,\sigma_r \quad \sigma_{r\theta} \quad -\mathrm{i}\sigma_{rz}\,]^T = \mathbf{T}_n\, \mathfrak{s}\, e^{-\mathrm{i}k_z z}$$

$$\mathfrak{s}(r, n, k_z, \omega) = \left\{ \begin{array}{c} \tilde{\sigma}_r \\ \tilde{\sigma}_{r\theta} \\ -\mathrm{i}\tilde{\sigma}_{rz} \end{array} \right\} = \mathbf{F}_n^{(1)}\, \mathbf{a}_1 + \mathbf{F}_n^{(2)}\, \mathbf{a}_2$$

$$\mathbf{F}_n^{(1)} = \mathbf{F}_n\left(H_n^{(1)}(k_\alpha r),\, H_n^{(1)}(k_\beta r) \right), \quad \mathbf{F}_n^{(2)} = \mathbf{F}_n\left(H_n^{(2)}(k_\alpha r),\, H_n^{(2)}(k_\beta r) \right)$$

$$\mathbf{F}_n = \left\{ \begin{array}{ccc} f_{11} & f_{12} & f_{13} \\ f_{21} & f_{22} & f_{23} \\ f_{31} & f_{32} & f_{33} \end{array} \right\}$$

$$(8.156)$$

which are defined in terms of a generic matrix $\mathbf{F}_n = \{\,f_{ij}\,\}$ with element functions $H_{\alpha n}$, $H_{\beta n}$ that are shorthand for the Bessel functions $H_n^{(1)}(k_\alpha r),\, H_n^{(1)}(k_\beta r)$, or $H_n^{(2)}(k_\alpha r),\, H_n^{(2)}(k_\beta r)$ (alternatively, $J_n(k_\alpha r),\, J_n(k_\beta r)$ and $Y_n(k_\alpha r),\, Y_n(k_\beta r)$ might be used). The elements of this matrix are

$$f_{11} = -k_\alpha\left\{ \lambda\left[1 + \left(\frac{k_z}{k_\alpha}\right)^2 \right] H_{\alpha n} + 2\mu\left[\frac{H'_{\alpha n}}{k_\alpha r} + \left(1 - \left(\frac{n}{k_\alpha r}\right)^2 \right) H_{\alpha n} \right] \right\}$$

$$f_{12} = \frac{2n\mu}{r}\left(H'_{\beta n} - \frac{H_{\beta n}}{k_\beta r} \right)$$

$$f_{13} = -2k_z\mu\left\{ \frac{H'_{\beta n}}{k_\beta r} + \left[1 - \left(\frac{n}{k_\beta r}\right)^2 \right] H_{\beta n} \right\}$$

$$f_{21} = \frac{2n\mu}{r}\left[H'_{\alpha n} - \frac{H_{\alpha n}}{k_\alpha r} \right]$$

$$f_{22} = -k_\beta \, \mu \left[2 \frac{H'_{\beta n}}{k_\beta r} + \left(1 - 2 \left(\frac{n}{k_\beta r} \right)^2 \right) H_{\beta n} \right]$$

$$f_{23} = k_z \frac{2n \, \mu}{k_\beta r} \left[H'_{\beta n} - \frac{H_{\beta n}}{k_\beta r} \right]$$

$$f_{31} = -2 \, k_z \, \mu \, H'_{\alpha n}$$

$$f_{32} = -k_z n \, \mu \frac{H_{\beta n}}{k_\beta r}$$

$$f_{33} = k_\beta \mu \left[1 - \left(\frac{k_z}{k_\beta} \right)^2 \right] H'_{\beta n}$$

(8.157)

8.8 Scalar Helmholtz equation in spherical coordinates

Consider the Helmholtz equation involving a scalar function Φ:

$$\nabla^2 \Phi + k^2 \Phi = 0 \tag{8.158}$$

which in spherical coordinates is expressed as

$$\frac{\partial^2 \Phi}{\partial R^2} + \frac{2}{R} \frac{\partial \Phi}{\partial R} + \frac{1}{R^2} \frac{\partial^2 \Phi}{\partial \phi^2} + \frac{\cot \phi}{R^2} \frac{\partial \Phi}{\partial \phi} + \frac{1}{R^2 \sin^2 \phi} \frac{\partial^2 \Phi}{\partial \theta^2} + k^2 \Phi = 0 \tag{8.159}$$

where $\Phi(R, \phi, \theta)$ is a scalar function of the spherical coordinates, R is the radius, ϕ the polar angle or co-latitude, and θ is the azimuth or longitude. To solve this problem, we use separation of variables, and begin by assuming a harmonic variation in the azimuth of the form

$$\Phi(R, \phi, \theta) = F(R) \, G(\phi) \, (c_1 \cos n\theta + c_2 \sin n\theta) \tag{8.160}$$

where c_1, c_2 are constants that may depend on the parameter n. To satisfy continuity in the azimuth at the transition from 2π to 0, n must be an integer, non-negative number. This implies

$$\frac{\partial^2 \Phi}{\partial \theta^2} = -n^2 \Phi \tag{8.161}$$

Substituting into the differential equation, multiplying by R^2, dividing by FG, and taking the terms in G to the right-hand side, we obtain

$$\frac{R^2}{F} \left[\frac{\partial^2 F}{\partial R^2} + \frac{2}{R} \frac{\partial F}{\partial R} + k^2 F \right] = -\frac{1}{G} \left[\frac{\partial^2 G}{\partial \phi^2} + \cot \phi \frac{\partial G}{\partial \phi} - \frac{n^2 G}{\sin^2 \phi} \right] \tag{8.162}$$

The left-hand side is a function of R only; the right-hand side is a function of ϕ only. Hence, each side must equal a constant, which we choose to be the parameter $m(m+1)$. It follows that

$$\frac{\partial^2 F}{\partial R^2} + \frac{2}{R} \frac{\partial F}{\partial R} + \left[k^2 - \frac{m(m+1)}{R^2} \right] F = 0 \qquad \text{(spherical Bessel equation)} \tag{8.163}$$

$$\frac{\partial^2 G}{\partial \phi^2} + \cot \phi \frac{\partial G}{\partial \phi} + \left[m(m+1) - \frac{n^2}{\sin^2 \phi} \right] G = 0 \quad \text{(assoc. Legendre equation)} \quad (8.164)$$

The solution to these two equations is

$$F(R) = c_3 \, h_m^{(1)}(kR) + c_4 \, h_m^{(2)}(kR) \quad \text{(spherical Hankel functions)} \quad (8.165)$$

or

$$F(R) = c_3 \, j_m(kR) + c_4 \, y_m(kR) \quad \text{(spherical Bessel functions)} \quad (8.166)$$

and

$$G(\phi) = c_5 \, P_m^n + c_6 \, Q_m^n \quad \text{(assoc. Legendre functions)} \quad (8.167)$$

in which again c_3, c_4, c_5, c_6 are constants that may depend on the parameters m, n It follows that in spherical coordinates, *particular* solutions to the Helmholtz equation exist that are the form

$$\boxed{\Phi(R, \phi, \theta) = (c_1 \cos n\theta + c_2 \sin n\theta) \left(c_3 \, h_m^{(1)}(kR) + c_4 \, h_m^{(2)}(kR) \right) (c_5 \, P_m^n + c_6 \, Q_m^n)}$$

$$(8.168)$$

In most cases, c_6 will be zero, because Q_m^n is singular at $\phi = 0, \pi$ (i.e., at the north and south poles), so for problems that include the full sphere, it must be excluded on physical grounds. For infinite media, c_3 may also be zero, to satisfy the radiation condition at $R = \infty$, whereas media that include the origin may need to be formulated in terms of the conventional spherical Bessel functions j_n, to avoid the singularity of the spherical Hankel functions (or y_n) at $R = 0$.

8.9 Vector Helmholtz equation in spherical coordinates

Consider the vector Helmholtz equation

$$\nabla^2 \mathbf{\Psi} + k^2 \mathbf{\Psi} = \mathbf{0} \quad (8.169)$$

or more precisely

$$\nabla \cdot \nabla \mathbf{\Psi} + k^2 \mathbf{\Psi} = \mathbf{0} \quad (8.170)$$

which involves the vector $\mathbf{\Psi} = \psi_r \hat{\mathbf{r}} + \psi_\phi \hat{\mathbf{s}} + \psi_\theta \hat{\mathbf{t}}$. We shall assume $\mathbf{\Psi}$ to be *solenoidal*, which means that it satisfies the *gauge condition* $\nabla \cdot \mathbf{\Psi} = 0$. Hence, only two of the three components of $\mathbf{\Psi}$ are independent, so the solution can only contain two independent functions. However,

$$\nabla \cdot \nabla \mathbf{\Psi} = \nabla(\nabla \cdot \mathbf{\Psi}) - \nabla \times \nabla \times \mathbf{\Psi}$$
$$= -\nabla \times \nabla \times \mathbf{\Psi} \quad (8.171)$$

Hence

$$\nabla \times \nabla \times \mathbf{\Psi} = k^2 \mathbf{\Psi} \quad (8.172)$$

This equation admits a solution of the form

$$\boxed{\begin{aligned}\boldsymbol{\Psi} &= \boldsymbol{\Psi}_1 + \boldsymbol{\Psi}_2 \\ &= R\psi\hat{\mathbf{r}} + \frac{1}{k^2}\nabla\left(\frac{\partial R\psi}{\partial R}\right) + \frac{1}{k}\nabla\times(R\chi\hat{\mathbf{r}})\end{aligned}}$$

(8.173)

in which

$$\boxed{\nabla^2\psi + k^2\psi = 0} \quad \text{and} \quad \boxed{\nabla^2\chi + k^2\chi = 0}$$

(8.174)

To prove that this is indeed the solution, we make use of some well-known vector operation identities, namely $\nabla\times\nabla f = 0$, $\nabla\cdot\nabla\times\mathbf{f} = 0$, which are true for any scalar and vector functions f and $\mathbf{f}$, respectively. Consider the first term:

$$\nabla\times\boldsymbol{\Psi}_1 = \nabla\times(R\psi\hat{\mathbf{r}}) + \frac{1}{k^2}\nabla\times\nabla\left(\frac{\partial R\psi}{\partial R}\right) = \nabla\times(R\psi\hat{\mathbf{r}})$$

$$= \frac{1}{\sin\phi}\frac{\partial\psi}{\partial\theta}\hat{\mathbf{s}} - \frac{\partial\psi}{\partial\phi}\hat{\mathbf{t}}$$

(8.175)

$$\nabla\times\nabla\times\boldsymbol{\Psi}_1 = \nabla\times\left[\frac{1}{\sin\phi}\frac{\partial\psi}{\partial\theta}\hat{\mathbf{s}} - \frac{\partial\psi}{\partial\phi}\hat{\mathbf{t}}\right]$$

$$= \nabla\left(\frac{\partial R\psi}{\partial R}\right) - R\hat{\mathbf{r}}\nabla^2\psi$$

(8.176)

and because $\boldsymbol{\Psi}_1$ must satisfy $\nabla\times\nabla\times\boldsymbol{\Psi}_1 = k^2\boldsymbol{\Psi}_1$, then

$$\nabla\left(\frac{\partial R\psi}{\partial R}\right) - R\hat{\mathbf{r}}\nabla^2\psi = k^2\left[R\psi\hat{\mathbf{r}} + \frac{1}{k^2}\nabla\left(\frac{\partial R\psi}{\partial R}\right)\right]$$

(8.177)

which results in the condition $\nabla^2\psi + k^2\psi = 0$. Similarly, the second term gives

$$k\nabla\times\boldsymbol{\Psi}_2 = \nabla\left(\frac{\partial R\chi}{\partial R}\right) - R\hat{\mathbf{r}}\nabla^2\chi$$

(8.178)

so

$$k\nabla\times\nabla\times\boldsymbol{\Psi}_2 = \nabla\times\nabla\left(\frac{\partial R\chi}{\partial R}\right) - \nabla\times(R\hat{\mathbf{r}}\nabla^2\chi) = k^3\boldsymbol{\Psi}_2$$

$$= -\nabla\times(R\hat{\mathbf{r}}\nabla^2\chi) = k^2\nabla\times(R\hat{\mathbf{r}}\chi)$$

(8.179)

$$\nabla\times\left[R\hat{\mathbf{r}}\left(\nabla^2\chi + k^2\chi\right)\right] = \mathbf{0}$$

(8.180)

which implies $\nabla^2\chi + k^2\chi = 0$. Thus, the solution provided does indeed satisfy the vector Helmholtz equation. It remains to verify the gauge condition:

$$\nabla\cdot\boldsymbol{\Psi} = \nabla\cdot(R\psi\hat{\mathbf{r}}) + \frac{1}{k^2}\nabla\cdot\nabla\left(\frac{\partial R\psi}{\partial R}\right) + \frac{1}{k}\nabla\cdot\nabla\times(R\chi\hat{\mathbf{r}})$$

$$= \nabla\cdot(R\psi\hat{\mathbf{r}}) + \frac{1}{k^2}\nabla\cdot\left[\nabla\times\nabla\times(R\psi\hat{\mathbf{r}}) + R\hat{\mathbf{r}}\nabla^2\psi\right]$$

$$= \nabla\cdot(R\psi\hat{\mathbf{r}}) - \nabla\cdot(R\psi\hat{\mathbf{r}}) = 0$$

(8.181)

Hence, the solution to the vector Helmholtz equation can now be formed with the solutions to the two scalar Helmholtz equations for ψ and χ, using for this purpose the expressions provided in the previous section:

$$\psi(R, \phi, \theta) = (c_1 \cos n\theta + c_2 \sin n\theta)\left(c_3 h_m^{(1)}(kR) + c_4 h_m^{(2)}(kR)\right)\left(c_5 P_m^n + c_6 Q_m^n\right)$$

(8.182)

$$\chi(R, \phi, \theta) = (d_1 \cos n\theta + d_2 \sin n\theta)\left(d_3 h_m^{(1)}(kR) + d_4 h_m^{(2)}(kR)\right)\left(d_5 P_m^n + d_6 Q_m^n\right)$$

(8.183)

8.10 Elastic wave equation in spherical coordinates

Consider the elastic wave equation for an isotropic medium

$$(\lambda + 2\mu) \nabla\nabla \cdot \mathbf{u} - \mu\nabla \times \nabla \times \mathbf{u} = \rho\ddot{\mathbf{u}}$$

(8.184)

Defining

$$\varepsilon = \nabla \cdot \mathbf{u} = \text{volumetric strain}$$

(8.185)

$$\mathbf{\Omega} = \tfrac{1}{2}\nabla \times \mathbf{u} = \text{rotation vector, which satisfies} \quad \nabla \cdot \mathbf{\Omega} = 0$$

(8.186)

we can write the elastic wave equation as

$$(\lambda + 2\mu) \nabla\varepsilon - 2\mu\nabla \times \mathbf{\Omega} = \rho\ddot{\mathbf{u}}$$

(8.187)

Applying in turn a gradient and then a curl to the above equation (but not both), we obtain after straightforward algebra

$$(\lambda + 2\mu) \nabla^2\varepsilon = \rho\ddot{\varepsilon} \quad \text{and} \quad \mu\nabla^2\mathbf{\Omega} = \rho\ddot{\mathbf{\Omega}}$$

(8.188)

These constitute a scalar and a vector wave equation. The latter consists of three scalar wave equations, of which one depends on the other two (because of the condition $\nabla \cdot \mathbf{\Omega} = 0$). Hence, for harmonic motion, these reduce to the two Helmholtz equations

$$\nabla^2\varepsilon + k_P^2\,\varepsilon = 0, \qquad k_P = \omega/\alpha, \quad \alpha = \sqrt{(\lambda + 2\mu)/\rho} \quad \Rightarrow \quad \text{P waves}$$

(8.189)

$$\nabla^2\mathbf{\Omega} + k_S^2\mathbf{\Omega} = \mathbf{0}, \qquad k_S = \omega/\beta, \quad \beta = \sqrt{\mu/\rho} \quad \Rightarrow \quad \text{S waves}$$

(8.190)

From the previous sections, the solutions to these two equations are

$$\varepsilon = \Phi, \qquad \nabla^2\Phi + k^2\Phi = 0, \qquad \nabla^2\psi + k^2\psi = 0, \qquad \nabla^2\chi + k^2\chi = 0$$

(8.191)

$$2\mathbf{\Omega} \equiv \mathbf{\Psi} = R\psi\hat{\mathbf{r}} + \frac{1}{k_S^2}\nabla\left(\frac{\partial R\psi}{\partial R}\right) + \frac{1}{k_S}\nabla \times (R\chi\hat{\mathbf{r}})$$

(8.192)

Also, for harmonic motion, the wave equation can be written as

$$(\lambda + 2\mu)\,\nabla\varepsilon - 2\mu\nabla \times \Omega = -\rho\omega^2\,\mathbf{u} \tag{8.193}$$

that is,

$$\mathbf{u} = \frac{\nabla \times \Psi}{k_S^2} - \frac{\nabla\Phi}{k_P^2}$$

$$= \frac{1}{k_S^2}\left[\nabla \times (R\psi\hat{\mathbf{r}}) + \frac{1}{k_S}\nabla \times \nabla \times (R\chi\hat{\mathbf{r}})\right] - \frac{1}{k_P^2}\nabla\Phi \tag{8.194}$$

and using the developments in the previous section

$$\mathbf{u} = \frac{1}{k_S^2}\left\{\left[k_S R\chi + \frac{1}{k_S}\frac{\partial^2(R\chi)}{\partial R^2}\right]\hat{\mathbf{r}} + \left[\frac{1}{\sin\phi}\frac{\partial\psi}{\partial\theta} + \frac{1}{k_S R}\frac{\partial^2(R\chi)}{\partial R\partial\phi}\right]\hat{\mathbf{s}}\right.$$

$$\left. + \left[\frac{1}{k_S R\sin\phi}\frac{\partial^2(R\chi)}{\partial R\partial\theta} - \frac{\partial\psi}{\partial\phi}\right]\hat{\mathbf{t}}\right\} - \frac{1}{k_P^2}\nabla\Phi \tag{8.195}$$

This expression shows that the displacement field consists of a longitudinal irrotational field consisting of primary waves (pressure waves, P-waves) eliciting particle motions that are normal to the P-wave front (the surface $\Phi = $ constant, whose normal is $\nabla\Phi$), and two transverse solenoidal fields consisting of secondary waves (shear waves, S-waves) eliciting particle motions that are contained in the tangent plane to the S-wave front (the surfaces $\psi = $ constant and/or $\chi = $ constant). The shear motion component in horizontal planes is referred to as the SH wave; the shear motion component in vertical planes, as the SV wave.

Since the solutions for ψ, χ, and Φ contain arbitrary constants of integration, we can incorporate the divisors k_P^2, k_S^2 into them, and write the displacement vector in matrix form as

$$\mathbf{u} = \left[\frac{\partial\Phi}{k_P\partial R} + \frac{1}{k_S}\left(\frac{\partial^2(R\chi)}{\partial R^2} + k_S^2 R\chi\right)\right]\hat{\mathbf{r}}$$

$$+ \left[\frac{1}{k_P R}\frac{\partial\Phi}{\partial\phi} + \frac{1}{k_S R}\frac{\partial^2(R\chi)}{\partial R\partial\phi} + \frac{1}{\sin\phi}\frac{\partial\psi}{\partial\theta}\right]\hat{\mathbf{s}}$$

$$+ \left[\frac{1}{k_P R\sin\phi}\frac{\partial\Phi}{\partial\theta} + \frac{1}{k_S R\sin\phi}\frac{\partial^2(R\chi)}{\partial R\partial\theta} - \frac{\partial\psi}{\partial\phi}\right]\hat{\mathbf{t}} \tag{8.196}$$

Observe that $k_S^2 R\chi = -R\nabla^2\chi$, which vanishes at zero frequency. Now, from the section on the scalar Helmholtz equation, the solutions for the scalar potentials Φ, χ, and ψ can be written in compact form as

$$\Phi = \begin{pmatrix}\cos n\theta \\ \sin n\theta\end{pmatrix} P_m^n\, h_{Pm}\, A_\Phi \tag{8.197}$$

$$\chi = \begin{pmatrix}\cos n\theta \\ \sin n\theta\end{pmatrix} P_m^n\, h_{Sm}\, A_\chi \tag{8.198}$$

$$\psi = \begin{pmatrix}-\sin n\theta \\ \cos n\theta\end{pmatrix} P_m^n\, h_{Sm}\, (-A_\psi) \tag{8.199}$$

in which A_Φ, A_χ, A_ψ are integration constants; $P_m^n(\phi)$ are the associated Legendre functions of order m and rank n (the Q_m^n functions are excluded on physical grounds); h_{Pm} (h_{Sm}) is shorthand for any of the spherical Bessel functions of order m and argument $k_P R$ ($k_S R$), but especially for the second spherical Hankel functions $h_m^{(2)}(k_P R)$, ($h_m^{(2)}(k_S R)$); and either the upper or lower trigonometric factors in parenthesis are used. The upper factors are called for in situations where the displacement field is known to be symmetric with respect to the vertical x, z plane (say, because of the spatial symmetry of the loads) the lower ones are for cases where this field is known to be antisymmetric with respect to this plane. Nonsymmetric fields can be obtained by linear combinations of these two cases. Introducing these solutions into the expression for **u**, carrying out the differentiation operations, and adding an overbar to remind us that this is a *particular* solution for given indices m, n, we obtain

$$\bar{u}_R = \begin{pmatrix} \cos n\theta \\ \sin n\theta \end{pmatrix} \left\{ P_m^n \frac{dh_{Pm}}{k_P\, dR} A_\Phi + \frac{m(m+1)}{k_S R} P_m^n h_{Sm} A_\chi \right\} \tag{8.200}$$

$$\bar{u}_\phi = \begin{pmatrix} \cos n\theta \\ \sin n\theta \end{pmatrix} \left\{ \frac{dP_m^n}{d\phi} \frac{h_{Pm}}{k_P R} A_\Phi + \frac{dP_m^n}{d\phi} \frac{1}{k_S R} \frac{d(Rh_{Sm})}{dR} A_\chi + \frac{n}{\sin\phi} P_m^n h_{Sm} A_\psi \right\} \tag{8.201}$$

$$\bar{u}_\theta = \begin{pmatrix} -\sin n\theta \\ \cos n\theta \end{pmatrix} \left\{ \frac{n}{k_P R \, \sin\phi} P_m^n h_{Pm} A_\Phi + \frac{n}{k_S R \sin\phi} P_m^n \frac{d(Rh_{Sm})}{dR} A_\chi + \frac{dP_m^n}{d\phi} h_{Sm} A_\psi \right\} \tag{8.202}$$

Defining also

$$\mathbf{T}_n = \text{diag}\left\{ \begin{pmatrix} \cos n\theta \\ \sin n\theta \end{pmatrix} \quad \begin{pmatrix} \cos n\theta \\ \sin n\theta \end{pmatrix} \quad \begin{pmatrix} -\sin n\theta \\ \cos n\theta \end{pmatrix} \right\} \tag{8.203}$$

we obtain

$$\begin{Bmatrix} \bar{u}_R \\ \bar{u}_\phi \\ \bar{u}_\theta \end{Bmatrix} = \mathbf{T}_n \begin{Bmatrix} P_m^n \dfrac{dh_{Pm}}{k_P\, dR} & m(m+1)P_m^n\dfrac{h_{Sm}}{k_S R} & 0 \\ \dfrac{dP_m^n}{d\phi}\dfrac{h_{Pm}}{k_P R} & \dfrac{dP_m^n}{d\phi}\dfrac{1}{k_S R}\dfrac{d(Rh_{Sm})}{dR} & \dfrac{n\,P_m^n}{\sin\phi}h_{Sm} \\ \dfrac{n\,P_m^n}{\sin\phi}\dfrac{h_{Pm}}{k_P R} & \dfrac{n\,P_m^n}{\sin\phi}\dfrac{1}{k_S R}\dfrac{d(Rh_{Sm})}{dR} & \dfrac{dP_m^n}{d\phi}h_{Sm} \end{Bmatrix} \begin{Bmatrix} A_\Phi \\ A_\chi \\ A_\psi \end{Bmatrix} \tag{8.204}$$

which can be factored into

$$\begin{Bmatrix} \bar{u}_R \\ \bar{u}_\phi \\ \bar{u}_\theta \end{Bmatrix} = \mathbf{T}_n \begin{Bmatrix} P_m^n & 0 & 0 \\ 0 & \dfrac{dP_m^n}{d\phi} & \dfrac{n\,P_m^n}{\sin\phi} \\ 0 & \dfrac{n\,P_m^n}{\sin\phi} & \dfrac{dP_m^n}{d\phi} \end{Bmatrix} \begin{Bmatrix} \dfrac{dh_{Pm}}{k_P\, dR} & m(m+1)\dfrac{h_{Sm}}{k_S R} & 0 \\ \dfrac{h_{Pm}}{k_P R} & \dfrac{1}{k_S R}\dfrac{d(Rh_{Sm})}{dR} & 0 \\ 0 & 0 & h_{Sm} \end{Bmatrix} \begin{Bmatrix} A_\Phi \\ A_\chi \\ A_\psi \end{Bmatrix} \tag{8.205}$$

Hence, a *particular* solution to the elastic wave equation in spherical coordinates is as given in the box below.

$$\bar{\mathbf{u}}(R, \phi, \theta, \omega) = \mathbf{T}_n \mathbf{L}_m^n \mathbf{H}_m \mathbf{a}, \qquad n \leq m$$

$\mathbf{a} = $ vector of integration constants;

$$\mathbf{T}_n = \text{diag}\left[\cos n\theta \quad \cos n\theta \quad -\sin n\theta\right]$$

or

$$\mathbf{T}_n = \text{diag}\left[\sin n\theta \quad \sin n\theta \quad \cos n\theta\right]$$

$$\mathbf{L}_m^n = \left\{ \begin{matrix} P_m^n & 0 & 0 \\[2mm] 0 & \dfrac{dP_m^n}{d\phi} & \dfrac{n\,P_m^n}{\sin\phi} \\[3mm] 0 & \dfrac{n\,P_m^n}{\sin\phi} & \dfrac{dP_m^n}{d\phi} \end{matrix} \right\}$$

$$\mathbf{H}_m = \left\{ \begin{matrix} \dfrac{dh_{Pm}}{dz_P} & m(m+1)\dfrac{h_{Sm}}{z_S} & 0 \\[3mm] \dfrac{h_{Pm}}{z_P} & \dfrac{1}{z_S}\dfrac{d(z_S h_{Sm})}{dz_S} & 0 \\[3mm] 0 & 0 & h_{Sm} \end{matrix} \right\}$$

(8.206)

in which $z_P = k_P R$, $z_S = k_S R$, $h_{Pm} = h_m(z_P)$, $h_{Sm} = h_m(z_S)$, h_m is any of the spherical Bessel functions, and P_m^n are the associated Legendre functions of order m and rank n. Table 10.8 shows a list of the first few spheroidal (or co-latitude) matrices $\mathbf{L}$. Thus, we have succeeded in separating the variations of the displacements in the radial direction from those in longitude and co-latitude. On the other hand, from Section 1.4.3, the stress in spherical surfaces associated with a displacement field $\bar{\mathbf{u}}$ is

$$\bar{\mathbf{s}}_R = \left[\sigma_R \quad \sigma_{R\phi} \quad \sigma_{R\theta}\right]^T = \mathbf{L}_R^T \bar{\mathbf{u}}$$

$$= \left[\mathbf{D}_{RR}\frac{\partial}{\partial R} + \mathbf{D}_{R\phi}\frac{1}{R}\frac{\partial}{\partial\phi} + \mathbf{D}_{R\theta}\frac{1}{R\sin\phi}\frac{\partial}{\partial\theta} + \mathbf{D}_{R1}\frac{1}{r} + \mathbf{D}_{R2}\frac{\cot\phi}{R}\right]\mathbf{T}_n \mathbf{L}_m^n \mathbf{H}_m \mathbf{a}$$

(8.207)

Introducing into this expression the material matrices for an isotropic medium given in Section 1.4.3, and carrying out the required differentiation operations, we obtain the stresses in spherical surfaces. It should be mentioned that these operations require tedious algebraic manipulations – including taking into account the differential equations for the spherical Bessel and associated Legendre functions. The final result is

$$
\bar{\mathbf{s}}_R = \begin{bmatrix} \sigma_R & \sigma_{R\phi} & \sigma_{R\theta} \end{bmatrix}^T = \mathbf{T}_n \mathbf{L}_m^n \mathbf{F}_m \mathbf{a},
$$

$$
\mathbf{F}_m = \left\{ \begin{matrix} f_{11} & f_{12} & 0 \\ f_{21} & f_{22} & 0 \\ 0 & 0 & f_{33} \end{matrix} \right\}
$$

$$
f_{11} = -k_P \left[(\lambda + 2\mu) h_m - \frac{2\mu}{z_P} \left(2h_{P,m+1} + m(m-1)\frac{h_{Pm}}{z_P} \right) \right]
$$

$$
f_{22} = -k_S \left[\mu h_m - \frac{2\mu}{z_S} \left(h_{S,m+1} - (m+1)(m-1)\frac{h_{Sm}}{z_S} \right) \right]
$$

$$
f_{12} = -k_S\, m(m+1)\frac{2\mu}{z_S} \left(h_{S,m+1} - (m-1)\frac{h_{Sm}}{z_S} \right)
$$

$$
f_{21} = -k_P \frac{2\mu}{z_P} \left(h_{P,m+1} - (m-1)\frac{h_{Pm}}{z_P} \right)
$$

$$
f_{33} = -k_S \mu \left(h_{S,m+1} - (m-1)\frac{h_{Sm}}{z_S} \right)
$$

$$(8.208)$$

9 Integral transform method

The integral transform method provides the most general framework for the analytical and numerical treatment of elastodynamic problems in unbounded continua, provided that the media exhibit specific geometric and material regularities. In particular, it can be used to solve problems of sources in unbounded, homogeneous, and layered media formulated in Cartesian, cylindrical, or spherical coordinates. We provide a brief introduction to this method while picking up the fundamental tools needed for the powerful *stiffness matrix method* described in the next chapter, and we illustrate these concepts by means of various examples.

In a nutshell, the method consists in carrying out an appropriate integral transform on the vector wave equation – including the source term – which changes the problem from a set of partial differential equations in the space-time domain to a system of coupled linear equations in the frequency–wavenumber domain. After solving the latter for the displacements, an inverse integral transformation is applied, which returns the sought-after displacements (i.e., the wave field) in space–time.

In principle, the method is exact, but only in the simplest of problems (e.g., Pekeris's or Chao's problem) are the inverse transforms amenable to exact evaluation via contour integration. In most other (more complicated) cases, the inversion must be carried out numerically.

9.1 Cartesian coordinates

Consider a horizontally layered, laterally unbounded system subjected to a source (or body load) **b** acting at some location. In addition, the system may be either finite or infinitely deep in the vertical direction. As we have seen in Section 1.4.1, the wave equation in Cartesian coordinates for such a system is given by

$$
\mathbf{D}_{xx}\frac{\partial^2 \mathbf{u}}{\partial x^2} + \mathbf{D}_{yy}\frac{\partial^2 \mathbf{u}}{\partial y^2} + \mathbf{D}_{zz}\frac{\partial^2 \mathbf{u}}{\partial z^2} + (\mathbf{D}_{xy} + \mathbf{D}_{yx})\frac{\partial^2 \mathbf{u}}{\partial x\,\partial y}
$$

$$
+ (\mathbf{D}_{yz} + \mathbf{D}_{zy})\frac{\partial^2 \mathbf{u}}{\partial y\,\partial z} + (\mathbf{D}_{xz} + \mathbf{D}_{zx})\frac{\partial^2 \mathbf{u}}{\partial x\,\partial z} + \mathbf{b} = \rho\ddot{\mathbf{u}} \tag{9.1}
$$

in which the $\mathbf{D}_{\alpha\beta}$ do not depend on x, y. Applying a Fourier transform in x, y, we obtain the wave equation in the wavenumber domain:

$$
\mathbf{D}_{zz}\frac{\partial^2\tilde{\mathbf{u}}}{\partial z^2} - i\left[k_x(\mathbf{D}_{xz} + \mathbf{D}_{zx}) + k_y(\mathbf{D}_{yz} + \mathbf{D}_{zy})\right]\frac{\partial\tilde{\mathbf{u}}}{\partial z}
$$

$$
- \left[k_x^2\,\mathbf{D}_{xx} + k_x k_y(\mathbf{D}_{xy} + \mathbf{D}_{yx}) + k_y^2\,\mathbf{D}_{yy}\right]\tilde{\mathbf{u}} + \tilde{\mathbf{b}} = \rho\ddot{\tilde{\mathbf{u}}} \tag{9.2}
$$

in which $\tilde{\mathbf{u}} = \tilde{\mathbf{u}}(k_x, k_y, z, t)$ and $\tilde{\mathbf{b}} = \tilde{\mathbf{b}}(k_x, k_y, z, t)$ are the spatial Fourier transforms of $\mathbf{u}$ and $\mathbf{b}$, and k_x, k_y are the two horizontal wavenumbers. After solving this one-dimensional wave propagation problem in $\tilde{\mathbf{u}}(z, t)$ for a dense set of horizontal wavenumbers, we proceed to obtain the displacements in space–time from the inverse Fourier transformation

$$
\mathbf{u}(x, y, z, t) = \left(\frac{1}{2\pi}\right)^2 \int_{-\infty}^{+\infty} \int_{-\infty}^{+\infty} \tilde{\mathbf{u}}(k_x, k_y, z, t)e^{-i(k_x x + k_y y)}\,dk_x\,dk_y \tag{9.3}
$$

For reasons to be seen later, it is convenient at this point to introduce the modified, transformed displacement and load vectors

$$
\mathfrak{u} = \begin{Bmatrix} 1 \\ & 1 \\ & & -i \end{Bmatrix} \begin{Bmatrix} \tilde{u}_x \\ \tilde{u}_y \\ \tilde{u}_z \end{Bmatrix} = \begin{Bmatrix} \tilde{u}_x \\ \tilde{u}_y \\ -i\tilde{u}_z \end{Bmatrix}, \tag{9.4a}
$$

$$
\mathfrak{b} = \begin{Bmatrix} 1 \\ & 1 \\ & & -i \end{Bmatrix} \begin{Bmatrix} \tilde{b}_x \\ \tilde{b}_y \\ \tilde{b}_z \end{Bmatrix} = \begin{Bmatrix} \tilde{b}_x \\ \tilde{b}_y \\ -i\tilde{b}_z \end{Bmatrix} \tag{9.4b}
$$

The Fourier-transformed wave equation is then

$$
(k_x^2\,\mathbf{D}_{xx} + k_x k_y\mathbf{B}_{xy} + k_y^2\,\mathbf{D}_{yy})\,\mathfrak{u} + (k_x\mathbf{B}_{xz} + k_y\mathbf{B}_{yz})\frac{\partial\mathfrak{u}}{\partial z} - \mathbf{D}_{zz}\frac{\partial^2\mathfrak{u}}{\partial z^2} + \rho\ddot{\mathfrak{u}} = \mathfrak{b} \tag{9.5}
$$

in which

$$
\mathbf{B}_{xy} = \begin{Bmatrix} 0 & \lambda + \mu & 0 \\ \lambda + \mu & 0 & 0 \\ 0 & 0 & 0 \end{Bmatrix}, \tag{9.6a}
$$

$$
\mathbf{B}_{xz} = \begin{Bmatrix} 0 & 0 & -(\lambda + \mu) \\ 0 & 0 & 0 \\ \lambda + \mu & 0 & 0 \end{Bmatrix}, \tag{9.6b}
$$

$$
\mathbf{B}_{yz} = \begin{Bmatrix} 0 & 0 & 0 \\ 0 & 0 & -(\lambda + \mu) \\ 0 & \lambda + \mu & 0 \end{Bmatrix} \tag{9.6c}
$$

In particular, for a problem in plane strain (i.e., no variation in the y direction), the Fourier-transformed wave equation reduces to

$$\mathfrak{b} - k_x^2\, \mathbf{D}_{xx}\mathbf{u} - k_x \mathbf{B}_{xz}\frac{\partial \mathbf{u}}{\partial z} + \mathbf{D}_{zz}\frac{\partial^2 \mathbf{u}}{\partial z^2} = \rho \ddot{\mathbf{u}} \tag{9.7}$$

which involves only real coefficient matrices. Additionally, it will be found that this equation is exactly the same as in cylindrical coordinates, and that it leads to symmetric stiffness matrices for layered systems.

The above equation involves two uncoupled systems that should be solved separately, namely, one for SV-P waves and one for SH waves. Their equations are

$$\begin{Bmatrix} \mu & 0 \\ 0 & \lambda+2\mu \end{Bmatrix} \frac{\partial^2}{\partial z^2} \begin{Bmatrix} \tilde{u}_x \\ -i\tilde{u}_z \end{Bmatrix} - k_x \begin{Bmatrix} 0 & -(\lambda+\mu) \\ \lambda+\mu & 0 \end{Bmatrix} \frac{\partial}{\partial z} \begin{Bmatrix} \tilde{u}_x \\ -i\tilde{u}_z \end{Bmatrix}$$

$$-k_x^2 \begin{Bmatrix} \lambda+2\mu & 0 \\ 0 & \mu \end{Bmatrix} \begin{Bmatrix} \tilde{u}_x \\ -i\tilde{u}_z \end{Bmatrix} + \begin{Bmatrix} \tilde{b}_x \\ -i\tilde{b}_z \end{Bmatrix} = \rho \frac{\partial^2}{\partial t^2} \begin{Bmatrix} \tilde{u}_x \\ -i\tilde{u}_z \end{Bmatrix}$$

$$\tilde{b}_y + \mu \frac{\partial^2 \tilde{u}_y}{\partial z^2} - \mu k_x^2 \tilde{u}_y = \rho \frac{\partial^2 \tilde{u}_y}{\partial t^2}$$

$$\tag{9.8}$$

For harmonic motions $\ddot{\mathbf{u}} = -\omega^2 \mathbf{u}$ and in the absence of sources ($\mathfrak{b} = \mathbf{0}$, i.e., a homogeneous equation), the transformed equation of motion admits solutions of the form

$$\mathbf{u}(k, z, \omega) = \mathbf{R}_1\, \mathbf{E}_z^{-1}\mathbf{a}_1 + \mathbf{R}_2\, \mathbf{E}_z\mathbf{a}_2, \qquad (\mathbf{a}_1, \mathbf{a}_2 \text{ } 3 \times 1 \text{ vectors of constants}) \tag{9.9}$$

$$\mathbf{E}_z = \mathbf{diag} \begin{Bmatrix} e^{kpz} & e^{ksz} & e^{ksz} \end{Bmatrix}$$

$$p = \sqrt{1 - \left(\frac{\omega}{\alpha k}\right)^2} = \sqrt{1 - \left(\frac{c}{\alpha}\right)^2}, \quad s = \sqrt{1 - \left(\frac{\omega}{\beta k}\right)^2} = \sqrt{1 - \left(\frac{c}{\beta}\right)^2}$$

($k \equiv k_x$ = horizontal wavenumber, $\quad c$ = phase velocity) $\tag{9.10}$

$$\mathbf{R}_1 = \begin{Bmatrix} 1 & 0 & -s \\ 0 & 1 & 0 \\ -p & 0 & 1 \end{Bmatrix}, \qquad \mathbf{R}_2 = \begin{Bmatrix} 1 & 0 & s \\ 0 & 1 & 0 \\ p & 0 & 1 \end{Bmatrix} \tag{9.11}$$

which corresponds to displacements of the form (overbar a reminder of factor $-i$ in u_z)

$$\bar{\mathbf{u}}(x, z, \omega) = \mathbf{u}\, e^{-ikx} = \left(\mathbf{R}_1\, \mathbf{E}_z^{-1}\mathbf{a}_1 + \mathbf{R}_2\, \mathbf{E}_z\mathbf{a}_2\right) e^{-ikx} \tag{9.12}$$

This solution agrees perfectly with the solution obtained earlier by separation of variables via the solution to the Helmholtz equation.

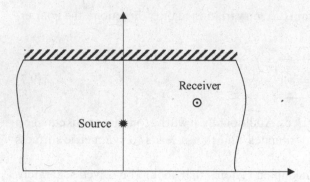

Example 9.1: Homogeneous stratum subjected to a harmonic, anti-plane line load

Consider a homogeneous stratum of total thickness h whose lower and upper surfaces are free and fixed, respectively (Fig. 9.1). This (admittedly upside down) stratum is subjected to a harmonic anti-plane (SH) line load of frequency ω at $x = 0$ and some arbitrary elevation z'. Taking the origin of coordinates at the lower (free) surface, this implies a body load of the form

$$b_y(x, z, \omega) = \delta(x)\,\delta(z - z') \tag{9.13}$$

whose Fourier transform in x is simply $\tilde{b}_y = \delta(z - z')$. As we have already observed, the structure of the matrices $\mathbf{D}_{xx}, \mathbf{D}_{zz}, \mathbf{B}_{xz}$ is such that the equation for transverse displacement $\tilde{u}_y$ is uncoupled from the in-plane degrees of freedom $\tilde{u}_x, \tilde{u}_z$. Hence, the governing wave equation in the frequency–wavenumber domain is

$$\left(\mu k^2 - \rho\omega^2\right)\tilde{u}_y - \mu\frac{\partial^2 \tilde{u}_y}{\partial z^2} = \delta(z - z') \tag{9.14}$$

where for notational simplicity we have written the horizontal wavenumber simply as $k_x \equiv k$. To solve this equation together with the boundary conditions at the two external surfaces, we proceed to divide it by the shear modulus and consider first the *homogeneous* problem in which there is no load term (i.e., source). To distinguish this special case, we write its solution as $\tilde{u}_y = \phi(z)$, so the differential equation and its boundary conditions are

$$\frac{\partial^2 \phi}{\partial z^2} + k_\beta^2\,\phi = 0, \qquad \left.\frac{\partial \phi}{\partial z}\right|_{z=0} = 0, \qquad \phi|_{z=h} = 0 \tag{9.15}$$

with $\beta = \sqrt{\mu/\rho}$ the shear wave velocity, and $k_\beta^2 = \omega^2/\beta^2 - k^2$. This is an eigenvalue problem in k_β whose solution is

$$\phi = A\cos k_\beta z + B\sin k_\beta z \tag{9.16}$$

in which A, B are constants. From the boundary conditions, we deduce

$$B = 0 \quad \text{and} \quad \cos k_\beta h = 0 \;\Rightarrow\; k_\beta h = \frac{\pi}{2}(2j - 1), \quad j = 1, 2, 3, \ldots \tag{9.17}$$

Writing k_j for the values of k that satisfy the characteristic equation for k_β, the solution of the *homogeneous* equation is found to be

$$\phi_j = A_j \cos \frac{(2j-1)\pi z}{2h} \tag{9.18}$$

$$k_j h = \pm \sqrt{\left(\frac{\omega h}{\beta}\right)^2 - \left(\frac{\pi(2j-1)}{2}\right)^2}$$

$$= \pm \frac{\pi}{2}\sqrt{\left(\frac{\omega}{\omega_1}\right)^2 - (2j-1)^2}, \quad j = 1, 2, 3, \dots \tag{9.19}$$

in which $\omega_1 = \pi\beta/2h$ is the fundamental (or first cutoff) frequency of the stratum, and the $\phi_j(z)$ are the normal propagation modes of the stratum in anti-plane shear with characteristic wavenumbers k_j. At each frequency there exist two solutions, of which we choose the one whose imaginary part is negative if k_j is purely imaginary, or whose real part is positive if k_j is purely real. This ensures that the wave field satisfies the radiation and boundedness conditions at infinity (for positive x). Observe that for any given frequency ω, there exist only a finite number of real eigenvalues k_j that represent propagating modes, while new modes emerge as ω rises above ω_1 in ratios 1:3:5:7....

We now return to our original problem. To solve the *inhomogeneous* wave propagation problem, we make use of modal superposition and express the displacement in terms of the normal modes:

$$\tilde{u}_y(k, z, \omega) = \sum_{j=1}^{\infty} A_j \phi_j(z) \tag{9.20}$$

with as yet undetermined *participation factors* A_j. Substituting this expansion into the original differential equation, we obtain

$$\sum_{j=1}^{\infty} A_j \left[\left(\mu k^2 - \rho\omega^2\right)\phi_j - \mu \frac{\partial^2 \phi_j}{\partial z^2} \right] = \delta(z - z') \tag{9.21}$$

which, after carrying out the differentiation and using the definition of the modal wavenumber k_j, reduces to

$$\mu \sum_{j=1}^{\infty} A_j \left(k^2 - k_j^2\right)\phi_j = \delta(z - z') \tag{9.22}$$

We next multiply both sides by another distinct mode $\phi_k(z)$ and integrate over the thickness of the stratum, which gives

$$\mu \sum_{j=1}^{\infty} A_j \left(k^2 - k_j^2\right) \int_0^h \phi_j \phi_k \, dz = \int_0^h \phi_k \delta(z - z') \, dz = \phi_k(z') \tag{9.23}$$

but from the orthogonality of the modes

$$\int_0^h \phi_j \phi_k \, dz = \frac{1}{2}h\delta_{jk} \tag{9.24}$$

so that

$$A_j = \frac{2\,\phi_k(z')}{\mu h(k^2 - k_j^2)} = \frac{2\cos\frac{\pi(2j-1)z'}{2h}}{\mu h(k^2 - k_j^2)} \tag{9.25}$$

Hence, the solution to our original problem follows from the inverse Fourier transform

$$u_y(x, \omega) = \frac{2}{\mu h}\sum_{j=1}^{\infty}\phi_j(z')\,\phi_j(z)\frac{1}{2\pi}\int_{-\infty}^{+\infty}\frac{e^{-ikx}}{k^2 - k_j^2}dk \tag{9.26}$$

Carrying out this transformation by means of contour integration, we obtain finally

$$u_y = \frac{1}{i\mu}\sum_{j=1}^{\infty}\frac{\cos\frac{\pi(2j-1)z'}{2h}\cos\frac{\pi(2j-1)z}{2h}}{k_j h}e^{-ik_j|x|} \tag{9.27}$$

This is the *exact* solution to this problem. Again, for any given frequency there exists only a finite number of modes with real wavenumbers; the remainder are evanescent modes that quickly attenuate with distance from the source. Hence, the above formula converges fast to the correct value, except of course, at $x = 0$, where there is a singularity. Observe also that at each cutoff frequency, one more modal wavenumber changes from imaginary (i.e., evanescent) to real (i.e., propagating). At this frequency, the modal wavenumber passes through zero (or attains a small value, if the medium has damping), a situation that represents a resonant condition. Hence, in the neighborhood of the cutoff frequencies, the last mode to become real contributes the most to the response because its denominator in the summation is small.

9.2 Cylindrical coordinates

From Section 1.4.2, the wave equation in cylindrical coordinates is

$$\mathbf{b} - \rho\ddot{\mathbf{u}} + \mathbf{D}_{rr}\frac{\partial^2\mathbf{u}}{\partial r^2} + \mathbf{D}_{\theta\theta}\frac{1}{r^2}\frac{\partial^2\mathbf{u}}{\partial\theta^2} + \mathbf{D}_{zz}\frac{\partial^2\mathbf{u}}{\partial z^2} + (\mathbf{D}_{r\theta} + \mathbf{D}_{\theta r})\frac{1}{r}\frac{\partial^2\mathbf{u}}{\partial r\partial\theta}$$

$$+ (\mathbf{D}_{rz} + \mathbf{D}_{zr})\frac{\partial^2\mathbf{u}}{\partial r\partial z} + (\mathbf{D}_{\theta z} + \mathbf{D}_{z\theta})\frac{1}{r}\frac{\partial^2\mathbf{u}}{\partial z\partial\theta} + \mathbf{D}_{rr}\frac{1}{r}\frac{\partial\mathbf{u}}{\partial r}$$

$$+ (\mathbf{D}_{\theta1} - \mathbf{D}_{1\theta})\frac{1}{r^2}\frac{\partial\mathbf{u}}{\partial\theta} + (\mathbf{D}_{rz} + \mathbf{D}_{z1} - \mathbf{D}_{1z})\frac{1}{r}\frac{\partial\mathbf{u}}{\partial z} - \mathbf{D}_{11}\frac{\mathbf{u}}{r^2} = \mathbf{0} \tag{9.28}$$

We now must distinguish between systems of flat layers (half-space, stratum, etc.) and systems of cylindrical layers (e.g., laminated cylinders), which involve distinct integral transforms. In the first case, the spatial transform is in horizontal planes, while in the latter, the transform is over the axial coordinate. In either case, the medium is assumed to be homogeneous and unbounded in the transformed direction.

9.2.1 Horizontally stratified media

To obtain the wave equation in the frequency–wavenumber domain, we first apply a Fourier transform in time, which changes $\ddot{\mathbf{u}} \to -\omega^2\mathbf{u}$, and then apply a Hankel transform in r and a

Fourier transform in θ (i.e., a Fourier–Bessel transform). This corresponds to the following operations (boxed equations are those commonly used in a solution):

1) Carry out a forward Fourier–Bessel transform on the source term,

$$\boxed{\tilde{\mathbf{b}}_n(k, z, \omega) = a_n \int_0^\infty r\, \mathbf{C}_n \int_0^{2\pi} \mathbf{T}_n \int_{-\infty}^{+\infty} \mathbf{b}(r, \theta, z, t)\, e^{-i\omega t}\, dt\, d\theta\, dr} \tag{9.29}$$

which admits the formal inversion

$$\mathbf{b}(r, \theta, z, t) = \frac{1}{2\pi} \int_{-\infty}^{+\infty} e^{i\omega t} \sum_{n=0}^{\infty} \mathbf{T}_n \int_0^\infty k\, \mathbf{C}_n \tilde{\mathbf{b}}_n(k, z, \omega)\, dk\, d\omega \tag{9.30}$$

2) Compute the solution in the frequency–wavenumber domain for the variation of displacements in z by solving the uncoupled equations for SV-P and SH waves.

3) Carry out an inverse Fourier–Bessel transform, and obtain the solution in the space–time domain (or the frequency domain, in which case the inverse Fourier transform over frequencies can be dispensed with):

$$\boxed{\mathbf{u}(r, \theta, z, t) = \frac{1}{2\pi} \int_{-\infty}^{+\infty} e^{i\omega t} \sum_{n=0}^{\infty} \mathbf{T}_n \int_0^\infty k\, \mathbf{C}_n \tilde{\mathbf{u}}_n(k, z, \omega)\, dk\, d\omega} \tag{9.31}$$

which again admits the formal inversion

$$\tilde{\mathbf{u}}_n(k, z, \omega) = a_n \int_0^\infty r\, \mathbf{C}_n \int_0^{2\pi} \mathbf{T}_n \int_{-\infty}^{+\infty} \mathbf{u}(r, \theta, z, t)\, e^{-i\omega t}\, dt\, d\theta\, dr \tag{9.32}$$

In these expressions

$$a_n = \begin{cases} \dfrac{1}{2\pi} & n = 0 \\ \dfrac{1}{\pi} & n \neq 0 \end{cases} \tag{9.33}$$

$$\mathbf{T}_n = \mathrm{diag}\left\{ \begin{pmatrix} \cos n\theta \\ \sin n\theta \end{pmatrix}, \begin{pmatrix} -\sin n\theta \\ \cos n\theta \end{pmatrix}, \begin{pmatrix} \cos n\theta \\ \sin n\theta \end{pmatrix} \right\} \tag{9.34}$$

$$\mathbf{C}_n = \begin{Bmatrix} J_n' & \frac{n}{kr} J_n & 0 \\ \frac{n}{kr} J_n & J_n' & 0 \\ 0 & 0 & J_n \end{Bmatrix} \tag{9.35}$$

The diagonal matrix $\mathbf{T}_n$ is formed with either the upper or the lower set of functions. The upper one is used when the loads, and therefore the displacements, are symmetric with respect to the x axis; the lower one is used when loads and displacements are antisymmetric with respect to this axis. For example, a point load in the x direction calls for the upper set with $n = 1$, whereas a load in the y direction involves the lower set with $n = 1$. By contrast, a vertical load requires the upper set with $n = 0$, whereas a torsional load about a vertical axis calls for the lower set with $n = 0$. More general, non-symmetric loads would call for a combination of these sets together with a summation over many (or even infinitely many) n. On the other hand, the elements $J_n = J_n(kr)$ of the matrix

C_n are the cylindrical Bessel functions of the first kind and order n. Observe that the differentiation of the Bessel functions is carried out with respect to the product kr, and not simply r. A brief summary of the fundamental properties of these functions is given in the Appendix.

To the uninitiated reader, it may not be at all obvious that the set of integral transform equations given above for $\mathbf{u}$ and $\tilde{\mathbf{u}}$ indeed constitute a direct and inverse Fourier–Bessel transform pair. The proof of the inversion is based on the fact that for arbitrary functions $f_j(r)$, the forward transform

$$\begin{Bmatrix} F_1 \\ F_2 \\ F_3 \end{Bmatrix} = \int_0^\infty r \begin{Bmatrix} J_n' & \frac{n}{kr}J_n & 0 \\ \frac{n}{kr}J_n & J_n' & 0 \\ 0 & 0 & J_n \end{Bmatrix} \begin{Bmatrix} f_1 \\ f_2 \\ f_3 \end{Bmatrix} dr \tag{9.36}$$

can be written in the fully diagonal form

$$\begin{Bmatrix} \frac{1}{2}(F_1+F_2) \\ \frac{1}{2}(F_2-F_1) \\ F_3 \end{Bmatrix} = \int_0^\infty r \begin{Bmatrix} J_{n-1} & 0 & 0 \\ 0 & J_{n+1} & 0 \\ 0 & 0 & J_n \end{Bmatrix} \begin{Bmatrix} \frac{1}{2}(f_1+f_2) \\ \frac{1}{2}(f_2-f_1) \\ f_3 \end{Bmatrix} dr \tag{9.37}$$

which is a self-reciprocating Hankel transform, i.e.,

$$\begin{Bmatrix} \frac{1}{2}(f_1+f_2) \\ \frac{1}{2}(f_2-f_1) \\ f_3 \end{Bmatrix} = \int_0^\infty k \begin{Bmatrix} J_{n-1} & 0 & 0 \\ 0 & J_{n+1} & 0 \\ 0 & 0 & J_n \end{Bmatrix} \begin{Bmatrix} \frac{1}{2}(F_1+F_2) \\ \frac{1}{2}(F_2-F_1) \\ F_3 \end{Bmatrix} dk \tag{9.38}$$

Also, the orthogonality of the Fourier series in θ must be used to complete the proof of inversion. This involves the identity

$$\int_0^{2\pi} \mathbf{T}_n \mathbf{T}_j \, d\theta = \pi \, \delta_{(nj)}(1 + \delta_{n0}\delta_{j0}) = \delta_{(n)j}/a_n. \tag{9.39}$$

Transformed wave equation
Applying a *spatial* Fourier–Bessel transform to the wave equation, or equivalently, assuming that $\mathbf{u}$ and $\mathbf{b}$ are of the form

$$\mathbf{u}(r, \theta, z, t) = \mathbf{T}_n \mathbf{C}_n \, \mathbf{f}(k, n, z, t), \qquad \mathbf{b}(r, \theta, z, t) = \mathbf{T}_n \mathbf{C}_n \, \mathbf{q}(k, n, z, t) \tag{9.40}$$

in which $\mathbf{f} = [f_1 \quad f_2 \quad f_3]^T$, one obtains the three expressions (after lengthy and tedious algebra, which requires also consideration of the differential equation for the Bessel functions)

$$\nabla\nabla{\cdot}\mathbf{u} = \mathbf{T}_n \mathbf{C}_n \begin{Bmatrix} -k\left(k\,f_1 - \frac{\partial}{\partial z}f_3\right) \\ 0 \\ \frac{\partial}{\partial z}\left(-k\,f_1 + \frac{\partial}{\partial z}f_3\right) \end{Bmatrix}, \quad \nabla\cdot\nabla\mathbf{u} = \mathbf{T}_n \mathbf{C}_n \left(\frac{\partial^2}{\partial z^2}\mathbf{f} - k^2\mathbf{f}\right) \tag{9.41}$$

$$\ddot{\mathbf{u}} = \mathbf{T}_n \mathbf{C}_n \ddot{\mathbf{f}} \tag{9.42}$$

Hence, the wave equation $(\lambda + \mu)\nabla\nabla \cdot \mathbf{u} + \mu\nabla \cdot \nabla\mathbf{u} + \mathbf{b} = \rho\ddot{\mathbf{u}}$ in cylindrical coordinates reduces to

$$\mathbf{T}_n\mathbf{C}_n\left\{\begin{bmatrix} q_1 \\ q_2 \\ q_3 \end{bmatrix} + (\lambda + \mu)\begin{bmatrix} -k\left(k\,f_1 - \frac{\partial}{\partial z}f_3\right) \\ 0 \\ \frac{\partial}{\partial z}\left(-k\,f_1 + \frac{\partial}{\partial z}f_3\right) \end{bmatrix} + \mu\left(\frac{\partial^2}{\partial z^2} - k^2 - \rho\frac{\partial^2}{\partial t^2}\right)\begin{bmatrix} f_1 \\ f_2 \\ f_3 \end{bmatrix}\right\} = \mathbf{0}$$

(9.43)

which implies

$$\begin{bmatrix} q_1 \\ q_2 \\ q_3 \end{bmatrix} + \begin{bmatrix} \mu & 0 & 0 \\ 0 & \mu & 0 \\ 0 & 0 & \lambda + 2\mu \end{bmatrix}\begin{bmatrix} f_1'' \\ f_2'' \\ f_3'' \end{bmatrix} - k\begin{bmatrix} 0 & 0 & -(\lambda+\mu) \\ 0 & 0 & 0 \\ \lambda+\mu & 0 & 0 \end{bmatrix}\begin{bmatrix} f_1' \\ f_2' \\ f_3' \end{bmatrix}$$

$$- k^2\begin{bmatrix} \lambda+2\mu & 0 & 0 \\ 0 & \mu & 0 \\ 0 & 0 & \mu \end{bmatrix}\begin{bmatrix} f_1 \\ f_2 \\ f_3 \end{bmatrix} = \rho\begin{bmatrix} \ddot{f}_1 \\ \ddot{f}_2 \\ \ddot{f}_3 \end{bmatrix}$$

(9.44)

This equation in turn separates into a pair of uncoupled equations – a 2-vector equation for the SV-P degrees of freedom, and a scalar equation for the SH degree of freedom – but these are no longer plane waves. Writing these two equations separately, we obtain

$$\begin{bmatrix} q_1 \\ q_3 \end{bmatrix} + \begin{bmatrix} \mu & 0 \\ 0 & \lambda+2\mu \end{bmatrix}\frac{\partial^2}{\partial z^2}\begin{bmatrix} f_1 \\ f_3 \end{bmatrix} - k\begin{bmatrix} 0 & -(\lambda+\mu) \\ \lambda+\mu & 0 \end{bmatrix}\frac{\partial}{\partial z}\begin{bmatrix} f_1 \\ f_3 \end{bmatrix}$$

$$- k^2\begin{bmatrix} \lambda+2\mu & 0 \\ 0 & \mu \end{bmatrix}\begin{bmatrix} f_1 \\ f_3 \end{bmatrix} = \rho\frac{\partial^2}{\partial t^2}\begin{bmatrix} f_1 \\ f_3 \end{bmatrix}$$

$$q_2 + \mu\left(\frac{\partial^2}{\partial z^2} - k^2\right)f_2 = \rho\frac{\partial^2 f_2}{\partial t^2}$$

(9.45)

These are the equations of motion in cylindrical coordinates expressed in the radial–azimuthal wavenumber domain. Observe that they are identical to the equation of motion for a plane-strain system in Cartesian coordinates expressed in the horizontal wavenumber domain – provided the latter is modified by application of the factor $-i = -\sqrt{-1}$ to the vertical components of displacements and loads. Hence, the same solution strategies used to find solutions to $\mathbf{f}(z, t)$ in a Cartesian system can be used for a cylindrical system, as will be illustrated later on.

Plane layers in cylindrical vs. Cartesian coordinates

The *transformed* equations in frequency–wavenumber space for horizontally homogeneous systems were found to be exactly the same in cylindrical and in Cartesian coordinates. This implies, among other things, that the solution in the frequency–wavenumber domain is identical in both, and so also are the solutions of the eigenvalue problems for guided waves in horizontally layered media. Furthermore, these wave propagation modes are *independent* of the azimuthal index n. For example, Love waves elicited by anti-plane line loads have the same dispersion and wave propagation characteristics as torsional waves in that same medium, except that the latter decay as they spread out – a result of the inverse

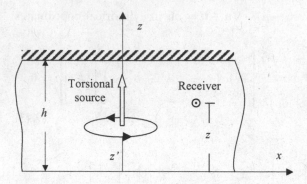

Figure 9.2: Stratum subjected to torsional ring source.

Fourier–Bessel transform – as will be shown in the example that follows. A corollary is that it is possible – at least in principle – to obtain the solution for point loads in cylindrical coordinates from the plane-strain solutions for line loads in that same layered medium, a property that is sometimes referred to as the *inversion of the descent of dimensions*. This could be achieved by Fourier-transforming the 2-D plane-strain solutions (both for SV-P and SH line loads) into frequency–wavenumber space, and inverting back into 3-D space via a Fourier–Bessel transformation.

Example 9.2: Homogeneous stratum subjected to a torsional ring load

Consider a homogeneous stratum of total thickness h whose lower and upper surfaces are free and fixed, respectively. This (inverted) stratum is subjected to a harmonic torsional ring load of radius R and frequency ω applied at some arbitrary elevation z'. Taking the origin of coordinates at the lower (free) surface, this implies a body load in the tangential direction of the form

$$b_\theta(r, \omega) = \frac{M_t}{2\pi R^2}\delta(z - z')\,\delta(r - R) \tag{9.46}$$

When this body load is integrated over a volume enclosing the load, it produces a net torsional moment M_t. Recognizing that this load is constant with θ and also antisymmetric with respect to the x axis, we conclude that the transform involves only the $n = 0$ azimuthal term, and that the lower component in the matrix $\mathbf{T}_n$ must be used, so $\mathbf{T}_0 = \mathrm{diag}\{0, 1, 0\}$. Hence, integration over θ simply produces a factor 2π that cancels out with the coefficient a_0 in front of the transform integral. The Fourier–Bessel transform for this load reduces then to

$$\tilde{\mathbf{b}}_n = \frac{M_t}{2\pi R^2}\delta(z - z')\int_0^\infty r \left\{\begin{matrix} J_0' & & \\ & J_0' & \\ & & J_0 \end{matrix}\right\}\left\{\begin{matrix} 0 \\ \delta(r - R) \\ 0 \end{matrix}\right\} dr$$

$$= \frac{M_t}{2\pi R}\delta(z - z')\left\{\begin{matrix} 0 \\ J_0'(kR) \\ 0 \end{matrix}\right\} = -\frac{M_t}{2\pi R}J_1(kR)\,\delta(z - z')\left\{\begin{matrix} 0 \\ 1 \\ 0 \end{matrix}\right\} \tag{9.47}$$

Observing the structure of the matrices $\mathbf{D}_{rr}$, $\mathbf{D}_{zz}$, $\mathbf{B}_{rz}$, we see that the equation for tangential displacement $\tilde{u}_\theta$ is uncoupled from the in-plane degrees of freedom $\tilde{u}_r$, $\tilde{u}_z$. Hence, the governing wave equation in the frequency–wavenumber domain is

$$\left(\mu k^2 - \rho\omega^2\right)\tilde{u}_\theta - \mu\frac{\partial^2\tilde{u}_\theta}{\partial z^2} = -\frac{M_t}{2\pi R}J_1(kR)\,\delta(z - z') \tag{9.48}$$

Other than the scaling factor before $\delta(z - z')$ on the right-hand side, this equation is identical to that in the example of an anti-plane line load given previously, and so also are the boundary conditions. Hence, the solution in the frequency–wavenumber domain, which is based on the homogeneous solution for normal modes, is identical:

$$\phi_j = A_j \cos \frac{(2j - 1)\pi z}{2h} \tag{9.49}$$

$$k_j h = \pm \sqrt{\left(\frac{\omega h}{\beta}\right)^2 - \left(\frac{\pi(2j - 1)}{2}\right)^2}$$

$$= \pm \frac{\pi}{2} \sqrt{\left(\frac{\omega}{\omega_1}\right)^2 - (2j - 1)^2}, \quad j = 1, 2, 3, \ldots \tag{9.50}$$

$$\tilde{u}_\theta(k, z, \omega) = \sum_{j=1}^{\infty} A_j \phi_j(z) \tag{9.51}$$

$$A_j = -\frac{M_t}{2\pi R} J_1(kR) \frac{2\phi_k(z')}{\mu h(k^2 - k_j^2)} \tag{9.52}$$

where again the roots k_j are chosen so that the imaginary part is negative, or the real part is positive if purely real. Hence, the solution to the torsional ring load follows from the inverse Hankel transform

$$u_\theta(x, \omega) = -\frac{M_t}{\pi R \mu h} \sum_{j=1}^{\infty} \phi_j(z') \phi_j(z) \int_0^{\infty} k \frac{J_1(kR) J_0'(kr)}{k^2 - k_j^2} dk \tag{9.53}$$

that is

$$u_\theta(x, \omega) = \frac{M_t}{\pi R \mu h} \sum_{j=1}^{\infty} \phi_j(z') \phi_j(z) \int_{-\infty}^{+\infty} \frac{k J_1(kR) J_1(kr)}{k^2 - k_j^2} dk \tag{9.54}$$

Now, assuming that the soil has a small amount of damping (which can be arbitrarily small), then $\mathrm{Im}(k_j) < 0$ for any frequency, which allows evaluating the Hankel transform as (see Appendix)

$$\int_0^{\infty} \frac{k J_1(kR) J_1(kr)}{k^2 - k_j^2} dk = \begin{cases} \frac{\pi}{2i} H_1^{(2)}(k_j R) J_1(k_j r) & r \leq R \\ \frac{\pi}{2i} J_1(k_j R) H_1^{(2)}(k_j r) & r \geq R \end{cases} \quad \mathrm{Im}(k_j) < 0 \tag{9.55}$$

with which we have obtained the complete solution. In particular, in the limit of a torsional point load $R \to 0$, the integral divided by R converges to

$$\lim_{R \to 0} \frac{1}{R} \int_0^{\infty} \frac{k J_1(kR) J_1(kr)}{k^2 - k_j^2} dk = \frac{\pi k_j}{4i} H_1^{(2)}(k_j r) \tag{9.56}$$

Finally

$$u_\theta(x, \omega) = \frac{M_t}{4i\mu h} \sum_{j=1}^{\infty} \cos \frac{\pi(2j - 1)z'}{2h} \cos \frac{\pi(2j - 1)z}{2h} k_j H_1^{(2)}(k_j r) \tag{9.57}$$

which is an *exact* solution for the wave field caused by a torsional point load applied in the interior of an elastic stratum.

9.2.2 Cylindrically stratified media

The starting point in this case is again the wave equation in cylindrical coordinates, but now we perform instead a Fourier transform in the axial and azimuthal directions:

$$\boxed{\tilde{\mathbf{b}}_n = a_n \int_0^{2\pi} \mathbf{T}_n \int_{-\infty}^{+\infty} \mathbf{b}(r,\theta,z,\omega)\,e^{-\mathrm{i}k_z z}\,dz\,d\theta}\,, \qquad a_n = \begin{cases} \dfrac{1}{2\pi} & n = 0 \\[2mm] \dfrac{1}{\pi} & n \neq 0 \end{cases} \tag{9.58}$$

$$\mathbf{b}(r,\theta,z,\omega) = \frac{1}{2\pi}\int_{-\infty}^{+\infty}\left\{\sum_{n=0}^{\infty}\mathbf{T}_n\tilde{\mathbf{b}}_n\right\}e^{-\mathrm{i}k_z z}\,dk_z \tag{9.59}$$

$$\tilde{\mathbf{u}}_n = a_n \int_0^{2\pi}\mathbf{T}_n\int_{-\infty}^{+\infty}\mathbf{u}(r,\theta,z,\omega)\,e^{-\mathrm{i}k_z z}\,dz\,d\theta \tag{9.60}$$

$$\boxed{\mathbf{u}(r,\theta,z,\omega) = \frac{1}{2\pi}\int_{-\infty}^{+\infty}\left\{\sum_{n=0}^{\infty}\mathbf{T}_n\tilde{\mathbf{u}}_n\right\}e^{-\mathrm{i}k_z z}\,dk_z} \tag{9.61}$$

in which again the boxed equations are those commonly used. Applying the stated Fourier transforms in θ, z, and t to the wave equation, collecting terms, and applying factors $-\mathrm{i} = -\sqrt{-1}$ to each of the vertical components so that $\mathbf{u} = \begin{bmatrix}\tilde{u}_r & \tilde{u}_\theta & -\mathrm{i}\tilde{u}_z\end{bmatrix}$ and $\mathbf{b} = \begin{bmatrix}\tilde{b}_r & \tilde{b}_\theta & -\mathrm{i}\tilde{b}_z\end{bmatrix}$, we obtain the one-dimensional system of differential equations in the radial direction,

$$\mathbf{b} + \left\{\rho\omega^2\mathbf{I} + \begin{bmatrix}\lambda+2\mu & 0 & 0 \\ 0 & \mu & 0 \\ 0 & 0 & \mu\end{bmatrix}\frac{\partial^2}{\partial r^2} + \begin{bmatrix}\lambda+2\mu & -n(\lambda+\mu) & 0 \\ n(\lambda+\mu) & \mu & 0 \\ 0 & 0 & \mu\end{bmatrix}\frac{1}{r}\frac{\partial}{\partial r}\right.$$

$$- \begin{bmatrix}\lambda+2\mu+n^2\mu & -n(\lambda+3\mu) & 0 \\ -n(\lambda+3\mu) & \mu+n^2(\lambda+2\mu) & 0 \\ 0 & 0 & n^2\mu\end{bmatrix}\frac{1}{r^2} - \begin{bmatrix}\mu & 0 & 0 \\ 0 & \mu & 0 \\ 0 & 0 & \lambda+2\mu\end{bmatrix}k_z^2$$

$$\left.+ \begin{bmatrix}0 & 0 & 1 \\ 0 & 0 & 0 \\ -1 & 0 & 0\end{bmatrix}(\lambda+\mu)k_z\frac{\partial}{\partial r} - \begin{bmatrix}0 & 0 & 0 \\ 0 & 0 & n \\ 1 & n & 0\end{bmatrix}(\lambda+\mu)\frac{k_z}{r}\right\}\mathbf{u} = \mathbf{0}\,. \tag{9.62}$$

In the absence of body loads (i.e., $\mathbf{b} = \mathbf{0}$), this equation can be shown to admit solutions of the form

$$\mathbf{u}(r,k_z,\omega) = \mathbf{H}_n^{(1)}\mathbf{a}_1 + \mathbf{H}_n^{(1)}\mathbf{a}_2 \quad (\mathbf{a}_1,\mathbf{a}_2 : 3\times 1 \text{ vectors of arbitrary constants}) \tag{9.63}$$

$$\mathbf{H}_n^{(1)} = \left\{\begin{matrix}(H_{\alpha n}^{(1)})' & n\dfrac{H_{\beta n}^{(1)}}{k_\beta r} & \dfrac{k_z}{k_\beta}(H_{\beta n}^{(1)})' \\[3mm] n\dfrac{H_{\alpha n}^{(1)}}{k_\alpha r} & (H_{\beta n}^{(1)})' & n\dfrac{k_z}{k_\beta}\dfrac{H_{\beta n}^{(1)}}{k_\beta r} \\[3mm] -\dfrac{k_z}{k_\alpha}H_{\alpha n}^{(1)} & 0 & H_{\beta n}^{(1)}\end{matrix}\right\}, \quad \mathbf{H}_n^{(2)} = \left\{\begin{matrix}(H_{\alpha n}^{(2)})' & n\dfrac{H_{\beta n}^{(2)}}{k_\beta r} & \dfrac{k_z}{k_\beta}(H_{\beta n}^{(2)})' \\[3mm] n\dfrac{H_{\alpha n}^{(2)}}{k_\alpha r} & (H_{\beta n}^{(2)})' & n\dfrac{k_z}{k_\beta}\dfrac{H_{\beta n}^{(2)}}{k_\beta r} \\[3mm] -\dfrac{k_z}{k_\alpha}H_{\alpha n}^{(2)} & 0 & H_{\beta n}^{(2)}\end{matrix}\right\},$$

$$\tag{9.64}$$

$$k_\alpha = \sqrt{k_P^2 - k_z^2}, \quad k_\beta = \sqrt{k_S^2 - k_z^2}, \qquad k_z = \text{axial wavenumber} \tag{9.65}$$

$$H_{\alpha n}^{(1)} = H_n^{(1)}(k_\alpha r), \quad H_{\alpha n}^{(2)} = H_n^{(2)}(k_\alpha r), \quad \left(H_{\alpha n}^{(1)}\right)' = \frac{d H_{\alpha n}^{(1)}}{d(k_\alpha r)}, \quad \left(H_{\alpha n}^{(2)}\right)' = \frac{d H_{\alpha n}^{(2)}}{d(k_\alpha r)} \tag{9.66}$$

$$H_{\beta n}^{(1)} = H_n^{(1)}(k_\beta r), \quad H_{\beta n}^{(2)} = H_n^{(2)}(k_\beta r), \quad \left(H_{\beta n}^{(1)}\right)' = \frac{d H_{\beta n}^{(1)}}{d(k_\beta r)}, \quad \left(H_{\beta n}^{(2)}\right)' = \frac{d H_{\beta n}^{(2)}}{d(k_\beta r)} \tag{9.67}$$

in which $H_{\alpha n}^{(1)}$, $H_{\alpha n}^{(2)}$ are the Hankel functions of the first and second kind of order n. The added sub-indices α, β indicate the argument type. Clearly, by appropriate choices for the constants $\mathbf{a}_1$, $\mathbf{a}_2$, the Hankel functions can also be changed into Bessel functions, except that these are not convenient because they become quasi-linearly dependent when the argument is complex, and especially imaginary. The above solution corresponds to displacements of the form

$$\bar{\mathbf{u}}(r, \theta, z, \omega) = \mathbf{T}_n \, \mathbf{u} \, e^{-ik_z z} = \mathbf{T}_n \left(\mathbf{H}_n^{(1)} \mathbf{a}_1 + \mathbf{H}_n^{(2)} \mathbf{a}_2\right) e^{-ik_z z} \tag{9.68}$$

This solution agrees with the results obtained earlier by different means in Section 8.7.2. Thus, when internal stresses in the transformed frequency–wavenumber space are computed, it will be found that they too coincide with those in Section 8.7.2, so they need not be repeated. In the interest of brevity and to avoid repetition, we postpone examples of application until the next chapter on the stiffness matrix method in Section 10.3.

9.3 Spherical coordinates

From Section 1.4.3, the wave equation for isotropic media in spherical coordinates is

$$
\begin{aligned}
\rho \ddot{\mathbf{u}} = \mathbf{b} + &\left\{ \begin{bmatrix} \lambda + 2\mu & \cdot & \cdot \\ \cdot & \mu & \cdot \\ \cdot & \cdot & \mu \end{bmatrix} \left(\frac{\partial^2}{\partial R^2} + \frac{2}{R} \frac{\partial}{\partial R} - \frac{2}{R^2} \right) \right. \\
&+ \begin{bmatrix} \mu & \cdot & \cdot \\ \cdot & \lambda + 2\mu & \cdot \\ \cdot & \cdot & \mu \end{bmatrix} \left(\frac{1}{R^2} \frac{\partial^2}{\partial \phi^2} + \frac{\cot\phi}{R^2} \frac{\partial}{\partial \phi} \right) \\
&+ \begin{bmatrix} \mu & \cdot & \cdot \\ \cdot & \mu & \cdot \\ \cdot & \cdot & \lambda + 2\mu \end{bmatrix} \frac{1}{R^2 \sin^2\phi} \frac{\partial^2}{\partial \theta^2} + \begin{bmatrix} 0 & \lambda+\mu & \cdot \\ \lambda+\mu & 0 & \cdot \\ \cdot & \cdot & 0 \end{bmatrix} \frac{1}{R} \frac{\partial^2}{\partial R \partial \phi} \\
&+ \begin{bmatrix} 0 & \cdot & \lambda+\mu \\ \cdot & 0 & \cdot \\ \lambda+\mu & \cdot & 0 \end{bmatrix} \frac{1}{R \sin\phi} \frac{\partial^2}{\partial R \partial \theta} + \begin{bmatrix} 0 & \cdot & \cdot \\ \cdot & 0 & \lambda+\mu \\ \cdot & \lambda+\mu & 0 \end{bmatrix} \frac{1}{R^2 \sin\phi} \frac{\partial^2}{\partial \phi \partial \theta} \\
&+ \begin{bmatrix} 0 & \lambda+\mu & \cdot \\ \cdot & 0 & \cdot \\ \cdot & \cdot & 0 \end{bmatrix} \frac{\cot\phi}{R} \frac{\partial}{\partial R} + \begin{bmatrix} 0 & -(\lambda+3\mu) & \cdot \\ 2(\lambda+2\mu) & 0 & \cdot \\ \cdot & \cdot & 0 \end{bmatrix} \frac{1}{R^2} \frac{\partial}{\partial \phi}
\end{aligned}
$$

(formula continued on next page)

(*continued*)

$$
+ \begin{bmatrix} 0 & \cdot & -(\lambda + 3\mu) \\ \cdot & 0 & \cdot \\ 2(\lambda + 2\mu) & \cdot & 0 \end{bmatrix} \frac{1}{R^2 \sin\phi} \frac{\partial}{\partial\theta}
$$

$$
+ \begin{bmatrix} 0 & \cdot & \cdot \\ \cdot & 0 & -(\lambda + 3\mu) \\ \cdot & \lambda + 3\mu & 0 \end{bmatrix} \frac{\cot\phi}{R^2 \sin\phi} \frac{\partial}{\partial\theta} + \begin{bmatrix} 0 & -(\lambda+3\mu) & \cdot \\ \cdot & 0 & \cdot \\ \cdot & \cdot & 0 \end{bmatrix} \frac{\cot\phi}{R^2}
$$

$$
+ \begin{bmatrix} 0 & \cdot & \cdot \\ \cdot & -\lambda & \cdot \\ \cdot & \cdot & \mu \end{bmatrix} \frac{1}{R^2 \sin^2\phi} + \begin{bmatrix} 0 & \cdot & \cdot \\ \cdot & -2\mu & \cdot \\ \cdot & \cdot & -2\mu \end{bmatrix} \frac{\cot^2\phi}{R^2} \Bigg\} \mathbf{u} \tag{9.69}
$$

This rather complicated looking system of differential equations can also be solved by means of transforms, except that because each spherical surface is finite in size, we employ series solutions in place of the integrals of the previous sections. However, instead of attempting to re-derive the relevant equations in the transformed domain for this case, we shall simply rely on the results already obtained in Section 8.10. Thus, the solution of this equation follows along the following formal steps (the matrices involved are defined in Section 8.10, and tabulated in Section 10.4):

1) Express the source in terms of spherical harmonics:

$$
\boxed{ \tilde{\mathbf{b}}_{mn}(R, \omega) = \mathbf{J}^{-1} \int_0^\pi \sin\phi \, \mathbf{L}_m^n \int_0^{2\pi} \mathbf{T}_n \, \mathbf{b}(R, \phi, \theta, \omega) \, d\theta \, d\phi } \tag{9.70}
$$

which admits the formal inversion

$$
\mathbf{b} = \sum_{m=0}^\infty \sum_{n=0}^m \mathbf{T}_n \, \mathbf{L}_m^n \tilde{\mathbf{b}}_{mn} \tag{9.71}
$$

2) For each m, n, and ω, obtain the solution $\tilde{\mathbf{u}}_{mn}(R)$ in the frequency–wavenumber domain by solving the one-dimensional system of equations in the radial coordinate R.

3) Obtain the solution in the space domain from the series solution

$$
\boxed{ \mathbf{u}(R, \phi, \theta, \omega) = \sum_{m=0}^\infty \sum_{n=0}^m \mathbf{T}_n \, \mathbf{L}_m^n \tilde{\mathbf{u}}_{mn} } \tag{9.72}
$$

which again admits the formal inversion

$$
\tilde{\mathbf{u}}_{mn}(R, \omega) = \mathbf{J}^{-1} \int_0^\pi \sin\phi \, \mathbf{L}_{m^n} \int_0^{2\pi} \mathbf{T}_n \mathbf{u}(R, \phi, \theta, \omega) \, d\theta \, d\phi \tag{9.73}
$$

The proof of these inversion formulas is obtained by considering the orthogonality properties of Fourier series in the azimuth and of the spheroidal matrix in co-latitude, namely,

$$
\int_0^{2\pi} \mathbf{T}_n \mathbf{T}_j \, d\theta = \pi \, \delta_{(nj)}(1 + \delta_{n0}\delta_{j0}) \tag{9.74}
$$

$$\int_0^\pi \mathbf{L}_m^n \mathbf{L}_k^n \sin\phi \, d\phi = \delta_{(mk)} \frac{(n+m)!}{(m+\frac{1}{2})(m-n)!} \text{diag}\{1 \quad m(m+1) \quad m(m+1)\} \qquad (9.75)$$

Hence

$$\int_0^\pi \sin\phi \, \mathbf{L}_k^n \int_0^{2\pi} \mathbf{T}_j \mathbf{u} \, d\theta \, d\phi = \sum_{m=0}^\infty \sum_{n=0}^m \left\{ \int_0^\pi \sin\phi \, \mathbf{L}_k^n \left[\int_0^{2\pi} \mathbf{T}_j \mathbf{T}_n d\theta \right] \mathbf{L}_m^n \, d\phi \right\} \tilde{\mathbf{u}}_{mn}$$

$$= \pi(1+\delta_{n0}) \sum_{m=0}^\infty \left[\int_0^\pi \sin\phi \, \mathbf{L}_k^n \mathbf{L}_m^n \, d\phi \right] \tilde{\mathbf{u}}_{mn}$$

$$= \frac{\pi(1+\delta_{n0})(n+m)!}{(m+\frac{1}{2})(m-n)!} \text{diag}\{1 \quad m(m+1)m(m+1)\} \tilde{\mathbf{u}}_{mn}$$

$$(9.76)$$

Defining the diagonal matrix

$$\mathbf{J} = \frac{\pi(1+\delta_{n0})(n+m)!}{(m+\frac{1}{2})(m-n)!} \begin{Bmatrix} 1 & 0 & 0 \\ 0 & m(m+1) & 0 \\ 0 & 0 & m(m+1) \end{Bmatrix} \qquad (9.77)$$

we obtain the result given above for $\tilde{\mathbf{u}}_{mn}$. For examples of application, see Section 10.4.

10 Stiffness matrix method for layered media

Closed-form solutions for layered media, or for homogeneous plates and strata with arbitrary boundary conditions, do not exist. Indeed, even the free–free plate – the so-called Mindlin plate – is ultimately intractable by purely analytical means. Thus, such problems must be solved with the aid of numerical tools. Among these, a widely used scheme is the *propagator matrix* or transfer matrix method of Haskell[1] and Thomson.[2] Nonetheless, we choose instead to present herein the related *stiffness* or *impedance matrix method*, which has the advantages over the propagator matrix method that are listed below and, at least in our judgment, no disadvantages in comparison with the latter. Thus, readers familiar with the propagator method are strongly encouraged to familiarize themselves with and switch to this superior method. Among the advantages of the stiffness matrix method are:

- Stiffness matrices are symmetric, while propagator matrices are not.
- Stiffness matrices involve half as many degrees of freedom as propagator matrices, so their bandwidth is only half as large. The former involve only displacements, whereas the state vector in propagator matrices contains both stresses and displacements.
- On account of the two previous items, the stiffness matrix method is nearly an order of magnitude faster than the propagator matrix method: The computational effort is smaller by a factor 2 due to symmetry, and a factor of more than 4 on account of bandwidth, which gives a total reduction of more than 8. Since the computations must be repeated for each frequency and each wavenumber, the savings are considerable.
- Stiffness matrices remain robust and stable for thick layers and/or high frequencies. In these situations, layer interfaces decouple naturally as a result of coupling terms tending to zero. Propagator matrices, on the other hand, contain terms of exponential growth that require special treatment.
- Stiffness matrices lead naturally to the solutions for normal modes (eigenvalue problems), source problems, and wave amplification problems, all without the need for

[1] Haskell, N. A., 1964, The dispersion of surface waves on multilayered media, *Bulletin of the Seismological Society of America*, Vol. 43, pp. 17–34.
[2] Thomson, W. T., 1950, Transmission of elastic waves through a stratified soil medium, *Journal of Applied Physics*, Vol. 21, pp. 98–93.

special manipulations or treatment for each class of problem. Indeed, one can readily solve simultaneously, and with virtually no added effort, for various sources.

- Stiffness matrices lend themselves to the application of substructuring techniques.
- Discrete versions of the stiffness matrices based on the finite element method – the so-called *thin-layer method* – can readily be obtained, with the advantage that the matrix formalism remains exactly the same. These discrete matrices allow finding the normal modes – both propagating and evanescent modes – from the solution of a standard eigenvalue problem, and thus avoid the use of search techniques. The reader is referred to the literature for further details on the thin-layer method.

Since its inception in 1981,[3] the stiffness matrix method has found wide applications in elastodynamics, especially in the dynamics of laminated plates and layered soils. A closely related, but not identical, method was also proposed much earlier by M. Biot.[4] In the following, we present the stiffness matrix method for Cartesian, cylindrical, and spherical coordinates. In a nutshell, we shall show that sources and displacements in a layered medium are related by an equation of the form $\mathbf{p} = \mathbf{Ku}$, in which $\mathbf{K}$ is a symmetric stiffness matrix.

10.1 Summary of method

The stiffness or impedance matrix method is a tool for the analysis of wave propagation problems in elastic media, namely *source* problems, *normal mode* problems, and *wave amplification* problems. In principle, it is an *exact* method in the sense that one obtains mathematical expressions for displacements that are free from approximations or discretization errors. However, the resulting expressions are generally intractable by purely analytical means and must ultimately be evaluated numerically. That such computation is fraught with difficulties and may thus introduce substantial errors if not carried out properly goes without saying, but a detailed discussion of these is beyond the scope of this book. The method can be applied to a large class of continua, such as beams and plates, but we restrict our presentation to the following three problems involving isotropic media formulated in Cartesian, cylindrical, and spherical coordinates:

- Laterally infinite, horizontally layered strata and/or half-spaces (plane strain).
- Infinitely long layered cylinders of finite or infinite radius, solid or hollow.
- Layered spheres of finite or infinite radius, solid or hollow.

The method is based on the use of integral transforms, and consists of the following steps:

- Transform the source(s) (if any), which are modeled as external tractions, from the space–time domain into the frequency–wavenumber domain. This produces a source vector $\mathbf{p}$, usually in closed form.
- For each frequency and wavenumber, determine the stiffness matrix of each layer and, by appropriate superposition, the stiffness matrix $\mathbf{K}$ of the complete layered system.

---

[3] Kausel, E. and Röesset, J. M., 1981, Stiffness matrices for layered soils, *Bulletin of the Seismological Society of America*, Vol. 71, pp. 1743–1761.

[4] Biot, M., 1963, Continuum dynamics of elastic plates and multilayered solids, *Journal of Mathematics and Mechanics*, Vol. 12, pp. 793–810.

This matrix is block-tridiagonal (i.e., narrowly banded) and symmetric, and in general its elements are complex. Damping is incorporated via complex moduli.

- Solve the system of equations $\mathbf{p} = \mathbf{K}\mathbf{u}$ by standard methods, and obtain the displacements in the frequency–wavenumber domain.
- Carry out an inverse transform into the spatial–temporal domain, which yields the desired response.

Omitting the inverse Fourier transform over frequencies, the requisite inverse integral transforms needed to obtain the displacement vector $\mathbf{u}$ in the spatial domain are of the following form:

- Cartesian coordinates, horizontal layers:

$$\mathbf{u}(x, z, \omega) = \frac{1}{2\pi} \int_{-\infty}^{+\infty} \tilde{\mathbf{u}}(k, z, \omega)\, e^{-ikx} dk \qquad (10.1)$$

- Cylindrical coordinates, horizontal layers:

$$\mathbf{u}(r, \theta, z, \omega) = \sum_{n=0}^{\infty} \mathbf{T}_n \int_0^{\infty} k\, \mathbf{C}_n \tilde{\mathbf{u}}_n(k, z, \omega)\, dk \qquad (10.2)$$

- Cylindrical coordinates, cylindrical layers:

$$\mathbf{u}(r, \theta, z, \omega) = \frac{1}{2\pi} \int_{-\infty}^{+\infty} \left\{ \sum_{n=0}^{\infty} \mathbf{T}_n\, \tilde{\mathbf{u}}_n(r, k_z, \omega) \right\} e^{-ik_z z}\, dk_z \qquad (10.3)$$

- Spherical coordinates, spherical layers:

$$\mathbf{u}(R, \phi, \theta, \omega) = \sum_{m=0}^{\infty} \sum_{n=0}^{m} \mathbf{T}_n\, \mathbf{L}_m^n \tilde{\mathbf{u}}_{mn}(R, \omega) \qquad (10.4)$$

Details and examples are given in the pages that follow.

10.2 Stiffness matrix method in Cartesian coordinates

Consider a homogeneous medium subjected to waves in the x, z plane, i.e., SV-P and SH waves in plane strain. In the absence of external sources (i.e., $\mathbf{b} = \mathbf{0}$), the elastic wave equation in the transformed frequency– wavenumber space ω, k is (see Section 9.1)

$$k^2 \mathbf{D}_{xx}\mathbf{u} + k\, \mathbf{B}_{xz} \frac{\partial \mathbf{u}}{\partial z} - \mathbf{D}_{zz} \frac{\partial^2 \mathbf{u}}{\partial z^2} - \rho\omega^2 \mathbf{u} = \mathbf{0} \qquad (10.5)$$

where for simplicity we have written the horizontal wavenumber as $k_x \equiv k$. Making an ansatz $\mathbf{u}(k, \omega, z) = \mathbf{v}(k, \omega)\, e^{nz}$ and substituting it into the above expression, we obtain

$$\left(k^2 \mathbf{D}_{xx} - n^2 \mathbf{D}_{zz} + n\, k\mathbf{B}_{xz} - \rho\omega^2 \mathbf{I}\right) \mathbf{v} = \mathbf{0} \qquad (10.6)$$

This equation is an eigenvalue problem in n for plane waves with vertical wavenumber $k_z = \pm i\, n$ and eigenvectors $\mathbf{v}$. Nontrivial solutions exist if the determinant of the 3×3 matrix in parenthesis vanishes, a condition that leads to a sixth order equation, which for an isotropic medium reduces to a quadratic and a biquadratic equation for SH and SV-P waves, respectively. Solving these equations and finding the eigenvectors $\mathbf{v}$, the resulting displacement field can be shown to be given by

$$\mathbf{u} = \left\{ \begin{array}{c} u_x \\ u_y \\ -i\,u_z \end{array} \right\} = \left\{ \mathbf{R}_1\,\mathbf{E}_z^{-1} \quad \mathbf{R}_2\,\mathbf{E}_z \right\} \left\{ \begin{array}{c} \mathbf{a}_1 \\ \mathbf{a}_2 \end{array} \right\} \tag{10.7}$$

with

$$\mathbf{R}_1 = \left\{ \begin{array}{ccc} 1 & 0 & -s \\ 0 & 1 & 0 \\ -p & 0 & 1 \end{array} \right\}, \qquad \mathbf{R}_2 = \left\{ \begin{array}{ccc} 1 & 0 & s \\ 0 & 1 & 0 \\ p & 0 & 1 \end{array} \right\} \tag{10.8}$$

$$\mathbf{E}_z = \text{diag}\left\{ e^{kpz} \quad e^{ksz} \quad e^{ksz} \right\}$$

Here $\mathbf{a}_1$, $\mathbf{a}_2$ are vectors of arbitrary constants (wave amplitudes), and

$$p = \sqrt{1 - \left(\frac{\omega}{k\alpha}\right)^2}, \qquad \alpha = \sqrt{\frac{\lambda + 2\mu}{\rho}} = \text{P-wave velocity} \tag{10.9}$$

$$s = \sqrt{1 - \left(\frac{\omega}{k\beta}\right)^2}, \qquad \beta = \sqrt{\frac{\mu}{\rho}} = \text{S-wave velocity} \tag{10.10}$$

Observe that $n_{1,4} = \pm kp, n_{2,5} = \pm ks, n_{3,6} = \pm ks$ are the six roots of the eigenvalue problem in n, and the six columns of $\left\{ \mathbf{R}_1 \quad \mathbf{R}_2 \right\}$ are the eigenvectors. The first and fourth columns in the solution for $\mathbf{u}$ above are P waves, the second and fifth are SH waves, and the third and sixth are SV waves. Depending on the values of the wavenumber k and frequency ω (i.e., on real or imaginary values for p, s), these solutions represent either plane body waves, or guided waves that evanesce in the vertical z direction. The terms in $\mathbf{a}_1$ are plane waves that propagate or decay in the positive z direction (i.e., upwards), whereas those in $\mathbf{a}_2$ propagate or decay in the negative z direction (i.e., downwards). Observe that SH waves are uncoupled from SV-P waves.

The stresses in horizontal planes associated with the solution found above – after we modify these by an imaginary unit factor applied to the vertical component – can be shown to be given by

$$\mathfrak{s} = \left\{ \begin{array}{c} \sigma_{xz} \\ \sigma_{yz} \\ -i\,\sigma_{zz} \end{array} \right\} = k\mu\left\{ -\mathbf{Q}_1\,\mathbf{E}_z^{-1} \quad \mathbf{Q}_2\,\mathbf{E}_z \right\} \left\{ \begin{array}{c} \mathbf{a}_1 \\ \mathbf{a}_2 \end{array} \right\} \tag{10.11}$$

with

$$\mathbf{Q}_1 = \left\{ \begin{array}{ccc} 2p & 0 & -(1+s^2) \\ 0 & s & 0 \\ -(1+s^2) & 0 & 2s \end{array} \right\}, \qquad \mathbf{Q}_2 = \left\{ \begin{array}{ccc} 2p & 0 & 1+s^2 \\ 0 & s & 0 \\ 1+s^2 & 0 & 2s \end{array} \right\} \tag{10.12}$$

Consider next a homogeneous, horizontal, elastic layer of arbitrary thickness h as a free body in space, and assume that it is subjected to the spatially and temporally harmonic wave field determined previously. To maintain dynamic equilibrium, we apply appropriate *tractions* $\mathfrak{p}_1 = \mathfrak{s}_1$, $\mathfrak{p}_2 = -\mathfrak{s}_2$ at the upper and lower external interfaces, respectively, which completely balance the internal stresses at these two locations. We also choose the origin of coordinates at the center of the layer and temporarily assign the suffixes 1 and 2, respectively, to the upper and lower external surfaces. Hence, if $\mathbf{u}_1$, $\mathbf{u}_2$ are the displacements at the two interfaces, we have

$$\mathbf{u}_1 = \left\{ \mathbf{R}_1 \, \mathbf{E}_{h/2}^{-1} \quad \mathbf{R}_2 \, \mathbf{E}_{h/2} \right\} \left\{ \begin{matrix} \mathbf{a}_1 \\ \mathbf{a}_2 \end{matrix} \right\}, \quad \mathbf{u}_2 = \left\{ \mathbf{R}_1 \, \mathbf{E}_{-h/2}^{-1} \quad \mathbf{R}_2 \, \mathbf{E}_{-h/2} \right\} \left\{ \begin{matrix} \mathbf{a}_1 \\ \mathbf{a}_2 \end{matrix} \right\} \tag{10.13}$$

which can be combined into

$$\left\{ \begin{matrix} \mathbf{u}_1 \\ \mathbf{u}_2 \end{matrix} \right\} = \left\{ \begin{matrix} \mathbf{R}_1 \, \mathbf{E}_{h/2}^{-1} & \mathbf{R}_2 \, \mathbf{E}_{h/2} \\ \mathbf{R}_1 \, \mathbf{E}_{-h/2}^{-1} & \mathbf{R}_2 \, \mathbf{E}_{-h/2} \end{matrix} \right\} \left\{ \begin{matrix} \mathbf{a}_1 \\ \mathbf{a}_2 \end{matrix} \right\} = \left\{ \begin{matrix} \mathbf{R}_1 \, \mathbf{E}_{h/2}^{-1} & \mathbf{R}_2 \, \mathbf{E}_{h/2} \\ \mathbf{R}_1 \, \mathbf{E}_{h/2} & \mathbf{R}_2 \, \mathbf{E}_{h/2}^{-1} \end{matrix} \right\} \left\{ \begin{matrix} \mathbf{a}_1 \\ \mathbf{a}_2 \end{matrix} \right\} \tag{10.14}$$

Thus, the amplitude vector is

$$\left\{ \begin{matrix} \mathbf{a}_1 \\ \mathbf{a}_2 \end{matrix} \right\} = \left\{ \begin{matrix} \mathbf{R}_1 \, \mathbf{E}_{h/2}^{-1} & \mathbf{R}_2 \, \mathbf{E}_{h/2} \\ \mathbf{R}_1 \, \mathbf{E}_{h/2} & \mathbf{R}_2 \, \mathbf{E}_{h/2}^{-1} \end{matrix} \right\}^{-1} \left\{ \begin{matrix} \mathbf{u}_1 \\ \mathbf{u}_2 \end{matrix} \right\} \tag{10.15}$$

Also, the combined traction vectors are

$$\left\{ \begin{matrix} \mathfrak{p}_1 \\ \mathfrak{p}_2 \end{matrix} \right\} = \left\{ \begin{matrix} \mathfrak{s}_1 \\ -\mathfrak{s}_2 \end{matrix} \right\} = k\mu \left\{ \begin{matrix} -\mathbf{Q}_1 \, \mathbf{E}_{h/2}^{-1} & \mathbf{Q}_2 \, \mathbf{E}_{h/2} \\ \mathbf{Q}_1 \, \mathbf{E}_{-h/2}^{-1} & -\mathbf{Q}_2 \, \mathbf{E}_{-h/2} \end{matrix} \right\} \left\{ \begin{matrix} \mathbf{a}_1 \\ \mathbf{a}_2 \end{matrix} \right\}$$

$$= k\mu \left\{ \begin{matrix} -\mathbf{Q}_1 \, \mathbf{E}_{h/2}^{-1} & \mathbf{Q}_2 \, \mathbf{E}_{h/2} \\ \mathbf{Q}_1 \, \mathbf{E}_{h/2} & -\mathbf{Q}_2 \, \mathbf{E}_{h/2}^{-1} \end{matrix} \right\} \left\{ \begin{matrix} \mathbf{a}_1 \\ \mathbf{a}_2 \end{matrix} \right\} \tag{10.16}$$

Eliminating the amplitude vector between these two equations, we obtain

$$\left\{ \begin{matrix} \mathfrak{p}_1 \\ \mathfrak{p}_2 \end{matrix} \right\} = k\mu \left\{ \begin{matrix} -\mathbf{Q}_1 \, \mathbf{E}_{h/2}^{-1} & \mathbf{Q}_2 \, \mathbf{E}_{h/2} \\ \mathbf{Q}_1 \, \mathbf{E}_{h/2} & -\mathbf{Q}_2 \, \mathbf{E}_{h/2}^{-1} \end{matrix} \right\} \left\{ \begin{matrix} \mathbf{R}_1 \, \mathbf{E}_{h/2}^{-1} & \mathbf{R}_2 \, \mathbf{E}_{h/2} \\ \mathbf{R}_1 \, \mathbf{E}_{h/2} & \mathbf{R}_2 \, \mathbf{E}_{h/2}^{-1} \end{matrix} \right\}^{-1} \left\{ \begin{matrix} \mathbf{u}_1 \\ \mathbf{u}_2 \end{matrix} \right\} \tag{10.17}$$

or more compactly

$$\boxed{ \left\{ \begin{matrix} \mathfrak{p}_1 \\ \mathfrak{p}_2 \end{matrix} \right\} = \left\{ \begin{matrix} \mathbf{K}_{11} & \mathbf{K}_{12} \\ \mathbf{K}_{21} & \mathbf{K}_{22} \end{matrix} \right\} \left\{ \begin{matrix} \mathbf{u}_1 \\ \mathbf{u}_2 \end{matrix} \right\} } \tag{10.18}$$

which relates the two interface tractions with the observed interface displacements. The matrix

$$\mathbf{K} = \left\{ \begin{matrix} \mathbf{K}_{11} & \mathbf{K}_{12} \\ \mathbf{K}_{21} & \mathbf{K}_{22} \end{matrix} \right\} \tag{10.19}$$

is the *stiffness* or *impedance matrix* of the layer, which can be shown to be *symmetric*. Some authors refer to it also as the *spectral element* matrix. Because the SH components (second and fifth degree of freedom) are uncoupled from the SV-P degrees of freedom (1,3,4,6), it is possible to separate the above equation into two separate matrix equations: a 4×4 for SV-P waves, and a 2×2 for SH waves. Although somewhat tedious, the previous matrix operations can be carried out by hand and closed-form expressions obtained for the elements of $\mathbf{K}$. We give these in Table 10.1 separately for SV-P and SH waves. Elastic half-spaces (i.e., $h \to \infty$) can also be considered by setting $\mathbf{a}_1 = 0$, that is, by discarding the waves that propagate or decay upwards. This leads to the impedance matrices for

Table 10.1. Stiffness matrices, Cartesian coordinates, plane strain

$$kp = \sqrt{k^2 - (\omega/\alpha)^2}, \qquad ks = \sqrt{k^2 - (\omega/\beta)^2}$$

Note 1: In the first Riemann sheet, $\text{Im}(kp) \geq 0$, $\text{Im}(kp) \geq 0$, and $\text{Re}(ks) \geq 0$, $\text{Im}(ks) \geq 0$, whether or not the medium has material damping (attenuation), that is, the products kp, ks are complex numbers in the first quadrant, no matter what the sign of k should be. Thus, the numbers p, s themselves are in the first quadrant if $k > 0$, and in the third if $k < 0$. If material (hysteretic) damping is present, the complex wavenumbers for P and S waves are of the form

$$\frac{\omega}{\alpha_c} = \frac{\omega}{\alpha}\Big(1 - \mathrm{i}\,\xi_P \text{sgn}(\omega)\Big), \qquad \frac{\omega}{\beta_c} = \frac{\omega}{\beta}\Big(1 - \mathrm{i}\,\xi_S \text{sgn}(\omega)\Big) \; .$$

in which α_c, β_c are the complex wave velocities. This is based on the (very close) approximation

$$\beta_c = \sqrt{\frac{\mu_c}{\rho}} = \sqrt{\frac{\mu}{\rho}\Big(1 + 2\mathrm{i}\,\xi_S \text{sgn}(\omega)\Big)} \approx \frac{1}{1 - \mathrm{i}\,\xi_S \text{sgn}(\omega)}\sqrt{\frac{\mu}{\rho}}$$

and a similar expression for α_c. These approximations greatly simplify exponential terms such as

$$\exp\Big(-\frac{\mathrm{i}\omega x}{\beta_c}\Big) = \exp\Big(-\frac{\mathrm{i}\omega x}{\beta}\Big) \exp\Big(-\frac{\xi_S \omega x}{\beta}\Big), \quad \xi_S = 1/(2Q_S), \quad Q_S = \text{quality factor}$$

Note 2: Given k, all SH and SV-P elements exhibit complex-conjugate symmetry with respect to $\pm\omega$. Given ω, the SV-P elements are checkerboard symmetric and antisymmetric with respect to $\pm k$.

1) SH waves

Layer		Half-space	
$\mathbf{K} = ks\mu \begin{Bmatrix} \coth ksh & -1/\sinh ksh \\ -1/\sinh ksh & \coth ksh \end{Bmatrix}$		$\mathbf{K} = ks\mu$	$k > 0, \; \omega > 0$
$\mathbf{K} = k\mu \begin{Bmatrix} \coth kh & -1/\sinh kh \\ -1/\sinh kh & \coth kh \end{Bmatrix}$		$\mathbf{K} = k\mu$	$k > 0, \; \omega = 0$
$\mathbf{K} = \rho\beta\omega \begin{Bmatrix} \cot\Omega_S & -1/\sin\Omega_S \\ -1/\sin\Omega_S & \cot\Omega_S \end{Bmatrix}, \; \Omega_S = \dfrac{\omega h}{\beta}$		$\mathbf{K} = \mathrm{i}\omega\rho\beta$	$k = 0, \; \omega > 0$
$\mathbf{K} = \dfrac{\mu}{h}\begin{Bmatrix} 1 & -1 \\ -1 & 1 \end{Bmatrix}$		$\mathbf{K} = 0$	$k = 0, \; \omega = 0$

2) SV-P waves

Note: The horizontal–vertical coupling terms in the matrices for SV-P waves given here have opposite sign from those in Kausel and Röesset (footnote 6). This is the result of a deliberate change here in the sign of the imaginary factor applied to the vertical components. Also, we replace r in said reference by p, to avoid confusion with the radius or range.

$$\mathbf{K} = \begin{Bmatrix} \mathbf{K}_{11} & \mathbf{K}_{12} \\ \mathbf{K}_{21} & \mathbf{K}_{22} \end{Bmatrix}$$

(continued)

Table 10.1. (*continued*)

a) Non-zero frequency, non-zero wavenumber: $k > 0$, $\omega > 0$

$$C_p = \cosh kph, \quad S_p = \sinh kph, \quad C_s = \cosh ksh, \quad S_s = \sinh ksh$$

$$D = 2(1 - C_p C_s) + \left(\frac{1}{ps} + ps\right) S_p S_s, \quad \Delta = ps - \left(\frac{1+s^2}{2}\right)^2 = \text{Rayleigh function}$$

$$\mathbf{K}_{11} = 2k\mu \left[\frac{1-s^2}{2D} \left\{ \begin{array}{cc} \frac{1}{s}(C_p S_s - ps\, C_s S_p) & 1 - C_p C_s + ps\, S_p S_s \\ 1 - C_p C_s + ps\, S_p S_s & \frac{1}{p}(C_s S_p - ps\, C_p S_s) \end{array} \right\} + \frac{1+s^2}{2} \left\{ \begin{array}{cc} 0 & 1 \\ 1 & 0 \end{array} \right\} \right]$$

$\mathbf{K}_{22}$ = same as $\mathbf{K}_{11}$, with off-diagonal signs reversed

$$\mathbf{K}_{12} = 2k\mu \left[\frac{1-s^2}{2D} \left\{ \begin{array}{cc} \frac{1}{s}(ps\, S_p - S_s) & C_p - C_s \\ -(C_p - C_s) & \frac{1}{p}(ps\, S_s - S_p) \end{array} \right\} \right], \quad \mathbf{K}_{21} = \mathbf{K}_{12}^T$$

$$\mathbf{K}_{12}^{-1} = \frac{1}{2k\mu} \left[\frac{2}{1-s^2} \left\{ \begin{array}{cc} \frac{1}{p}(ps\, S_s - S_p) & -(C_p - C_s) \\ C_p - C_s & \frac{1}{s}(ps\, S_p - S_s) \end{array} \right\} \right]$$

Lower half-space $z < 0$ (for an upper half-space, reverse the sign of the off-diagonal terms):

$$\mathbf{K} = 2k\mu \left[\frac{1-s^2}{2(1-ps)} \left\{ \begin{array}{cc} p & -1 \\ -1 & s \end{array} \right\} + \left\{ \begin{array}{cc} 0 & 1 \\ 1 & 0 \end{array} \right\} \right]$$

$$\mathbf{K}^{-1} = \frac{1}{2k\mu \, \Delta} \left\{ \begin{array}{cc} s\frac{1}{2}(1-s^2) & ps - \frac{1}{2}(1+s^2) \\ ps - \frac{1}{2}(1+s^2) & p\frac{1}{2}(1-s^2) \end{array} \right\}$$

b) Zero frequency, non-zero wavenumber: $k > 0$, $\omega = 0$

$$\kappa = kh, \quad C = \cosh \kappa, \quad S = \sinh \kappa, \quad a = \frac{\beta}{\alpha}$$
$$D = (1+a^2)^2 S^2 - \kappa^2 (1-a^2)^2$$

$$\mathbf{K}_{11} = 2k\mu \left[\frac{1}{D} \left\{ \begin{array}{cc} (1+a^2) SC - \kappa(1-a^2) & -(1+a^2) S^2 \\ -(1+a^2) S^2 & (1+a^2) SC + \kappa(1-a^2) \end{array} \right\} + \left\{ \begin{array}{cc} 0 & 1 \\ 1 & 0 \end{array} \right\} \right]$$

$\mathbf{K}_{22}$ = same as $\mathbf{K}_{11}$, with off-diagonal signs reversed

$$\mathbf{K}_{12} = \frac{2k\mu}{D} \left\{ \begin{array}{cc} \kappa(1-a^2) C - (1+a^2) S & \kappa(1-a^2) S \\ -\kappa(1-a^2) S & -\left[\kappa(1-a^2) C + (1+a^2) S\right] \end{array} \right\}, \quad \mathbf{K}_{21} = \mathbf{K}_{12}^T$$

$$\mathbf{K}_{12}^{-1} = \frac{1}{2k\mu} \left\{ \begin{array}{cc} -\left[\kappa(1-a^2) C + (1+a^2) S\right] & -\kappa(1-a^2) S \\ \kappa(1-a^2) S & \kappa(1-a^2) C - (1+a^2) S \end{array} \right\}$$

Lower half-space $z \leq 0$ (for upper half-space, reverse the sign of the off-diagonal terms):

$$\mathbf{K} = \frac{2k\mu}{1+a^2} \begin{Bmatrix} 1 & a^2 \\ a^2 & 1 \end{Bmatrix}, \quad \mathbf{K}^{-1} = \frac{1}{2k\mu(1-a^2)} \begin{Bmatrix} 1 & -a^2 \\ -a^2 & 1 \end{Bmatrix}$$

Note: $\dfrac{1}{2(1-a^2)} = 1 - \nu$

c) Non-zero frequency, zero wavenumber: $k = 0, \ \omega > 0$

$$\Omega_P = \frac{\omega h}{\alpha}, \quad \Omega_S = \frac{\omega h}{\beta}$$

$$\mathbf{K} = \rho\omega \begin{Bmatrix} \beta \cot \Omega_S & 0 & -\beta/\sin \Omega_S & 0 \\ 0 & \alpha \cot \Omega_P & 0 & -\alpha/\sin \Omega_P \\ -\beta/\sin \Omega_S & 0 & \beta \cot \Omega_S & 0 \\ 0 & -\alpha/\sin \Omega_P & 0 & \alpha \cot \Omega_P \end{Bmatrix}$$

Half-space $\mathbf{K} = i\omega\rho \begin{Bmatrix} \beta & 0 \\ 0 & \alpha \end{Bmatrix}$

d) Zero frequency, zero wavenumber: $k = 0, \ \omega = 0$

$$\mathbf{K} = \frac{\mu}{h} \begin{Bmatrix} 1 & 0 & -1 & 0 \\ 0 & a^{-2} & 0 & -a^{-2} \\ -1 & 0 & 1 & 0 \\ 0 & -a^{-2} & 0 & a^{-2} \end{Bmatrix}$$

Note: $\mu a^{-2} = \lambda + 2\mu$

Half-space $\mathbf{K} = \begin{Bmatrix} 0 & 0 \\ 0 & 0 \end{Bmatrix}$

half-spaces listed in the tables. When the frequency and/or layer thickness are large, the coupling elements $\mathbf{K}_{12}, \mathbf{K}_{21}$ tend to zero, while the main elements $\mathbf{K}_{11}, \mathbf{K}_{22}$ converge to the impedances of a lower and an upper half-space, respectively.

Having found the stiffness matrix for a given layer, we can move on and consider next a layered system with $N-1$ layers and N interfaces, each of which is characterized by a stiffness matrix that depends on the horizontal wavenumber, the frequency, and the material properties and thickness of the layer. Thus, we can assemble the system impedance matrix by overlapping appropriately the matrices for neighboring layers. The result is a block-tridiagonal matrix equation of the form

$$\begin{Bmatrix} \mathbf{p}_1 \\ \mathbf{p}_2 \\ \mathbf{p}_3 \\ \vdots \\ \mathbf{p}_N \end{Bmatrix} = \begin{Bmatrix} \mathbf{K}_{11} & \mathbf{K}_{12} & \mathbf{0} & \cdots & \mathbf{0} \\ \mathbf{K}_{21} & \mathbf{K}_{22} & \mathbf{K}_{23} & \cdots & \mathbf{0} \\ \mathbf{0} & \mathbf{K}_{32} & \mathbf{K}_{33} & \ddots & \vdots \\ \vdots & \vdots & \vdots & \ddots & \ddots & \mathbf{K}_{N-1,N} \\ \mathbf{0} & \mathbf{0} & \cdots & \mathbf{K}_{N,N-1} & \mathbf{K}_{NN} \end{Bmatrix} \begin{Bmatrix} \mathbf{u}_1 \\ \mathbf{u}_2 \\ \mathbf{u}_3 \\ \vdots \\ \mathbf{u}_N \end{Bmatrix} \qquad (10.20)$$

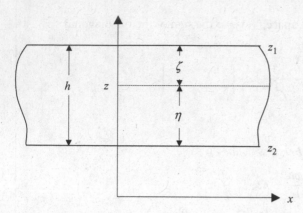

Figure 10.1: Analytic continuation in layer.

which can be written compactly as

$$\mathfrak{p} = \mathbf{K}\mathbf{u} \tag{10.21}$$

Here, $\mathfrak{p} = \mathfrak{p}(k, \omega)$ is the vector of external tractions applied at the interfaces, i.e., the source vector in the frequency–wavenumber domain, and $\mathbf{K}$ is the symmetric system stiffness matrix, which is generally composed of complex elements. Thus, solving this equation for the displacements and then carrying out appropriate integral transforms back into the spatial–temporal domain, we obtain the requisite solution for the layered medium. We shall demonstrate the use of these matrices later on in this chapter by means of various examples.

10.2.1 Analytic continuation in the layers

The stiffness matrix formulation described in the foregoing is based on the assumption that the external sources are applied solely in the form of tractions at the layer interfaces, and that there are no sources within the layer. This does not constitute a limitation of the method, however, since it is always possible to define an auxiliary interface at the location of an internal source and subdivide the physical layer into two sub-layers with identical material properties. On the other hand, this subdivision technique allows us also to obtain displacements within the layer in terms of the displacements at the interfaces by means of *analytic continuation*, as described next.

Consider an individual layer of finite thickness h, and subdivide it at elevation z it into two sub-layers of thicknesses ζ and η, respectively (so that $\zeta + \eta = h$) (see Fig. 10.1). Using super-indices ζ, η to identify the two sub-layers, the dynamic equilibrium equation for this system is

$$\left\{ \begin{array}{c} \mathfrak{p}_1 \\ \mathbf{0} \\ \mathfrak{p}_2 \end{array} \right\} = \left\{ \begin{array}{ccc} \mathbf{K}_{11}^{\zeta} & \mathbf{K}_{12}^{\zeta} & \mathbf{0} \\ \mathbf{K}_{21}^{\zeta} & \mathbf{K}_{22}^{\zeta} + \mathbf{K}_{11}^{\eta} & \mathbf{K}_{12}^{\eta} \\ \mathbf{0} & \mathbf{K}_{21}^{\eta} & \mathbf{K}_{22}^{\eta} \end{array} \right\} \left\{ \begin{array}{c} \mathbf{u}_1 \\ \mathbf{u} \\ \mathbf{u}_2 \end{array} \right\} \tag{10.22}$$

Condensation of the internal degree of freedom gives

$$\mathbf{u}(z) = - \left(\mathbf{K}_{22}^{\zeta} + \mathbf{K}_{11}^{\eta} \right)^{-1} \left(\mathbf{K}_{21}^{\zeta} \mathbf{u}_1 + \mathbf{K}_{12}^{\eta} \mathbf{u}_2 \right) \tag{10.23}$$

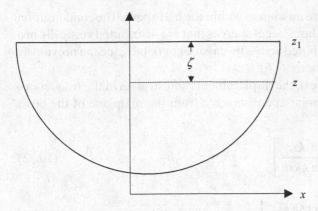

Figure 10.2: Analytic continuation in half-space.

On the other hand, after condensation, we recover the stiffness matrix for the complete layer. This implies the identities

$$\mathbf{K}_{11} = \mathbf{K}_{11}^{\zeta} - \mathbf{K}_{12}^{\zeta} \left(\mathbf{K}_{21}^{\eta}\right)^{-1} \mathbf{K}_{21}, \quad \mathbf{K}_{12} = -\mathbf{K}_{12}^{\zeta} \left(\mathbf{K}_{22}^{\zeta} + \mathbf{K}_{11}^{\eta}\right)^{-1} \mathbf{K}_{12}^{\eta} \tag{10.24}$$

$$\mathbf{K}_{21} = -\mathbf{K}_{21}^{\eta} \left(\mathbf{K}_{22}^{\zeta} + \mathbf{K}_{11}^{\eta}\right)^{-1} \mathbf{K}_{21}^{\zeta}, \quad \mathbf{K}_{22} = \mathbf{K}_{22}^{\eta} - \mathbf{K}_{21}^{\eta} \left(\mathbf{K}_{12}^{\zeta}\right)^{-1} \mathbf{K}_{12} \tag{10.25}$$

It follows that

$$\boxed{\mathbf{u}(z) = \left(\mathbf{K}_{21}^{\eta}\right)^{-1} \mathbf{K}_{21} \, \mathbf{u}_1 + \left(\mathbf{K}_{12}^{\zeta}\right)^{-1} \mathbf{K}_{12} \mathbf{u}_2} \quad (\text{layer} \quad z_2 \leq z \leq z_1) \tag{10.26}$$

which provides the displacements at interior points at elevation z in terms of the interface displacements. Of course, the latter must have been obtained first by solving the complete system. Table 10.1 gives explicit expression for the inverses needed above. Observe also that when either ζ or η becomes zero, the above expression returns the correct value at each interface. For example, when $z = z_2$, we have

$$\mathbf{K}_{21}^0 = \{\infty\} \quad \Rightarrow \quad \left(\mathbf{K}_{21}^0\right)^{-1} = \mathbf{0}, \quad \text{so} \quad \left(\mathbf{K}_{21}^h\right)^{-1} \mathbf{K}_{21} = \mathbf{I} \quad \text{and} \quad \mathbf{u} = \mathbf{u}_2. \tag{10.27}$$

In the special case of a half-space loaded at its surface, we have $z \leq z_1$, $\eta = \infty$, $\zeta = z_1 - z$, $\mathbf{K}_{11}^{\eta} \equiv \mathbf{K}_{\text{half}}$ (i.e., the half-space impedance), $\mathbf{K}_{12}^{\eta} = \mathbf{0}$, so that

$$\mathbf{u}(z) = -\left(\mathbf{K}_{22}^{\zeta} + \mathbf{K}_{\text{half}}\right)^{-1} \mathbf{K}_{21}^{\zeta} \, \mathbf{u}_1 \tag{10.28}$$

However, since condensation of the auxiliary interface must return a half-space, this implies

$$\mathbf{K}_{\text{half}} = \mathbf{K}_{11}^{\zeta} - \mathbf{K}_{12}^{\zeta} \left(\mathbf{K}_{22}^{\zeta} + \mathbf{K}_{\text{half}}\right)^{-1} \mathbf{K}_{21}^{\zeta} \tag{10.29}$$

so

$$\left(\mathbf{K}_{22}^{\zeta} + \mathbf{K}_{\text{half}}\right)^{-1} \mathbf{K}_{21}^{\zeta} = -\left(\mathbf{K}_{12}^{\zeta}\right)^{-1} \left(\mathbf{K}_{\text{half}} - \mathbf{K}_{11}^{\zeta}\right) \tag{10.30}$$

Hence

$$\boxed{\mathbf{u}(z) = \left(\mathbf{K}_{12}^{\zeta}\right)^{-1} \left(\mathbf{K}_{\text{half}} - \mathbf{K}_{11}^{\zeta}\right) \mathbf{u}_1} \quad (\text{half-space} \quad z \leq z_1) \tag{10.31}$$

which assumes, of course, that there are no sources within the half-space. The condensation strategy can be used also to evaluate the effect of sources that are distributed vertically in a continuous fashion, i.e., sources distributed across the layers. For details, see the previously cited work by Kausel and Röesset (footnote 6).

In the anti-plane case (i.e., SH waves), the displacements and stresses in the frequency–wavenumber domain at an interior point at a distance z from the midplane of the layer can be written explicitly as

$$u_y(z) = \frac{1}{2}\left[u_m \frac{\cosh ksz}{\cosh \frac{1}{2}ksh} + \Delta u \frac{\sinh ksz}{\sinh \frac{1}{2}ksh}\right], \quad -\tfrac{1}{2}h \leq z \leq \tfrac{1}{2}h \tag{10.32}$$

$$\tau_{yz}(z) = ks\mu \left\{ u_m \frac{\sinh ksz}{\cosh \frac{1}{2}ksh} + \Delta u \frac{\cosh ksz}{\sinh \frac{1}{2}ksh}\right\}, \quad -\tfrac{1}{2}h \leq z \leq \tfrac{1}{2}h$$

$$= \frac{1}{2}\left[\tau_m \frac{\cosh ksz}{\cosh \frac{1}{2}ksh} + \Delta\tau \frac{\sinh ksz}{\sinh \frac{1}{2}ksh}\right] \tag{10.33}$$

in which

$$u_m = \frac{u_{y1} + u_{y2}}{2}, \qquad \Delta u = \frac{u_{y1} - u_{y2}}{2}, \quad s = \sqrt{1 - (\omega/k\beta)^2} \tag{10.34}$$

$$\tau_m = \frac{\tau_{yz1} + \tau_{yz2}}{2}, \qquad \Delta\tau = \frac{\tau_{yz1} - \tau_{yz2}}{2} \tag{10.35}$$

These expressions for SH waves are based on placing the origin of coordinates at the center of the layer. Hence, they break down in the case of a half-space. In that case,

$$u_y(\zeta) = \frac{u_{y1}}{\sinh ks\zeta + \cosh ks\zeta}, \quad \zeta = z_1 - z \tag{10.36}$$

$$\tau_{yz}(\zeta) = \frac{\tau_{yz1}}{\sinh ks\zeta + \cosh ks\zeta} \tag{10.37}$$

10.2.2 Numerical computation of stiffness matrices

The elements of the stiffness matrices listed in Table 10.1 have terms containing hyperbolic functions of complex argument, which can attain very large numerical values even if the stiffness elements themselves will not. Thus, to avoid errors associated with severe cancellations of large numbers, it is essential to address this problem explicitly in the program routine that evaluates these functions. To illustrate the type of strategy needed, consider the following decomposition:

$$\begin{aligned} C_p = \cosh kph &= \cosh(a + ib) = \cosh a \cos b + i \ \sinh a \sin b \\ &= e^a \tfrac{1}{2}\left[(1 + e^{-2a})\cos b + i\,(1 - e^{-2a})\sin b\right] \\ &\underset{\text{def}}{=} e^a\, C_1 \end{aligned} \tag{10.38}$$

where we have assumed a to be a positive number (appropriate changes are needed if $a < 0$). The factor C_1 is now well behaved. Similarly,

$$C_s = \cosh ksh = \cosh(c + id) = e^c\, C_2, \quad S_p = e^a\, S_1, \quad S_s = e^c\, S_2 \tag{10.39}$$

For example, introducing this into the first term in the first element of $\mathbf{K}_{11}$, we obtain

$$
\frac{C_p S_s}{D} = \frac{e^{a+c} C_1 S_2}{2 \left(1 - e^{a+c} C_1 C_2\right) + \left(\frac{1}{ps} + ps\right) e^{a+c} S_1 S_2}
$$

$$
= \frac{C_1 S_2}{2 \left(e^{-(a+c)} - C_1 C_2\right) + \left(\frac{1}{ps} + ps\right) S_1 S_2} \tag{10.40}
$$

which is now numerically stable and well behaved.

10.2.3 Summary of computation

Without the imaginary unit factor applied to each of the vertical components of displacements and stresses (i.e., $-i\,u_z$, $-i\,\sigma_{zz}$), the stiffness matrices would not be symmetric. Moreover, if the soil had damping, they would not be Hermitian either. Thus, use of the imaginary factor is essential to preserve symmetry. On the other hand, though a positive sign for this factor (i.e., $+i\,u_z$, $+i\,\sigma_{zz}$) would also have led to symmetric matrices, we chose a negative sign instead so as to achieve matrices in cylindrical coordinates that are identical to those in Cartesian coordinates. Thus, after solving for the transformed displacements, the vertical components must be multiplied by $+i$ to recover the physical components. Formally, the solution is then obtained along the following lines:

1) Form the (physical) load vector $\tilde{\mathbf{p}}$ (the tilde refers to the frequency–wavenumber domain), $\quad \tilde{\mathbf{p}}(k, z, \omega) = \int_{-\infty}^{+\infty} \mathbf{p}(x, z, \omega)\, e^{ikx} dk.$

2) Multiply all axial components $\tilde{\mathbf{p}}$ of by $-i = -\sqrt{-1}$: $\tilde{\mathbf{p}} \to \mathbf{p}$.

3) Form the system matrix $\mathbf{K}$ and solve for $\mathbf{u}$: $\mathbf{u} = \mathbf{K}^{-1}\mathbf{p}$.

4) Multiply all *axial* (vertical) components of $\mathbf{u}$ by $+i = \sqrt{-1}$: $\mathbf{u} \to \tilde{\mathbf{u}}$.

5) Find the actual displacements at each interface with the inverse Fourier transform $\mathbf{u}(x, z, \omega) = \frac{1}{2\pi} \int_{-\infty}^{+\infty} \tilde{\mathbf{u}}(k, z, \omega)\, e^{-ikx} dk.$

Example 10.1: Rayleigh waves in a half-space

In the absence of external tractions at the surface of a half-space, the dynamic equilibrium equation is

$$
\mathbf{K}\mathbf{u} = \mathbf{0} \quad \Rightarrow \quad \det \mathbf{K} = 0 \tag{10.41}
$$

and from Table 10.1,

$$
\left| 2k\mu \left\{ \frac{1 - s^2}{2\,(1 - ps)} \begin{bmatrix} p & -1 \\ -1 & s \end{bmatrix} + \begin{bmatrix} 0 & 1 \\ 1 & 0 \end{bmatrix} \right\} \right| = 0 \quad \Rightarrow \quad \Delta = ps - \left(\frac{1 + s^2}{2} \right)^2 = 0
$$

$$
\tag{10.42}
$$

which when written in full can be recognized as the well-known Rayleigh function

$$\sqrt{1-\left(\frac{C_R}{\alpha}\right)^2}\sqrt{1-\left(\frac{C_R}{\beta}\right)^2}-\left(1-\frac{1}{2}\left(\frac{C_R}{\beta}\right)^2\right)^2=0, \quad C_R=\frac{\omega}{k}\equiv\frac{\omega}{k_R} \tag{10.43}$$

This equation can be changed into a bicubic equation with three roots, of which only one is physically meaningful and the other two are spurious roots that result from the rationalization process of squaring the two radicals. The single physical root is found to be independent of frequency, so Rayleigh waves are non-dispersive. A very tight approximation to the speed of Rayleigh waves as a function of Poisson's ratio is

$$\frac{C_R}{\beta}=0.874+0.197\nu-0.056\nu^2-0.0276\nu^3 \tag{10.44}$$

The relative amplitudes of Rayleigh waves at the surface follow from the eigenvectors, which are

$$\left\{\begin{matrix}u\\-iw\end{matrix}\right\}=\left\{\begin{matrix}1\\\frac{ap}{a-1}\end{matrix}\right\} \quad\Rightarrow\quad \left\{\begin{matrix}u\\w\end{matrix}\right\}=\left\{\begin{matrix}1\\\frac{iap}{a-1}\end{matrix}\right\} \quad\text{with}\quad a=\frac{1-s^2}{2(1-ps)} \tag{10.45}$$

Thus, the vertical displacements are 90° out of phase with the horizontal.

Example 10.2: **Love waves in a layer over an elastic half-space**

Consider an elastic layer of thickness h underlain by an elastic half-space. From Table 10.1, the stiffness matrices for the layer (1) and the half-space (2) are, respectively,

$$\mathbf{K}_1=\frac{ks_1\mu_1}{\sinh ks_1h}\left\{\begin{matrix}\cosh ks_1h & -1\\-1 & \cosh ks_1h\end{matrix}\right\}, \quad \mathbf{K}_2=ks_2\mu_2 \tag{10.46}$$

with $s_j=\sqrt{1-(\omega/k\beta_j)^2}$, $j=1,2$. The system equation is then

$$\mathbf{K}=ks_1\mu_1\left\{\begin{matrix}\coth ks_1h & -1/\sinh ks_1h\\-1/\sinh ks_1h & \coth ks_1h+\frac{s_2\mu_2}{s_1\mu_1}\end{matrix}\right\} \tag{10.47}$$

Free waves require the determinant of the system stiffness matrix to vanish, i.e.,

$$\coth^2 ks_1h+\frac{s_2\mu_2}{s_1\mu_1}\coth ks_1h-\frac{1}{\sinh^2 ks_1h}=0 \tag{10.48}$$

and after brief algebra

$$\tanh ks_1h+\frac{s_2\mu_2}{s_1\mu_1}=0, \quad\text{or in full,}\quad \tanh h\sqrt{k^2-k_1^2}+\frac{\mu_2\sqrt{k^2-k_2^2}}{\mu_1\sqrt{k^2-k_1^2}}=0 \tag{10.49}$$

with $k_j=\omega/\beta_j$. The above characteristic equation can also be written as

$$\tan h\sqrt{k_1^2-k^2}=\frac{\mu_2}{\mu_1}\frac{\sqrt{k^2-k_2^2}}{\sqrt{k_1^2-k^2}}=0 \tag{10.50}$$

which is the classical equation for Love waves. Once we have found the roots of this equation, we can obtain the displacement amplitudes at the interfaces as eigenvectors of the equation $\mathbf{Ku} = \mathbf{0}$. Thereafter, displacements caused by Love waves within the layer or the half-space can be obtained by means of the analytic continuation technique given previously.

Example 10.3: SH line source at some depth in a layer over an elastic half-space

The geometry is as in Example 10.2, but now the medium is subjected to an impulsive line source at some elevation z' and in direction y, which corresponds to a body load

$$b_y(x, z, t) = \delta(t)\,\delta(x)\,\delta(z - z') \tag{10.51}$$

Carrying out a Fourier transform in x and t, we obtain the source in the frequency–wavenumber domain as

$$\tilde{b}_y(k, z, \omega) = \delta(z - z') \tag{10.52}$$

which is equivalent to a surface traction $p_y = 1$ in a horizontal plane at elevation $z = z'$. If the source is in the upper layer at a depth $\zeta = z_1 - z'$ with complement $\eta = h - \zeta = z' - z_2$ and defining $\kappa = s_2\mu_2/s_1\mu_1$ then

$$ks_1\mu_1 \left\{ \begin{array}{ccc} \coth ks_1\zeta & -1/\sinh ks_1\zeta & 0 \\ -1/\sinh ks_1\zeta & \coth ks_1\zeta + \coth ks_1\eta & -1/\sinh ks_1\eta \\ 0 & -1/\sinh ks_1\eta & \coth ks_1\eta + \kappa \end{array} \right\} \left\{ \begin{array}{c} \tilde{u}_{y1} \\ \tilde{u}_y \\ \tilde{u}_{y2} \end{array} \right\} = \left\{ \begin{array}{c} 0 \\ 1 \\ 0 \end{array} \right\} \tag{10.53}$$

whereas if the source is in the half-space at $\zeta = z_2 - z'$ and defining again $\kappa = s_2\mu_2/s_1\mu_1$, then the system is

$$ks_1\mu_1 \left\{ \begin{array}{ccc} \coth ks_1 h & -1/\sinh ks_1 h & 0 \\ -1/\sinh ks_1 h & \coth ks_1 h + \kappa \coth ks_2\zeta & -\kappa/\sinh ks_2\zeta \\ 0 & -\kappa/\sinh ks_2\zeta & \kappa\,(\coth ks_2\zeta + 1) \end{array} \right\} \left\{ \begin{array}{c} \tilde{u}_{y1} \\ \tilde{u}_{y2} \\ \tilde{u}_y \end{array} \right\} = \left\{ \begin{array}{c} 0 \\ 0 \\ 1 \end{array} \right\} \tag{10.54}$$

We added tildes to the components to remind us that that the displacements in these expressions are cast in the frequency–wavenumber domain, i.e., $\tilde{u}_y = \tilde{u}_y(k, \omega)$. While the above systems could be solved analytically and the result simplified with the aid of basic trigonometric identities, in the normal use of the stiffness matrix method – especially for a system with several layers – this is done numerically for each frequency and wavenumber, as it is also in the propagator matrix method. After thus finding the response functions, we obtain the displacements in space–time by carrying out *numerically* an inverse Fourier transform over wavenumbers and, if needed, over frequencies, that is,

$$u_y(x, t) = \left(\frac{1}{2\pi}\right)^2 \int_{-\infty}^{+\infty} \int_{-\infty}^{+\infty} \tilde{u}_y\, e^{i(\omega t - kx)} dk\, d\omega \tag{10.55}$$

and similar expressions for the other components. If the medium has attenuation, the system has no singularities, so the numerical integrals are feasible. However, the kernels for layered media are likely to be wavy, and they will exhibit sharp peaks and undulations that increase in number with increasing frequency. These may affect the accuracy with which the transforms can be evaluated, and to minimize this undesirable effect, we strongly recommend the use of complex frequencies[5] (the so-called *complex exponential window method*[6]), which is a well-established technique in computational seismology.

If instead of the line source we had applied in this problem a *strip load* q_0 distributed uniformly over a horizontal area of width $2a$ with total intensity $2a\,q_0 = 1$, the wavenumber representation of the distributed load would have been

$$\tilde{p}_y = q_0 \int_{-a}^{+a} e^{ikx} dx = \frac{q_0}{i\,k}\left(e^{i\,ka} - e^{-i\,ka}\right) = 2a\,q_0\left(\frac{\sin ka}{ka}\right) = \frac{\sin ka}{ka} \tag{10.56}$$

Thus, the only difference from our previous problem is that the number 1 on the right-hand side is replaced by $(\sin ka)/ka$. More generally, any arbitrary external traction $p_y(x)$ distributed horizontally over an interface at elevation z' would lead to a load term of the form

$$\tilde{p}_y(k) = \int_{-\infty}^{+\infty} p_y(x)\,e^{ikx} dx \tag{10.57}$$

Alternatively, we could continue using a unit load on the right-hand side and, only after solving for the displacements, multiply these by $\tilde{p}_y(k)$. This would allow us to solve simultaneously for various load types, for example, a strip load of various widths, or a line load and, say, a single couple (whose Fourier transform is $-i\,k$).

Example 10.4: Mindlin plate subjected to horizontal and vertical line loads at upper surface

A homogeneous plate of arbitrary thickness h is subjected to either a horizontal *or* a vertical line load at the upper surface. In principle, this problem is similar to the previous one, except that the Mindlin plate calls for the use of the SV-P stiffness matrix (Tables, 10.1, 10.2), and the two independent line loads require replacing the load vector on the right-hand side by a two-column matrix. Thus, our problem would be of the general form **Ku= p**, or in full

$$\begin{Bmatrix} K_{11} & K_{12} & K_{13} & K_{14} \\ K_{21} & K_{22} & K_{23} & K_{24} \\ K_{31} & K_{32} & K_{33} & K_{34} \\ K_{41} & K_{42} & K_{43} & K_{44} \end{Bmatrix} \begin{Bmatrix} \tilde{u}_{1x} & \tilde{u}_{1z} \\ -i\,\tilde{w}_{1x} & -i\,\tilde{w}_{1z} \\ \tilde{u}_{2x} & \tilde{u}_{2z} \\ -i\,\tilde{w}_{2x} & -i\,\tilde{w}_{2z} \end{Bmatrix} = \begin{Bmatrix} 1 & 0 \\ 0 & -i \\ 0 & 0 \\ 0 & 0 \end{Bmatrix} \tag{10.58}$$

Observe the extra factor $-i$ in the second column on the right-hand side applied to the vertical load in the frequency–wavenumber domain ($\tilde{p}_{1z} = 1$). In principle, we first solve simultaneously for the two displacement vectors, then multiply the vertical components of these vectors by $+i$ (to remove the implied imaginary factors from these), and finally carry out inverse Fourier transformations back into the spatial domain. Of course, this succinct description glosses over many computational details that cannot be addressed herein, such as the Cholesky decomposition of **K**, the wavenumber step, the tails of integrals, and so on.

[5] Phinney, R. A., 1965, Theoretical calculation of the spectrum of first arrivals in a layered medium, *Journal of Geophysics Research*, Vol. 70, pp. 5107–5123.

[6] Kausel, E. and Roësset, J. M., 1992, Frequency domain analysis of undamped systems, *Journal of Engineering Mechanics*, Vol. 118, No. 4, pp. 721–734.

An alternative solution method could consist in carrying out a superposition of the so-called normal modes (or Lamb's modes) of the plate. These are obtained from the eigenvalue problem

$$\mathbf{K}\boldsymbol{\varphi} = \mathbf{0} \tag{10.59}$$

Because of the symmetry of the plate with respect to the midplane, this eigenvalue problem admits two type of modes: symmetric and antisymmetric. The displacement components for the upper and lower interfaces in these two types of modes satisfy the kinematic constraint

$$\boldsymbol{\varphi} = \left\{ \begin{array}{c} \boldsymbol{\varphi}_1 \\ \boldsymbol{\varphi}_2 \end{array} \right\}, \quad \boldsymbol{\varphi}_2 = \pm \mathbf{J}\,\boldsymbol{\varphi}_1, \quad \mathbf{J} = \left\{ \begin{array}{cc} 1 & 0 \\ 0 & -1 \end{array} \right\} \tag{10.60}$$

Hence, the 4×4 eigenvalue problem can be written as

$$\left\{ \begin{array}{cc} \mathbf{K}_{11} & \mathbf{K}_{12} \\ \mathbf{K}_{21} & \mathbf{K}_{22} \end{array} \right\} \left\{ \begin{array}{c} \boldsymbol{\varphi}_1 \\ \pm \mathbf{J}\,\boldsymbol{\varphi}_1 \end{array} \right\} = \left\{ \begin{array}{c} \mathbf{0} \\ \mathbf{0} \end{array} \right\} \Rightarrow \left\{ \begin{array}{cc} \mathbf{K}_{11} & \pm \mathbf{K}_{12}\,\mathbf{J} \\ \pm \mathbf{J}\,\mathbf{K}_{21} & \mathbf{J}\,\mathbf{K}_{22}\,\mathbf{J} \end{array} \right\} \left\{ \begin{array}{c} \boldsymbol{\varphi}_1 \\ \boldsymbol{\varphi}_1 \end{array} \right\} = \left\{ \begin{array}{c} \mathbf{0} \\ \mathbf{0} \end{array} \right\} \tag{10.61}$$

Because of the structure of the stiffness matrices, the second row is identical to the first. Hence, the reduced eigenvalue problem is simply

$$[\mathbf{K}_{11} \pm \mathbf{K}_{12}\,\mathbf{J}]\,\boldsymbol{\varphi}_1 = \mathbf{0} \tag{10.62}$$

The eigenvalues for the symmetric and antisymmetric modes are obtained by setting equal to zero the two determinants of this equation, one for each sign. Since this problem involves a single homogeneous layer, it is possible to work out the two determinants in closed form, although only after considerable and tedious algebra. They are

$$ps \tanh \tfrac{1}{2}kph - q^2 \tanh \tfrac{1}{2}ksh = 0 \tag{10.63a}$$

and

$$ps \coth \tfrac{1}{2}kph - q^2 \coth \tfrac{1}{2}ksh = 0 \tag{10.63b}$$

with

$$q = \frac{1 + s^2}{2} \tag{10.64}$$

$$p = \sqrt{1 - (\omega/k\alpha)^2} = \sqrt{1 - (c/\alpha)^2} \tag{10.65}$$

$$s = \sqrt{1 - (\omega/k\beta)^2} = \sqrt{1 - (c/\beta)^2} \tag{10.66}$$

in which c is the phase velocity (or *celerity*) of the modes. These are deceptively simple-looking transcendental eigenvalue equations, yet their full solution for complex wavenumbers k (i.e., evanescent modes) is exceedingly difficult[7,8] (see Fig. 10.3). For given values of the frequency ω, the real modes of propagation can be found using search techniques. Of course, the results should agree with those of a direct numerical search of the determinant of the stiffness matrix.

[7] Mindlin, R. D., 1957, Mathematical theory of elastic plates, *Proceedings 11th Annual Symposium on Frequency Control*, U.S. Army Signal Engineering Laboratory, Asbury Park, NJ.

[8] Mindlin, R. D. 1960, Waves and vibrations in isotropic, elastic plates, *Proceedings 1st Symposium on Naval Structural Mechanics, Stanford University, CA (Aug. 1958)*, Pergamon Press.

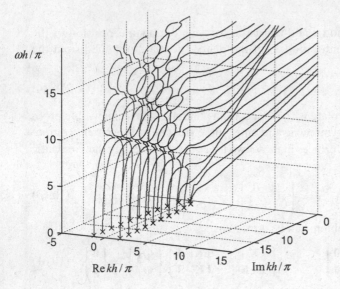

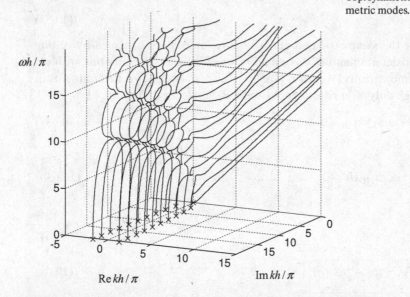

Figure 10.3: Wave spectrum for Mindlin plate, $\nu = 0.31$.
Top: symmetric modes. Bottom: antisymmetric modes.

By the way, observe that $\Delta = ps - q^2$ is the Rayleigh function, which is useful for showing that at high frequencies, the phase velocity of some of these modes converges to the Rayleigh wave velocity. After finding the modes, the displacements for the line load could be obtained by modal superposition, very much along the lines of what was done in the example in Section 9.1. In principle, such superposition must include both the propagating modes and the evanescent modes (real vs. complex eigenvalues or characteristic wavenumbers k). The exception is when we are interested only in the response at some distance from the source, in which case the propagating modes may suffice. The great advantage of modal solutions is that the inverse transforms back into the space domain can be carried out analytically.

A substantial simplification to this problem can be achieved by means of the thin-layer method[9] (TLM). In this alternative, the plate is subdivided into an adequate number of sub-layers that are thin in the finite element sense, the displacements are then discretized in the direction of layering via interpolation polynomials, and the method of weighted residuals is applied to obtain the discrete equations of motion. The TLM then changes the eigenvalue problem from transcendental to conventional – in essence, a quadratic eigenvalue problem in narrowly banded matrices that can be cast as a linear eigenvalue problem of double size. Hence, the propagation modes – propagating as well as evanescent – can readily be found.

Example 10.5: Amplification of plane SV-P waves in a layered medium

We demonstrate next the application of stiffness matrices to the wave amplification problem by recourse to substructuring techniques. Consider a stratified medium consisting of $N-1$ layers (N interfaces) underlain by an elastic half-space; see Fig. 10.4. This medium is subjected to plane P or SV waves that originate in the half-space with prescribed inclination θ with respect to the vertical. Hence, these waves have horizontal and vertical wavenumbers

$$k_x = \frac{\omega}{\alpha}\sin\theta, \quad k_z = \frac{\omega}{\alpha}\cos\theta, \quad k_x p_j = \frac{\omega}{\alpha}\sqrt{\sin^2\theta - \left(\frac{\alpha}{\alpha_j}\right)^2} \quad \text{for P waves} \quad (10.67)$$

$$k_x = \frac{\omega}{\beta}\sin\theta, \quad k_z = \frac{\omega}{\beta}\cos\theta, \quad k_x s_j = \frac{\omega}{\beta}\sqrt{\sin^2\theta - \left(\frac{\beta}{\beta_j}\right)^2} \quad \text{for SV waves} \quad (10.68)$$

in which the non-subscripted wave velocities are those of the half-space. In the absence of the upper layers, the response caused by these plane waves at the surface of the half-space is well known, and formulas can be found in many standard references. We refer to this response as the *outcropping motion*. Even simpler, if we replace the upper layers by an upper half-space with the same properties as the half-space underneath, so that together they form a homogeneous full space, then the motion is simply that of a plane wave. Either way, we distinguish the outcropping motion or full space with a superscript star (i.e. $\mathbf{u}^*$) and assume it to be known (see Fig. 10.4c). We shall show that to obtain the solution to the wave amplification problem with the stiffness matrix method, it suffices to apply a fictitious source at the half-space interface equal to $\mathbf{K}_{\text{half}}\mathbf{u}^*_{\text{outcrop}} = \mathbf{K}_{\text{full}}\mathbf{u}^*_{\text{full}}$ (the second of which is much simpler).

Consider Fig. 10.4b, which shows the layered system separated into two substructures or free bodies, namely the layers and a lower half-space. Compare this against Fig. 10.4c, where we consider a reference problem with *known* solution that consists either of a lower half-space by itself alone (i.e., a rock outcrop) or of a full space divided into a lower and an upper half-space, and in either case the seismic source is placed in the lower medium. Focusing attention on the lower substructure in each figure, we see that the motions at the surface of the two lower half-spaces differ from one

[9] Kausel, E., 1994, Thin layer method: formulation in the time domain, *International Journal for Numerical Methods in Engineering*, Vol. 37, pp. 927–941. Also, see references in that paper to earlier work on the subject.

a)

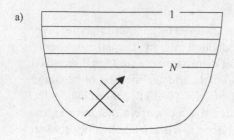

b)

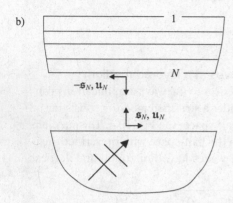

Figure 10.4: Amplification of plane SV-P waves using stiffness matrices. a) Actual layered system with seismic source. b) Layers and half-space as free bodies. c) Half-space alone (either a rock outcrop or a divided full space).

c)

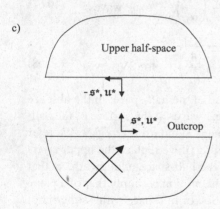

another, despite the fact that the lower substructures are identical and the source is exactly the same. This is ostensibly the result of the interface stresses acting between the layers and the half-space, so the difference in motions is solely the result of the differences in stresses, which act as secondary "sources" applied at the half-space interface. Since the stress imbalance and deviations in displacements are observed on the same horizon, they must satisfy the dynamic equilibrium equation

$$\mathfrak{s}_N - \mathfrak{s}^* = \mathbf{K}_{\text{half}}\left(\mathbf{u}_N - \mathbf{u}^*\right), \quad \text{so that} \quad -\mathfrak{s}_N = -\mathfrak{s}^* + \mathbf{K}_{\text{half}}\mathbf{u}^* - \mathbf{K}_{\text{half}}\mathbf{u}_N \quad (10.69)$$

In the case of a rock outcrop, $\mathfrak{s}^* = \mathbf{0}$ is the stress-free condition at the surface of the half-space, and $\mathbf{u}^*$ is the (presumably known) outcropping motion elicited by the source. For the full space, on the other hand, $\mathbf{u}^*$ and $\mathbf{s}^*$ are the displacements and stresses observed at the elevation that in the layered system constitutes the half-space interface. Taking into account the fact that there are no sources in the upper half of the full space, we conclude that the external stresses $-\mathbf{s}^*$ applied to it must satisfy the dynamic equilibrium equation

$$-\mathbf{s}^* = \mathbf{K}_{\text{upper\_half}}\mathbf{u}^*, \quad \text{so} \quad \mathbf{K}_{\text{half}}\mathbf{u}^* - \mathbf{s}^* \equiv \left(\mathbf{K}_{\text{lower\_half}} + \mathbf{K}_{\text{upper\_half}}\right)\mathbf{u}^* = \mathbf{K}_{\text{full}}\mathbf{u}^* \quad (10.70)$$

in which $\mathbf{K}_{\text{upper\_half}}$ is the impedance of the upper half-space (identical to $\mathbf{K}_{\text{lower\_half}} \equiv \mathbf{K}_{\text{half}}$, but with signs changed in checkerboard fashion), and $\mathbf{K}_{\text{full}}$ is the impedance of the full space. The latter is simply a *diagonal* matrix whose elements equal twice the diagonal elements of the stiffness matrix of the half-space. This second alternative has the advantage that there is no need for us to be concerned with reflections, critical angles, and the like.

Finally, applying the stresses $-\mathbf{s}_N$ as external tractions at the bottom of the upper layers in Fig. 10.4b, we obtain after brief algebra

$$\begin{Bmatrix} \mathbf{K}_{11} & \mathbf{K}_{12} & \cdots & & \mathbf{0} \\ \mathbf{K}_{21} & \mathbf{K}_{22} & \ddots & & \vdots \\ \mathbf{0} & \ddots & \ddots & & \mathbf{K}_{N\ 1,N} \\ \mathbf{0} & \cdots & \mathbf{K}_{N,N-1} & \mathbf{K}_{NN} + \mathbf{K}_{\text{half}} \end{Bmatrix} \begin{Bmatrix} \mathbf{u}_1 \\ \mathbf{u}_2 \\ \vdots \\ \mathbf{u}_N \end{Bmatrix} = \begin{Bmatrix} \mathbf{0} \\ \vdots \\ \mathbf{0} \\ \mathbf{K}_{\text{half}}\mathbf{u}^* - \mathbf{s}^* \end{Bmatrix} = \begin{Bmatrix} \mathbf{0} \\ \vdots \\ \mathbf{0} \\ \mathbf{K}_{\text{half}}\mathbf{u}^*_{\text{outcrop}} \end{Bmatrix} = \begin{Bmatrix} \mathbf{0} \\ \vdots \\ \mathbf{0} \\ \mathbf{K}_{\text{full}}\mathbf{u}^*_{\text{full}} \end{Bmatrix}$$

$$(10.71)$$

Clearly, the matrix on the left-hand side is that of the system of layers together with the half-space, and it must be assembled using the appropriate horizontal wavenumber $k = k_x$ given previously for either of the two wave types being considered. For example, in the case of a P wave with amplitude A propagating at some angle θ with the vertical, $\mathbf{u}^*$ is simply

$$\mathbf{u}^*_{\text{full}} = A\begin{Bmatrix} \sin\theta \\ -\mathrm{i}\,\cos\theta \end{Bmatrix} e^{-\mathrm{i}(k_x x + k_z z)}\bigg|_{z=0} \quad (10.72)$$

Observe that – as in all previous application of the stiffness matrix method in plane strain – the vertical components of $\mathbf{u}^*$ and $\mathbf{s}^*$ must carry a factor $-\mathrm{i}$. Thus, it suffices for us to multiply the motions $\mathbf{u}^*$ caused by a plane wave in the interior of a *full space* by the impedance of that full space, and apply these as external loads to the layered medium. Finally, observe that

$$\mathbf{K}_{\text{half}}\mathbf{u}^*_{\text{outcrop}} = \mathbf{K}_{\text{full}}\mathbf{u}^*_{\text{full}} \quad \text{implies} \quad \mathbf{u}^*_{\text{outcrop}} = \mathbf{K}^{-1}_{\text{half}}\mathbf{K}_{\text{full}}\mathbf{u}^*_{\text{full}} \quad (10.73)$$

which constitutes an easy way to compute the reflection of plane waves at the surface of an elastic half-space, without worrying about critical angles and the like.

10.3 Stiffness matrix method in cylindrical coordinates

Consider an elastic system subjected to dynamic sources whose material, geometric, and boundary conditions – but not necessarily the loads – exhibit cylindrical symmetry with respect to the vertical axis z. In the ensuing, we distinguish between the cases of horizontal,

Table 10.2. Flat layers in cylindrical coordinates

$$\tilde{\mathbf{u}}(k, z, \omega) = \left\{ \begin{array}{c} \tilde{u}_r \\ \tilde{u}_\theta \\ \tilde{u}_z \end{array} \right\},$$

$$= \frac{1}{2\pi} \int_{-\infty}^{+\infty} e^{i\omega t} \sum_{n=0}^{\infty} \mathbf{T}_n \int_0^{\infty} k \, \mathbf{C}_n \tilde{\mathbf{u}}(k, n, z, \omega) \, dk \, d\omega$$

$$\mathbf{T}_n = \mathrm{diag}\left[\begin{pmatrix} \cos n\theta \\ \sin n\theta \end{pmatrix} \begin{pmatrix} -\sin n\theta \\ \cos n\theta \end{pmatrix} \begin{pmatrix} \cos n\theta \\ \sin n\theta \end{pmatrix} \right] \quad \text{(upper or lower elements used)}$$

$$\mathbf{C}_n = \left\{ \begin{array}{ccc} J_n' & \dfrac{n}{kr} J_n & 0 \\ \dfrac{n}{kr} J_n & J_n' & 0 \\ 0 & 0 & J_n \end{array} \right\}, \qquad J_n' = \frac{d J_n(kr)}{d(kr)}$$

The stiffness matrix **K** is assembled with the elements of the plane strain matrix:

$$\mathbf{K} = \left\{ \begin{array}{ccc} K_{11}^{\mathrm{SVP}} & 0 & K_{12}^{\mathrm{SVP}} \\ 0 & K^{\mathrm{SH}} & 0 \\ K_{21}^{\mathrm{SVP}} & 0 & K_{22}^{\mathrm{SVP}} \end{array} \right\} \quad \text{from Table 10.1}$$

Form system matrix.

$$\tilde{\mathbf{p}}(k, n, z, \omega) = \frac{2 - \delta_{0n}}{2\pi} \int_0^{\infty} r \, \mathbf{C}_n \int_0^{2\pi} \mathbf{T}_n \int_{-\infty}^{+\infty} \mathbf{p}(r, \theta, z, t) \, e^{-i\omega t} dt \, d\theta \, dr$$

$\tilde{\mathbf{p}} = \mathbf{K}\tilde{\mathbf{u}}$. Solve separately for the SV-P and SH components, for they are uncoupled.

parallel layers that are perpendicular to the symmetry axis z, and of cylindrical layers that are concentric with it.

10.3.1 Horizontally layered system

As we saw in Section 9.2.1, the equations of motion in cylindrical coordinates for radially homogeneous media, when expressed in the frequency–wavenumber domain, are identical to those in plane strain. Hence, the stiffness matrix method for flat layers in cylindrical coordinates is virtually identical to that in plane strain. The only two differences are that a) the vertical components of stresses and displacements in cylindrical coordinates do *not* carry the implied imaginary unit factor $-i$, and b) a *Hankel transform* is used in place of the Fourier transform to obtain the field quantities in the spatial domain. The stiffness matrices for single or multiple layers are otherwise identical, and are handled in exactly the same way as in plane strain. Thus, it suffices to summarize the equations in Table 10.2 and illustrate the method by means of examples.

Example 10.6: Elastic half-space subjected to a tangential point load at its surface (Chao's problem)

A lower elastic half-space $z < 0$ is subjected at its surface to a harmonic, horizontal point load $b_x = \delta(x)\,\delta(y)\,\delta(z)$, which is equivalent to a surface traction in the horizontal plane $z = 0$ in direction x of the form $p_x = \delta(x)\,\delta(y)$. In cylindrical coordinates, this corresponds to a surface traction with azimuthal index $n = 1$ and components

$$\mathbf{p}_1 = \begin{Bmatrix} p_r \\ p_\theta \\ p_z \end{Bmatrix}_{(1)} = \frac{1}{2\pi} \begin{Bmatrix} \cos\theta \\ -\sin\theta \\ 0 \end{Bmatrix} \frac{\delta(r)}{r} = \begin{Bmatrix} \cos\theta & 0 & 0 \\ 0 & -\sin\theta & 0 \\ 0 & 0 & \cos\theta \end{Bmatrix} \begin{Bmatrix} 1 \\ 1 \\ 0 \end{Bmatrix} \frac{\delta(r)}{2\pi r}$$

$$= \mathbf{T}_1 \begin{Bmatrix} 1 \\ 1 \\ 0 \end{Bmatrix} \frac{\delta(r)}{2\pi r} \tag{10.74}$$

This is so because the radial and tangential components of the load, when projected in the x direction, added together, and integrated over a small circular area of radius R enclosing the origin, give a unit load in direction x:

$$\int_0^R \int_0^{2\pi} \left(\cos^2\theta + \sin^2\theta\right) \frac{\delta(r)}{2\pi r} r\, dr\, d\theta = \int_0^R \delta(r)\, dr - 1 \tag{10.75}$$

To solve this problem using stiffness matrices, we begin by casting the load in the frequency–wavenumber domain:

$$\tilde{\mathbf{p}}_n = a_n \int_0^\infty r\, \mathbf{C}_n \int_0^{2\pi} \mathbf{T}_n \mathbf{T}_1 \begin{Bmatrix} 1 \\ 1 \\ 0 \end{Bmatrix} \frac{\delta(r)}{2\pi r} d\theta\, dr, \qquad a_n = \frac{1}{\pi}\left(1 - \frac{1}{2}\delta_{n0}\right) \tag{10.76}$$

Because of orthogonality conditions satisfied by the integral in the azimuth θ, only the term $n = 1$ survives, so

$$\tilde{\mathbf{p}}_1 = \begin{Bmatrix} \tilde{p}_r \\ \tilde{p}_\theta \\ \tilde{p}_z \end{Bmatrix}_{(1)} = \int_0^\infty \begin{Bmatrix} J_1' & \frac{1}{kr}J_1 & 0 \\ \frac{1}{kr}J_1 & J_1' & 0 \\ 0 & 0 & J_1 \end{Bmatrix} \begin{Bmatrix} 1 \\ 1 \\ 0 \end{Bmatrix} \frac{\delta(r)}{2\pi} dr = \int_0^\infty \begin{Bmatrix} J_0 \\ J_0 \\ 0 \end{Bmatrix} \frac{\delta(r)}{2\pi} dr = \frac{1}{2\pi} \begin{Bmatrix} 1 \\ 1 \\ 0 \end{Bmatrix}$$

$$\tag{10.77}$$

This implies that of all azimuthal components, only $\tilde{\mathbf{u}}_1$ will exist. The displacement components in the frequency–wavenumber domain are then

$$\tilde{\mathbf{u}}_1 = \mathbf{K}^{-1}\tilde{\mathbf{p}}_1 \tag{10.78}$$

in which $\mathbf{K}$ is the stiffness matrix of the half-space, which is assembled with the elements given in Table 10.1 for SV-P waves *and* for SH waves. Formally, the stiffness matrix will be of the form

$$\mathbf{K} = \begin{Bmatrix} K_{11}^{SVP} & 0 & K_{12}^{SVP} \\ 0 & K^{SH} & 0 \\ K_{21}^{SVP} & 0 & K_{22}^{SVP} \end{Bmatrix} \tag{10.79}$$

but because the SV-P components are uncoupled from the SH components, the two should be assembled and solved separately – and this is especially true for layered media.

Therefore, from Table 10.1, we have

$$
\begin{Bmatrix} \tilde{u}_r \\ \tilde{u}_z \end{Bmatrix}_{(1)} = \begin{Bmatrix} K_{11}^{SVP} & K_{12}^{SVP} \\ K_{21}^{SVP} & K_{22}^{SVP} \end{Bmatrix}^{-1} \begin{Bmatrix} \tilde{p}_r \\ \tilde{p}_z \end{Bmatrix}_{(1)}
$$

$$
= \frac{1}{2k\mu\,\Delta} \begin{Bmatrix} s\,\frac{1}{2}(1-s^2) & ps - \frac{1}{2}(1+s^2) \\ ps - \frac{1}{2}(1+s^2) & p\,\frac{1}{2}(1-s^2) \end{Bmatrix} \frac{1}{2\pi} \begin{Bmatrix} 1 \\ 0 \end{Bmatrix}
$$

$$
= \frac{1}{4\pi\, k\mu\,\Delta} \begin{Bmatrix} s\,\frac{1}{2}(1-s^2) \\ ps - \frac{1}{2}(1+s^2) \end{Bmatrix} \tag{10.80}
$$

and

$$
\tilde{u}_{\theta(1)} = \frac{1}{K^{SH}}\,\tilde{p}_{\theta(1)}
$$

$$
= \left(\frac{1}{ks\,\mu}\right)\left(\frac{1}{2\pi}\right) = \frac{1}{2\pi\,ks\,\mu} \tag{10.81}
$$

Observe that, unlike the plane-strain case, the vertical components do not carry an imaginary factor. Finally, the displacements in the spatial domain are

$$
\mathbf{u} = \begin{Bmatrix} u_r \\ u_\theta \\ u_z \end{Bmatrix} = \mathbf{T}_1 \int_0^\infty \begin{Bmatrix} J_1' & \frac{1}{kr}J_1 & 0 \\ \frac{1}{kr}J_1 & J_1' & 0 \\ 0 & 0 & J_1 \end{Bmatrix} \begin{Bmatrix} \tilde{u}_r \\ \tilde{u}_\theta \\ \tilde{u}_z \end{Bmatrix}_{(1)} k\,dk \tag{10.82}
$$

that is

$$
u_r = (\cos\theta) \left[\frac{1}{4\pi\mu} \int_0^\infty \frac{s\,\frac{1}{2}(1-s^2)}{\Delta}\left(J_0(kr) - \frac{J_1(kr)}{kr}\right) dk + \frac{1}{2\pi\mu}\int_0^\infty \frac{1}{s}\frac{J_1(kr)}{kr}dk \right] \tag{10.83}
$$

$$
u_\theta = (-\sin\theta) \left[\frac{1}{4\pi\mu} \int_0^\infty \frac{s\,\frac{1}{2}(1-s^2)}{\Delta}\frac{J_1(kr)}{kr}\,dk + \frac{1}{2\pi\mu}\int_0^\infty \frac{1}{s}\left(J_0(kr) - \frac{J_1(kr)}{kr}\right) dk \right] \tag{10.84}
$$

$$
u_z = (\cos\theta) \left[\frac{1}{4\pi\mu} \int_0^\infty \frac{ps - \frac{1}{2}(1+s^2)}{\Delta}J_1(kr)\,dk \right] \tag{10.85}
$$

which can be evaluated by numerical integration or, as Chao does, by analytical means.

Example 10.7: Same as Example 10.6, but for a layer over an elastic half-space
The procedure is identical to that in Example 10.6. The only difference now is that in the frequency–wavenumber domain, we must first solve the two systems

$$
\begin{Bmatrix} \tilde{u}_{r1} \\ \tilde{u}_{z1} \\ \tilde{u}_{r2} \\ \tilde{u}_{z2} \end{Bmatrix}_{(1)} = (\mathbf{K}^{SVP})^{-1} \frac{1}{2\pi} \begin{Bmatrix} 1 \\ 0 \\ 0 \\ 0 \end{Bmatrix} \quad \text{and} \quad \begin{Bmatrix} \tilde{u}_{\theta1} \\ \tilde{u}_{\theta2} \end{Bmatrix}_{(1)} = (\mathbf{K}^{SH})^{-1} \frac{1}{2\pi} \begin{Bmatrix} 1 \\ 0 \end{Bmatrix} \tag{10.86}
$$

in which the matrices $\mathbf{K}^{SVP}$ and $\mathbf{K}^{SH}$ for the layered system are assembled with the elements in Table 10.1. In this case, it will be found that the transforms back into space for each elevation are intractable by purely analytical means, so they must be done numerically, using for *each* elevation the inversion formula

$$\mathbf{u}_j = \begin{Bmatrix} u_{rj} \\ u_{\theta j} \\ u_{zj} \end{Bmatrix} = \mathbf{T}_1 \int_0^\infty \begin{Bmatrix} J_1' & \frac{1}{kr}J_1 & 0 \\ \frac{1}{kr}J & J_1' & 0 \\ 0 & 0 & J_1 \end{Bmatrix} \begin{Bmatrix} \tilde{u}_{rj} \\ \tilde{u}_{\theta j} \\ \tilde{u}_{zj} \end{Bmatrix}_{(1)} k\,dk, \qquad j = 1, 2 \qquad (10.87)$$

Example 10.8: Elastic half-space subjected to a vertical point load at its surface (Pekeris's problem)

This is again very similar to Example 10.6, except that the vertical load in cylindrical coordinates corresponds to a surface traction with azimuthal index $n = 0$, and its components are

$$\mathbf{p}_0 = \begin{Bmatrix} p_r \\ p_\theta \\ p_z \end{Bmatrix}_{(0)} = \frac{1}{2\pi} \begin{Bmatrix} 0 \\ 0 \\ 1 \end{Bmatrix} \frac{\delta(r)}{r} \equiv \mathbf{T}_0 \begin{Bmatrix} 0 \\ 0 \\ 1 \end{Bmatrix} \frac{\delta(r)}{2\pi r}, \quad \mathbf{T}_0 = \mathrm{diag}\{1 \quad 0 \quad 1\} \qquad (10.88)$$

Hence, the load in the frequency–wavenumber domain is

$$\tilde{\mathbf{p}}_0 = \begin{Bmatrix} \tilde{p}_r \\ \tilde{p}_\theta \\ \tilde{p}_z \end{Bmatrix}_{(0)} = \int_0^\infty \begin{Bmatrix} J_0' & 0 & 0 \\ 0 & J_0' & 0 \\ 0 & 0 & J_0 \end{Bmatrix} \begin{Bmatrix} 0 \\ 0 \\ 1 \end{Bmatrix} \frac{\delta(r)}{2\pi} dr = \frac{1}{2\pi} \begin{Bmatrix} 0 \\ 0 \\ 1 \end{Bmatrix} \qquad (10.89)$$

The solution for the displacement vector in ω, k is then

$$\begin{Bmatrix} \tilde{u}_r \\ \tilde{u}_z \end{Bmatrix}_{(0)} = \begin{Bmatrix} K_{11}^{SVP} & K_{12}^{SVP} \\ K_{21}^{SVP} & K_{22}^{SVP} \end{Bmatrix}^{-1} \begin{Bmatrix} \tilde{p}_r \\ \tilde{p}_z \end{Bmatrix}_{(0)}$$

$$= \frac{1}{2k\mu\,\Delta} \begin{Bmatrix} s\frac{1}{2}(1-s^2) & ps - \frac{1}{2}(1+s^2) \\ ps - \frac{1}{2}(1+s^2) & p\frac{1}{2}(1-s^2) \end{Bmatrix} \frac{1}{2\pi} \begin{Bmatrix} 0 \\ 1 \end{Bmatrix}$$

$$= \frac{1}{4\pi\,k\mu\,\Delta} \begin{Bmatrix} ps - \frac{1}{2}(1+s^2) \\ p\frac{1}{2}(1-s^2) \end{Bmatrix} \qquad (10.90)$$

$$\tilde{u}_{\theta(0)} = \frac{1}{K^{SH}} \tilde{p}_{\theta(0)} = \left(\frac{1}{ks\mu}\right)(0) = 0 \qquad (10.91)$$

Finally, the solution in the space domain is

$$\mathbf{u} = \begin{Bmatrix} u_r \\ u_\theta \\ u_z \end{Bmatrix} = \mathbf{T}_0 \int_0^\infty \begin{Bmatrix} J_0' & 0 & 0 \\ 0 & J_0' & 0 \\ 0 & 0 & J_0 \end{Bmatrix} \begin{Bmatrix} \tilde{u}_r \\ \tilde{u}_\theta \\ \tilde{u}_z \end{Bmatrix}_{(0)} k\,dk \qquad (10.92)$$

or in full

$$u_r = -\frac{1}{4\pi\mu} \int_0^\infty \frac{ps - \frac{1}{2}(1+s^2)}{\Delta} J_1(kr)\,dk \qquad (10.93)$$

$$u_\theta = 0 \tag{10.94}$$

$$u_z = \frac{1}{4\pi\mu} \int_0^\infty \frac{s\,\frac{1}{2}(1-s^2)}{\Delta} J_0(kr)\,dk \tag{10.95}$$

10.3.2 Radially layered system

Consider a system of $N-1$ concentric cylindrical layers, each of arbitrary thickness, whose N interfaces we number from the outside to the inside. These layers could optionally be surrounded by an unboundedly large exterior region. We assume the material properties in each layer to be independent of the azimuth θ, and to be homogeneous within each cylindrical layer.

As shown in Section 9.2.2, for given axial (i.e., vertical) wavenumber k_z and frequency ω, a particular solution for the displacement vector at a point in the interior of an individual cylindrical layer is of the form

$$\bar{\mathbf{u}}(r,\theta,z,\omega) = \begin{bmatrix} \bar{u}_r & \bar{u}_\theta & -\mathrm{i}\,\bar{u}_z \end{bmatrix}^T = \mathbf{T}_n \left(\mathbf{H}_n^{(1)}\,\mathbf{a}_1 + \mathbf{H}_n^{(2)}\,\mathbf{a}_2 \right) e^{-\mathrm{i}k_z z} \equiv \mathbf{T}_n\,\mathfrak{u}\,e^{-\mathrm{i}k_z z} \tag{10.96}$$

$$\mathfrak{u} = \mathfrak{u}(r, k_z, \omega) = \begin{bmatrix} \tilde{u}_r & \tilde{u}_\theta & -\mathrm{i}\,\tilde{u}_z \end{bmatrix}^T = \mathbf{H}_n^{(1)}\,\mathbf{a}_1 + \mathbf{H}_n^{(2)}\,\mathbf{a}_2 \tag{10.97}$$

in which $\mathbf{a}_1$, $\mathbf{a}_2$ are arbitrary constants, and the matrices $\mathbf{T}_n$, $\mathbf{H}_n^{(1)}$, $\mathbf{H}_n^{(2)}$ are as listed in Table 10.3. The overbars are a reminder that this is a solution in the frequency–wavenumber domain with an added imaginary factor; the tildes indicate that the variations with z and θ have been separated. Also, the stresses in cylindrical surfaces are

$$\begin{bmatrix} \bar{\sigma}_r & \bar{\sigma}_{r\theta} & -\mathrm{i}\bar{\sigma}_{rz} \end{bmatrix}^T = \mathbf{T}_n \left(\mathbf{F}_n^{(1)}\,\mathbf{c}_1 + \mathbf{F}_n^{(2)}\,\mathbf{c}_2 \right) e^{-\mathrm{i}k_z z} = \mathbf{T}_n\,\mathfrak{s}\,e^{-\mathrm{i}k_z z} \tag{10.98}$$

$$\mathfrak{s} = \mathfrak{s}(r) = \begin{bmatrix} \tilde{\sigma}_r & \tilde{\sigma}_{r\theta} & -\mathrm{i}\tilde{\sigma}_{rz} \end{bmatrix}^T = \mathbf{F}_n^{(1)}\,\mathbf{a}_1 + \mathbf{F}_n^{(2)}\,\mathbf{a}_2 \tag{10.99}$$

in which $\mathbf{F}_n^{(1)}$, $\mathbf{F}_n^{(2)}$ are given in Table 10.4, and are constructed with first and second Hankel functions, respectively, as indicated by the superscript.

a) Single layer

Let's consider first the outermost layer as an isolated, free layer whose bounding outer and inner surfaces have radii r_1, r_2, respectively. Evaluating the displacements and stresses at these two surfaces, defining the tractions *per radian* as the products of the stresses and the interface radius, and writing both together in matrix form, we obtain

$$\begin{Bmatrix} \mathfrak{u}_1 \\ \mathfrak{u}_2 \end{Bmatrix} = \begin{Bmatrix} \mathbf{H}_{n1}^{(1)} & \mathbf{H}_{n1}^{(2)} \\ \mathbf{H}_{n2}^{(1)} & \mathbf{H}_{n2}^{(2)} \end{Bmatrix} \begin{Bmatrix} \mathbf{a}_1 \\ \mathbf{a}_2 \end{Bmatrix}, \quad \begin{Bmatrix} \mathfrak{p}_1 \\ \mathfrak{p}_2 \end{Bmatrix} = \begin{Bmatrix} r_1\,\mathfrak{s}_1 \\ -r_2\,\mathfrak{s}_2 \end{Bmatrix} = \begin{Bmatrix} r_1\,\mathbf{F}_{n1}^{(1)} & r_1\,\mathbf{F}_{n1}^{(2)} \\ -r_2\,\mathbf{F}_{n2}^{(1)} & -r_2\,\mathbf{F}_{n2}^{(2)} \end{Bmatrix} \begin{Bmatrix} \mathbf{a}_1 \\ \mathbf{a}_2 \end{Bmatrix}$$

$$\tag{10.100}$$

in which subscripts indicate the location (i.e., radius) at which the matrices are being evaluated, and superscripts indicate the kind of Hankel functions used. The negative sign

Table 10.3. Matrix for displacements in cylindrical layers

Obtain $\mathbf{u}$ by solving $\mathbf{u} = \mathbf{K}^{-1}\mathbf{p}$ for the system of layers, then for each interface:

$$\mathbf{u}(r, n, k_z, \omega) = \left\{ \begin{array}{c} \tilde{u}_r \\ \tilde{u}_\theta \\ -\mathrm{i}\,\tilde{u}_z \end{array} \right\} \quad \rightarrow \quad \tilde{\mathbf{u}} = \left\{ \begin{array}{c} \tilde{u}_r \\ \tilde{u}_\theta \\ \tilde{u}_z \end{array} \right\}$$

$$\mathbf{u}(r, \theta, z, \omega) = \left\{ \begin{array}{c} u_r \\ u_\theta \\ u_z \end{array} \right\} = \frac{1}{2\pi} \int_{-\infty}^{+\infty} \left\{ \sum_{n=0}^{\infty} \mathbf{T}_n\, \tilde{\mathbf{u}}_n \right\} e^{-\mathrm{i}k_z z}\, dk_z$$

$$\mathbf{T}_n = \mathrm{diag}[\, \cos n\theta \quad -\sin n\theta \quad \cos n\theta \,]$$

or

$$\mathbf{T}_n = \mathrm{diag}[\, \sin n\theta \quad \cos n\theta \quad \sin n\theta \,]$$

$$\mathbf{H}_n = \left\{ \begin{array}{ccc} H'_{\alpha n} & n\dfrac{H_{\beta n}}{k_\beta r} & \dfrac{k_z}{k_\beta} H'_{\beta n} \\[2ex] n\dfrac{H_{\alpha n}}{k_\alpha r} & H'_{\beta n} & n\dfrac{k_z}{k_\beta}\dfrac{H_{\beta n}}{k_\beta r} \\[2ex] -\dfrac{k_z}{k_\alpha} H_{\alpha n} & 0 & H_{\beta n} \end{array} \right\}, \quad \mathbf{H}_{nj}^{(i)} = \mathbf{H}_n\left(H_n^{(i)}(kr_j)\right)$$

in which

$$H_{\alpha n} = \left\{ \begin{array}{ll} H_n^{(1)}(k_\alpha r),\ H_n^{(2)}(k_\alpha r) & \text{layer} \\[1ex] H_n^{(2)}(k_\alpha r) & \text{exterior} \\[1ex] J_n(k_\alpha r) & \text{core} \end{array} \right.$$

$$H_{\beta n} = \left\{ \begin{array}{ll} H_n^{(1)}(k_\beta r),\ H_n^{(2)}(k_\beta r) & \text{layer} \\[1ex] H_n^{(2)}(k_\beta r) & \text{exterior} \\[1ex] J_n(k_\beta r) & \text{core} \end{array} \right.$$

$$H'_{\alpha n} = \frac{d H_{\alpha n}}{d(k_\alpha r)}, \qquad H'_{\beta n} = \frac{d H_{\beta n}}{d(k_\beta r)}$$

in the second row comes from the fact that external tractions are opposite in direction to the internal stresses at the inner surface. Eliminating the constants $\mathbf{a}_1$, $\mathbf{a}_2$ between these two matrices, we obtain

$$\left\{ \begin{array}{c} \mathbf{p}_1 \\ \mathbf{p}_2 \end{array} \right\} = \left\{ \begin{array}{cc} r_1\, \mathbf{F}_{n1}^{(1)} & r_1\, \mathbf{F}_{n1}^{(2)} \\[1ex] -r_2\, \mathbf{F}_{n2}^{(1)} & -r_2\, \mathbf{F}_{n2}^{(2)} \end{array} \right\} \left\{ \begin{array}{cc} \mathbf{H}_{n1}^{(1)} & \mathbf{H}_{n1}^{(2)} \\[1ex] \mathbf{H}_{n2}^{(1)} & \mathbf{H}_{n2}^{(2)} \end{array} \right\}^{-1} \left\{ \begin{array}{c} \mathbf{u}_1 \\ \mathbf{u}_2 \end{array} \right\}$$

$$= \left\{ \begin{array}{cc} \mathbf{K}_{11} & \mathbf{K}_{12} \\ \mathbf{K}_{21} & \mathbf{K}_{22} \end{array} \right\} \left\{ \begin{array}{c} \mathbf{u}_1 \\ \mathbf{u}_2 \end{array} \right\} \tag{10.101}$$

Table 10.4. Elements of matrix for stresses in cylindrical surfaces

$$\mathfrak{s}(r, n, k_z, \omega) = \left\{ \begin{array}{c} \tilde{\sigma}_r \\ \tilde{\sigma}_{r\theta} \\ -i\tilde{\sigma}_{rz} \end{array} \right\}, \quad \tilde{\mathbf{s}} = \left\{ \begin{array}{c} \tilde{\sigma}_r \\ \tilde{\sigma}_{r\theta} \\ \tilde{\sigma}_{rz} \end{array} \right\}, \quad \mathbf{F}_n = \left\{ \begin{array}{ccc} f_{11} & f_{12} & f_{13} \\ f_{21} & f_{22} & f_{23} \\ f_{31} & f_{32} & f_{33} \end{array} \right\}$$

$$\mathbf{F}_n = \left\{ f_{ij}(H_{\alpha n}, H_{\beta n}) \right\}, \quad \mathbf{F}_{nj}^{(i)} = \mathbf{F}_n \left(H_n^{(i)}(kr_j) \right)$$

in which $H_{\alpha n}$, etc., are the same as in Table 10.3. We list below the elements for a generic version of $\mathbf{F}_n$:

$$f_{11} = -k_\alpha \left\{ \lambda \left[1 + \left(\frac{k_z}{k_\alpha} \right)^2 \right] H_{\alpha n} + 2\mu \left[\frac{H'_{\alpha n}}{k_\alpha r} + \left(1 - \left(\frac{n}{k_\alpha r} \right)^2 \right) H_{\alpha n} \right] \right\}$$

$$f_{12} = \frac{2n\mu}{r} \left(H'_{\beta n} - \frac{H_{\beta n}}{k_\beta r} \right)$$

$$f_{13} = -2k_z\mu \left\{ \frac{H'_{\beta n}}{k_\beta r} + \left[1 - \left(\frac{n}{k_\beta r} \right)^2 \right] H_{\beta n} \right\}$$

$$f_{21} = \frac{2n\mu}{r} \left[H'_{\alpha n} - \frac{H_{\alpha n}}{k_\alpha r} \right]$$

$$f_{22} = -k_\beta \mu \left[2\frac{H'_{\beta n}}{k_\beta r} + \left(1 - 2 \left(\frac{n}{k_\beta r} \right)^2 \right) H_{\beta n} \right]$$

$$f_{23} = k_z \frac{2n\mu}{k_\beta r} \left[H'_{\beta n} - \frac{H_{\beta n}}{k_\beta r} \right]$$

$$f_{31} = -2k_z\mu H'_{\alpha n}$$

$$f_{32} = -k_z n\mu \frac{H_{\beta n}}{k_\beta r}$$

$$f_{33} = k_\beta\mu \left[1 - \left(\frac{k_z}{k_\beta} \right)^2 \right] H'_{\beta n}$$

in which

$$\mathbf{K} = \left\{ \begin{array}{cc} \mathbf{K}_{11} & \mathbf{K}_{12} \\ \mathbf{K}_{21} & \mathbf{K}_{22} \end{array} \right\} = \left\{ \begin{array}{cc} r_1 \mathbf{F}_{n1}^{(1)} & r_1 \mathbf{F}_{n1}^{(2)} \\ -r_2 \mathbf{F}_{n2}^{(1)} & -r_2 \mathbf{F}_{n2}^{(2)} \end{array} \right\} \left\{ \begin{array}{cc} \mathbf{H}_{n1}^{(1)} & \mathbf{H}_{n1}^{(2)} \\ \mathbf{H}_{n2}^{(1)} & \mathbf{H}_{n2}^{(2)} \end{array} \right\}^{-1} \qquad (10.102)$$

is the *symmetric* stiffness (or impedance) matrix of the cylindrical layer.

b) Unbounded external region

In the case of a cylindrical cavity within an unbounded homogeneous space, the external region can be regarded as a layer whose external radius is infinitely large. In this case, we must set $\mathbf{a}_1 = 0$ (to satisfy the radiation and boundedness conditions at infinity) and use thus only second Hankel functions. Hence, the 3×3 impedance matrix of the exterior region is

$$\mathbf{K}_{ext} = -r\, \mathbf{F}_n^{(2)} \left(\mathbf{H}_n^{(2)} \right)^{-1} \tag{10.103}$$

which is assembled with the matrices $\mathbf{H}_n$ and $\mathbf{F}_n$ in Tables 10.3 and 10.4, both of which must be constructed using second Hankel functions evaluated at the radius r of the external region.

c) Layered system

Finally, consider a layered system, which may optionally be surrounded by an unbounded exterior region. To construct the global (system) stiffness matrix, it suffices to overlap the layer matrices for each layer as appropriate, beginning with the impedance matrix of the exterior region. The result is a system of equations of the block-tridiagonal form

$$
\begin{Bmatrix} \mathbf{p}_1 \\ \mathbf{p}_2 \\ \mathbf{p}_3 \\ \vdots \\ \mathbf{p}_N \end{Bmatrix}
=
\begin{Bmatrix}
\mathbf{K}_{11} & \mathbf{K}_{12} & \mathbf{0} & \cdots & \mathbf{0} \\
\mathbf{K}_{21} & \mathbf{K}_{22} & \mathbf{K}_{23} & \cdots & \mathbf{0} \\
\mathbf{0} & \mathbf{K}_{32} & \mathbf{K}_{33} & \ddots & \vdots \\
\vdots & \vdots & \ddots & \ddots & \mathbf{K}_{N-1,N} \\
\mathbf{0} & \mathbf{0} & \cdots & \mathbf{K}_{N,N-1} & \mathbf{K}_{NN}
\end{Bmatrix}
\begin{Bmatrix} \mathbf{u}_1 \\ \mathbf{u}_2 \\ \mathbf{u}_3 \\ \vdots \\ \mathbf{u}_N \end{Bmatrix}
\tag{10.104}
$$

or more compactly

$$\mathbf{p} = \mathbf{K}\,\mathbf{u} \tag{10.105}$$

in which $\mathbf{p}$ is the vector of external sources (tractions per radian) in the frequency–wavenumber domain, which act as external tractions at the layer interfaces. Formally, the solution is then obtained along the following lines:

1) Form the (physical) load vector

$$\tilde{\mathbf{p}}(r, n, k_z, \omega) = \tfrac{2-\delta_{0n}}{2\pi} \int_0^{2\pi} \mathbf{T}_n \int_{-\infty}^{+\infty} \mathbf{p}(r, \theta, z, \omega)\, e^{-ik_z z}\, dz\, d\theta \tag{10.106}$$

(usually requires multiplying external pressures by r to obtain tractions per radian).

2) Multiply all axial components of $\tilde{\mathbf{p}}$ by $-i = -\sqrt{-1}$; $\tilde{\mathbf{p}} \to \mathbf{p}$.

3) Form system matrix $\mathbf{K}$ and solve for $\mathbf{u} = \mathbf{K}^{-1}\mathbf{p}$.

4) Multiply all *axial* (vertical) components of $\mathbf{u}$ by $+i = \sqrt{-1}$; $\mathbf{u} \to \tilde{\mathbf{u}}$.

5) Find the actual displacements at each interface from the inverse transform

$$\mathbf{u}(r, \theta, z, \omega) = \frac{1}{2\pi} \int_{-\infty}^{+\infty} \left\{ \sum_{n=0}^{\infty} \mathbf{T}_n \tilde{\mathbf{u}}_n \right\} e^{-ik_z z}\, dk_z \tag{10.107}$$

Table 10.5. Stiffness matrix for cylindrical layers

See Tables 10.3, 10.4 for matrices $\mathbf{H}_n$, $\mathbf{F}_n$ used herein
$1 \rightarrow$ outer surface, $2 \rightarrow$ inner surface

$$k_P = \omega/\alpha, \quad k_S = \omega/\beta, \quad k_z = \text{axial wavenumber}$$

$$k_\alpha = \sqrt{k_P^2 - k_z^2}, \qquad k_\beta = \sqrt{k_S^2 - k_z^2}$$

$$\mathbf{K} = \left\{ \begin{matrix} r_1 \mathbf{F}_{n1}^{(1)} & r_1 \mathbf{F}_{n1}^{(2)} \\ -r_2 \mathbf{F}_{n2}^{(1)} & -r_2 \mathbf{F}_{n2}^{(2)} \end{matrix} \right\} \left\{ \begin{matrix} \mathbf{H}_{n1}^{(1)} & \mathbf{H}_{n1}^{(2)} \\ \mathbf{H}_{n2}^{(1)} & \mathbf{H}_{n2}^{(2)} \end{matrix} \right\}^{-1} ; \quad \text{single layer,} \quad \left\{ \begin{matrix} \mathbf{H}_{ni}^{(j)} = \mathbf{H}_n \left(H_n^{(j)}(kr_i) \right) \\ \mathbf{F}_{ni}^{(j)} = \mathbf{F}_n \left(H_n^{(j)}(kr_i) \right) \end{matrix} \right.$$

$$\mathbf{K}_{ext} = -r \, \mathbf{F}_n^{(2)} \left(\mathbf{H}_n^{(2)} \right)^{-1}; \quad \text{unbounded exterior region,} \quad \left\{ \begin{matrix} \mathbf{H}_n^{(2)} = \mathbf{H}_n \left(H_{ni}^{(2)}(kr_i) \right) \\ \mathbf{F}_n^{(2)} = \mathbf{F}_n \left(H_{ni}^{(2)}(kr_i) \right) \end{matrix} \right.$$

$$\mathbf{K}_{core} = r \, \mathbf{F}_n \mathbf{H}_n^{-1}; \quad \text{solid core,} \quad \left\{ \begin{matrix} \mathbf{H}_n = \mathbf{H}_n \left(J_n(kr) \right) \\ \mathbf{F}_n = \mathbf{F}_n \left(J_n(kr) \right) \end{matrix} \right.$$

System of layers:

$$\mathfrak{p} = \mathbf{K}\,\mathbf{u}$$

Note: Expressing the matrices above as $\mathbf{F}_{ni}^{(j)} = \mathbf{F}_{ij}$, $\mathbf{H}_{ni}^{(j)} = \mathbf{H}_{ij}$, the single layer matrix can be shown to be given by

$$\mathbf{K} = \left\{ \begin{matrix} \mathbf{K}_{11} & \mathbf{K}_{12} \\ \mathbf{K}_{21} & \mathbf{K}_{22} \end{matrix} \right\} = \left\{ \begin{matrix} r_1 \left(\mathbf{F}_{11} - \mathbf{F}_{12}\mathbf{H}_{22}^{-1}\mathbf{H}_{21} \right) \mathbf{H}_1 & r_1 \left(\mathbf{F}_{12} - \mathbf{F}_{11}\mathbf{H}_{11}^{-1}\mathbf{H}_{12} \right) \mathbf{H}_2 \\ -r_2 \left(\mathbf{F}_{21} - \mathbf{F}_{22}\mathbf{H}_{22}^{-1}\mathbf{H}_{21} \right) \mathbf{H}_1 & -r_2 \left(\mathbf{F}_{22} - \mathbf{F}_{21}\mathbf{H}_{11}^{-1}\mathbf{H}_{12} \right) \mathbf{H}_2 \end{matrix} \right\}$$

in which

$$\mathbf{H}_1 = \left(\mathbf{H}_{11} - \mathbf{H}_{12}\mathbf{H}_{22}^{-1}\mathbf{H}_{21} \right)^{-1}, \quad \mathbf{H}_2 = \left(\mathbf{H}_{22} - \mathbf{H}_{21}\mathbf{H}_{11}^{-1}\mathbf{H}_{12} \right)^{-1}$$

This form is computationally more robust. Though it is not obvious from the above, the layer matrix is symmetric.

---

d) Solid core

As the inner radius of a thick tube approaches zero and forms a solid cylinder (or perhaps the core of a layered system), a singularity develops at the axis that decouples it from the exterior. To avoid the presence of sources on the axis that effectively vanish from the exterior, we need to carry out a condensation of the degrees of freedom on the axis prior to taking the limit $r_1 \rightarrow 0$. Using the simplified notation in Table 10.5, the force–displacement equation for the thick tube can be written as

$$\left\{ \begin{matrix} \mathbf{K}_{11} & \mathbf{K}_{12} \\ \mathbf{K}_{21} & \mathbf{K}_{22} \end{matrix} \right\} \left\{ \begin{matrix} \mathbf{u}_1 \\ \mathbf{u}_2 \end{matrix} \right\} = \left\{ \begin{matrix} \mathfrak{p}_1 \\ \mathfrak{p}_2 \end{matrix} \right\} \tag{10.108}$$

Condensation of the inner degrees of freedom yields

$$\left(\mathbf{K}_{22} - \mathbf{K}_{21}\mathbf{K}_{11}^{-1}\mathbf{K}_{12}\right)\mathbf{u}_2 = \mathbf{p}_2 - \mathbf{K}_{21}\mathbf{K}_{11}^{-1}\,\mathbf{p}_1 \tag{10.109}$$

As will be seen, when $r_1 \to 0$, the product $\mathbf{K}_{21}\mathbf{K}_{11}^{-1}$ on the right-hand side will tend to zero at the same rate as the singularly large forces $\mathbf{p}_1$ go to infinity, so that their product will remain finite.

We explore next the limit of the matrices when the inner radius goes to zero. For this purpose, we shall construct the stiffness matrices in Table 10.5 using Bessel and Neumann functions, namely

$$\mathbf{F}_{11}\left(Y_n(kr_1)\right),\ \mathbf{H}_{11}\left(Y_n(kr_1)\right) \tag{10.110}$$

$$\mathbf{F}_{12}\left(J_n(kr_1)\right),\ \mathbf{H}_{12}\left(J_n(kr_1)\right) \tag{10.111}$$

$$\mathbf{F}_{21}\left(Y_n(kr_2)\right),\ \mathbf{H}_{21}\left(Y_n(kr_2)\right) \tag{10.112}$$

$$\mathbf{F}_{22}\left(J_n(kr_2)\right),\ \mathbf{H}_{22}\left(J_n(kr_2)\right) \tag{10.113}$$

Hence, all elements of $\mathbf{F}_{11}, \mathbf{H}_{11}$ (but only these) will be singularly large when $r_1 \to 0$, and $\mathbf{H}_{11}^{-1} \to \mathbf{O}$ (the null matrix). This implies the following trends:

$$\mathbf{H}_1 = \left(\mathbf{H}_{11} - \mathbf{H}_{12}\mathbf{H}_{22}^{-1}\mathbf{H}_{21}\right)^{-1} \to \mathbf{H}_{11}^{-1}, \quad \mathbf{H}_2 = \left(\mathbf{H}_{22} - \mathbf{H}_{21}\mathbf{H}_{11}^{-1}\mathbf{H}_{12}\right)^{-1} \to \mathbf{H}_{22}^{-1} \tag{10.114}$$

$$\mathbf{K}_{11} = r_1\left(\mathbf{F}_{11} - \mathbf{F}_{12}\mathbf{H}_{22}^{-1}\mathbf{H}_{21}\right)\mathbf{H}_1 \to \lim_{r_1 \to 0}\left(r_1\mathbf{F}_{11}\mathbf{H}_{11}^{-1}\right) \tag{10.115}$$

$$\mathbf{K}_{12} = r_1\left(\mathbf{F}_{12} - \mathbf{F}_{11}\mathbf{H}_{11}^{-1}\mathbf{H}_{12}\right)\mathbf{H}_2 \to \mathbf{O} \tag{10.116}$$

$$\mathbf{K}_{21} = -r_2\left(\mathbf{F}_{21} - \mathbf{F}_{22}\mathbf{H}_{22}^{-1}\mathbf{H}_{21}\right)\mathbf{H}_1 \to \mathbf{O} \tag{10.117}$$

$$\mathbf{K}_{22} = -r_2\left(\mathbf{F}_{22} - \mathbf{F}_{21}\mathbf{H}_{11}^{-1}\mathbf{H}_{12}\right)\mathbf{H}_2 \to r_2\,\mathbf{F}_{22}\mathbf{H}_{22}^{-1} \tag{10.118}$$

When $r_1 \to 0$, the elements of $\mathbf{K}_{11}$ become either zero, finite, or infinite, but because the source on the axis is also infinitely large, the product of stiffness and displacement at this location is indeterminate. Thus, $\mathbf{K}_{11}$ is not particularly interesting to us. On the other hand, we observe also that the submatrix $\mathbf{K}_{22}$ converges to what we refer as the stiffness of the *solid core*, a 3×3 matrix that is constructed with conventional Bessel functions. On the other hand, the coupling forces are well behaved, as will be seen next. From Table 10.5, we obtain

$$\mathbf{K}_{21}\mathbf{K}_{11}^{-1} = -r_2\left(\mathbf{F}_{21} - \mathbf{F}_{22}\mathbf{H}_{22}^{-1}\mathbf{H}_{21}\right)\mathbf{H}_1\,\mathbf{H}_1^{-1}\left(r_1\mathbf{F}_{11} - r_1\mathbf{F}_{12}\mathbf{H}_{22}^{-1}\mathbf{H}_{21}\right)^{-1}$$

$$\to -r_2\left(\mathbf{F}_{21} - \mathbf{F}_{22}\mathbf{H}_{22}^{-1}\mathbf{H}_{21}\right)\left(r_1\,\mathbf{F}_{11}\right)^{-1} \tag{10.119}$$

so

$$\mathbf{K}_{22}\mathbf{u}_2 = \mathbf{p}_2 + r_2\left(\mathbf{F}_{21} - \mathbf{F}_{22}\mathbf{H}_{22}^{-1}\mathbf{H}_{21}\right)\left(r_1\mathbf{F}_{11}\right)^{-1}\mathbf{p}_1 \tag{10.120}$$

in which $\mathbf{K}_{22} = r_2 \mathbf{F}_{22}\mathbf{H}_{22}^{-1}$ is the stiffness of the cylindrical core. If the product $(r_1\mathbf{F}_{11})^{-1}\mathbf{p}_1$ remains finite as $\mathbf{F}_{11}^{-1} \to \mathbf{O}$ and $\mathbf{p}_1 \to \infty$, this system of equations has a finite solution. This shows that the source on the axis can be expressed in terms of a fictitious source applied at the exterior,

$$\mathbf{p}_{2eq} = \mathbf{p}_2 + r_2 \left(\mathbf{F}_{21} - \mathbf{F}_{22}\mathbf{H}_{22}^{-1}\mathbf{H}_{21} \right) \left[(r_1\mathbf{F}_{11})^{-1}\mathbf{p}_1 \right] \tag{10.121}$$

At this point, we may generalize the above by allowing $\mathbf{K}_{22}$ to receive stiffness (impedance) contributions from external cylindrical layers, if any. Thus, a point source on the axis can be expressed in terms of fictitious loads on the exterior surface of the core (or, for that matter, on any internal cylindrical surface, provided the core is appropriately subdivided into two sub-layers).

We proceed next to examine the limit $\lim_{r_1 \to 0} \mathbf{F}_{11}^{-1}$ when the inner radius converges to the axis. While all terms are zero in the limit, we shall give the expressions for the leading terms in $z_\beta = k_\beta r_1$, because the source itself will be singular to some order, as will be seen.

Defining $a = \beta/\alpha$, $z_\alpha = k_\alpha r_1$, $z_\beta = k_\beta r_1$, $k_\alpha = \sqrt{k_P^2 - k_z^2}$, $k_\beta = \sqrt{k_S^2 - k_z^2}$, $k_P = \omega/\alpha$, $k_S = \omega/\beta$ and using MATLAB's symbolic toolbox, we obtain the following:

$n = 0$:

$$\lim_{r_1 \to 0} (r_1\mathbf{F}_{11})^{-1} \to \frac{\pi}{4\mu\left(1+\left(\frac{k_z}{k_\beta}\right)^2\right)} \left\{ \begin{array}{ccc} -\frac{k_\alpha}{k_\beta}\left(1 - \left(\frac{k_z}{k_\beta}\right)^2\right)z_\beta & 0 & -2\frac{k_\alpha}{k_\beta}\frac{k_z}{k_\beta} \\ 0 & -\left(1 + \left(\frac{k_z}{k_\beta}\right)^2\right)z_\beta & 0 \\ -2\frac{k_z}{k_\beta}z_\beta & 0 & 2 \end{array} \right\} \tag{10.122}$$

$n = 1$:

$$\lim_{r_1 \to 0} (r_1\mathbf{F}_{11})^{-1} \to \frac{\pi}{8\mu\left[1+\left(\frac{k_z}{k_\beta}\right)^2\right]} \left\{ \begin{array}{ccc} \left(\frac{k_\alpha}{k_\beta}\right)^2 & \left(\frac{k_z}{k_\beta}\right)^2 & -3\left(\frac{k_\alpha}{k_\beta}\right)^2\frac{k_z}{k_\beta}z_\beta \\ 1 + \left(\frac{k_z}{k_\beta}\right)^2 & 1 + \left(\frac{k_z}{k_b}\right)^2 & -\left[1 + \left(\frac{k_z}{k_\beta}\right)^2\right]\frac{k_z}{k_\beta}z_\beta \\ \frac{k_z}{k_\beta} & \frac{k_z}{k_\beta} & \left[2 - \left(\frac{k_z}{k_\beta}\right)^2\right]z_\beta \end{array} \right\} \tag{10.123}$$

$n \geq 2$:

$$\lim_{r_1 \to 0} (r_1\mathbf{F}_{11})^{-1} \to \frac{\pi z_\beta^{n-1}}{2^n n!\mu\left(1-\left(\frac{k_\alpha}{k_\beta}\right)^2\right)} \left\{ \begin{array}{ccc} n\left(\frac{k_\alpha}{k_\beta}\right)^{n+1} & n\left(\frac{k_\alpha}{k_\beta}\right)^{n+1} & -\left(\frac{k_\alpha}{k_\beta}\right)^{n+1}\frac{k_z}{k_\beta}z_\beta \\ n\left[1 + \left(\frac{k_z}{k_\beta}\right)^2\right] & n\left[1 + \left(\frac{k_z}{k_\beta}\right)^2\right] & -\left[1 + \left(\frac{k_z}{k_\beta}\right)^2\right]a^2\frac{k_z}{k_\beta}z_\beta \\ n\frac{k_z}{k_\beta} & n\frac{k_z}{k_\beta} & \left[1 - a^2\left(1 + \left(\frac{k_z}{k_\beta}\right)^2\right)\right]z_\beta \end{array} \right\} \tag{10.124}$$

To illustrate matters, let's consider the three particular cases of a torsional point source, an axial point load, and a lateral point load acting on the axis of a solid cylinder, which have a spatial–temporal variation $\delta(z)\,\delta(t)$. When converted to the axial wavenumber–frequency domain, these terms give unity, so we shall omit them in the ensuing.

Torsional point source (n = 0)

A torsional point source of unit intensity can be obtained as the limit of a tangential ring load p_θ acting on a ring of radius $r_1 \to 0$, which elicits a torsional moment $M_t = 2\pi \, \tilde{p}_\theta \, r_1 = 1$ (observe that $\tilde{p}_\theta$ is the traction *per radian*, so the total ring load is 2π times larger, and multiplication by the radius gives the moment). Hence, the load vector in the frequency–axial-wavenumber domain for a torsional point source of unit intensity is

$$\mathbf{p}_1 = \left\{ \begin{array}{c} \tilde{p}_r \\ \tilde{p}_\theta \\ -\mathrm{i}\,\tilde{p}_z \end{array} \right\} = \frac{1}{2\pi r_1} \left\{ \begin{array}{c} 0 \\ 1 \\ 0 \end{array} \right\} \tag{10.125}$$

Hence,

$$\lim_{r_1 \to 0} (r_1 \mathbf{F}_{11})^{-1} \, \mathbf{p}_1 = \left(\frac{1}{2\pi r_1} \right) \frac{\pi}{4\mu \left(1 + \left(\frac{k_z}{k_\beta}\right)^2\right)} \left\{ \begin{array}{ccc} -\frac{k_\alpha}{k_\beta}\left(1 - \left(\frac{k_z}{k_\beta}\right)^2\right) z_\beta & 0 & -2\frac{k_\alpha}{k_\beta}\frac{k_z}{k_\beta} \\ 0 & -\left(1 + \left(\frac{k_z}{k_\beta}\right)^2\right) z_\beta & 0 \\ -2\frac{k_z}{k_\beta} z_\beta & 0 & 2 \end{array} \right\} \left\{ \begin{array}{c} 0 \\ 1 \\ 0 \end{array} \right\} \tag{10.126}$$

that is,

$$\lim_{r_1 \to 0} (r_1 \mathbf{F}_{11})^{-1} \, \mathbf{p}_1 = \frac{k_\beta}{8\mu} \left\{ \begin{array}{c} 0 \\ 1 \\ 0 \end{array} \right\} \quad \text{(use to form } \mathbf{p}_{2eq}) \tag{10.127}$$

Axial point load (n = 0)

This load can be visualized as the limit of an axial ring traction per radian, p_z, acting on a ring of radius $r_1 \to 0$, which implies an axial load $P_z = 2\pi \, \tilde{p}_z = 1$. Hence, the load vector in the frequency–axial-wavenumber domain for a unit axial load is

$$\mathbf{p}_1 = \left\{ \begin{array}{c} {}^\bullet\tilde{p}_r \\ \tilde{p}_\theta \\ -\mathrm{i}\,\tilde{p}_z \end{array} \right\} = \frac{-\mathrm{i}}{2\pi} \left\{ \begin{array}{c} 0 \\ 0 \\ 1 \end{array} \right\} \tag{10.128}$$

$$\lim_{r_1 \to 0} (r_1 \mathbf{F}_{11})^{-1} \, \mathbf{p}_1 = \left(\frac{-\mathrm{i}}{2\pi} \right) \frac{\pi}{4\mu \left(1 + \left(\frac{k_z}{k_\beta}\right)^2\right)}$$

$$\times \left\{ \begin{array}{ccc} -\frac{k_\alpha}{k_\beta}\left(1 - \left(\frac{k_z}{k_\beta}\right)^2\right) z_\beta & 0 & -2\frac{k_\alpha}{k_\beta}\frac{k_z}{k_\beta} \\ 0 & -\left(1 + \left(\frac{k_z}{k_\beta}\right)^2\right) z_\beta & 0 \\ -2\frac{k_z}{k_\beta} z_\beta & 0 & 2 \end{array} \right\} \left\{ \begin{array}{c} 0 \\ 0 \\ 1 \end{array} \right\} \tag{10.129}$$

that is,

$$\lim_{r_1 \to 0} (r_1 \mathbf{F}_{11})^{-1} \, \mathfrak{p}_1 = \frac{\mathrm{i}}{4\mu \left(1 + \left(\frac{k_z}{k_\beta}\right)^2\right)} \left\{ \begin{array}{c} \left(\frac{k_\alpha}{k_\beta}\right)^2 \frac{k_z}{k_\beta} \\ 0 \\ -1 \end{array} \right\} \quad \text{(use to form } \mathfrak{p}_{2eq}) \tag{10.130}$$

Lateral point load (n = 1)
A lateral load of unit intensity can be visualized as the combination of radial and tangential ring loads per radian of equal strength acting on a ring of radius $r_1 \to 0$:

$$\mathbf{p} = \left\{ \begin{array}{c} p_r \\ p_\theta \\ p_z \end{array} \right\} = \frac{1}{2\pi} \left\{ \begin{array}{ccc} \cos\theta & 0 & 0 \\ 0 & -\sin\theta & 0 \\ 0 & 0 & \cos\theta \end{array} \right\} \left\{ \begin{array}{c} 1 \\ 1 \\ 0 \end{array} \right\} = \frac{1}{2\pi} \mathbf{T}_1 \left\{ \begin{array}{c} 1 \\ 1 \\ 0 \end{array} \right\} \tag{10.131}$$

Projecting this onto the direction of the load and integrating over the ring, we obtain a unit load in the x direction:

$$\int_0^{2\pi} (p_r \cos\theta - p_\theta \sin\theta) \, r_1 \, d\theta = \frac{1}{2\pi r_1} \int_0^{2\pi} \left(\cos^2\theta + \sin^2\theta\right) r_1 \, d\theta = 1 \tag{10.132}$$

The load vector in the azimuthal wavenumber domain follows from the Fourier transform

$$\tilde{\mathbf{p}} = \frac{1}{\pi} \int_0^{2\pi} \mathbf{T}_1 \, \mathbf{p}(r_1, \theta, k_z, \omega) d\theta = \frac{1}{4\pi} \left\{ \begin{array}{c} 1 \\ 1 \\ 0 \end{array} \right\} \tag{10.133}$$

(the additional factor $1/2$ arises from the integral $\frac{1}{\pi} \int_0^{2\pi} \mathbf{T}_1 \mathbf{T}_1 \, d\theta = \frac{1}{2}\mathbf{I}$). Hence,

$$\mathfrak{p}_1 = \left\{ \begin{array}{c} \tilde{p}_r \\ \tilde{p}_\theta \\ -\mathrm{i}\,\tilde{p}_z \end{array} \right\} = \frac{1}{4\pi} \left\{ \begin{array}{c} 1 \\ 1 \\ 0 \end{array} \right\} \tag{10.134}$$

$$\lim_{r_1 \to 0} (r_1 \mathbf{F}_{11})^{-1} \, \mathfrak{p}_1 = \left(\frac{1}{4\pi}\right) \frac{\pi}{8\mu \left[1 + \left(\frac{k_z}{k_\beta}\right)^2\right]}$$

$$\times \left\{ \begin{array}{ccc} \left(\frac{k_\alpha}{k_\beta}\right)^2 & \left(\frac{k_\alpha}{k_\beta}\right)^2 & -3\left(\frac{k_\alpha}{k_\beta}\right)^2 \frac{k_z}{k_\beta} z_\beta \\ 1 + \left(\frac{k_z}{k_\beta}\right)^2 & 1 + \left(\frac{k_z}{k_\beta}\right)^2 & -\left[1 + \left(\frac{k_z}{k_\beta}\right)^2\right] \frac{k_z}{k_\beta} z_\beta \\ \frac{k_z}{k_\beta} & \frac{k_z}{k_\beta} & \left[2 - \left(\frac{k_z}{k_\beta}\right)^2\right] z_\beta \end{array} \right\} \left\{ \begin{array}{c} 1 \\ 1 \\ 0 \end{array} \right\} \tag{10.135}$$

that is,

$$\lim_{r_1 \to 0} (r_1 \mathbf{F}_{11})^{-1} \, \mathfrak{p}_1 = \frac{1}{16\mu \left[1 + \left(\frac{k_z}{k_\beta}\right)^2\right]} \left\{ \begin{array}{c} \left(\frac{k_\alpha}{k_\beta}\right)^2 \\ 1 + \left(\frac{k_z}{k_\beta}\right)^2 \\ \frac{k_z}{k_\beta} \end{array} \right\} \quad \text{(use to form } \mathfrak{p}_{2eq}) \tag{10.136}$$

In summary, a *solid core* of radius r is characterized by the following properties:

$$\mathbf{K}_{core} = r\,\mathbf{F}_n\mathbf{H}_n^{-1} \qquad \text{Stiffness matrix (using } J_n)$$

$$\mathbf{p}_{equiv} = \mathbf{p}_{core} + r\left(\mathbf{F}_{21} - \mathbf{F}_{22}\mathbf{H}_{22}^{-1}\mathbf{H}_{21}\right)\mathbf{q}_{axis} \qquad \text{Total load per radian at external surface}$$

$$\mathbf{q}_{axis} = \lim_{r_1 \to 0}\left(r_1\mathbf{F}_{11}\right)^{-1}\mathbf{p}_1 \qquad \text{Fictitious load due to source on axis}$$

$$\mathbf{q}_{axis} = \frac{k_\beta}{8\mu}\begin{Bmatrix} 0 \\ 1 \\ 0 \end{Bmatrix} \qquad \text{Torsional point source, } n = 0$$

$$\mathbf{q}_{axis} = \frac{\mathrm{i}}{4\mu\left(1+\left(\frac{k_z}{k_\beta}\right)^2\right)}\begin{Bmatrix} \left(\frac{k_\alpha}{k_\beta}\right)^2\frac{k_z}{k_\beta} \\ 0 \\ -1 \end{Bmatrix} \qquad \text{Axial point source, } n = 0$$

$$\mathbf{q}_{axis} = \frac{1}{16\mu\left[1+\left(\frac{k_z}{k_\beta}\right)^2\right]}\begin{Bmatrix} \left(\frac{k_\alpha}{k_\beta}\right)^2 \\ 1+\left(\frac{k_z}{k_\beta}\right)^2 \\ \frac{k_z}{k_\beta} \end{Bmatrix} \qquad \text{Lateral point source, } n = 1$$

$$\mathbf{H}_{21} = \mathbf{H}_n\left(Y_n(kr_2)\right), \qquad \mathbf{H}_{22} = \mathbf{H}_n\left(J_n(kr_2)\right) \qquad \text{(Table 10.3)}$$

$$\mathbf{F}_{21} = \mathbf{F}_n\left(Y_n(kr_2)\right), \qquad \mathbf{F}_{22} = \mathbf{F}_n\left(J_n(kr_2)\right) \qquad \text{(Table 10.4)}$$

$$\text{(10.137)}$$

Example 10.9: Normal modes of solid rod

The dynamic stiffness matrix of a solid rod of radius $r_1 = r$ is obtained by disregarding in the formulation all of the elements of the layer matrix associated with the inner radius, inasmuch as $r_2 = 0$ at the axis. Also, we replace the Hankel functions by Bessel functions, to avoid a singularity on the axis. Hence, the symmetric stiffness matrix is

$$\mathbf{K} = r\,\mathbf{F}_n\left(\mathbf{H}_n\right)^{-1} \tag{10.138}$$

which must be assembled with Bessel functions of the first kind, J_n. A free vibration problem is characterized by $\mathbf{K}\mathbf{u} = \mathbf{0}$, which is satisfied only if the determinant of the stiffness matrix vanishes, i.e.,

$$\det\mathbf{K} = \det\left(r\,\mathbf{F}_n(\mathbf{H}_n)^{-1}\right) = 0 \quad \Rightarrow \quad \det\mathbf{F}_n = 0 \tag{10.139}$$

the roots of which can be obtained by numerical search for any combination of frequency ω, axial wavenumber k_z, and azimuthal index n.

Example 10.10: Layered, solid rod subjected to load on axis

Consider a layered, solid rod composed of N material layers numbered from the outside to the core. As in Example 10.9, we begin by discarding the degrees of freedom associated

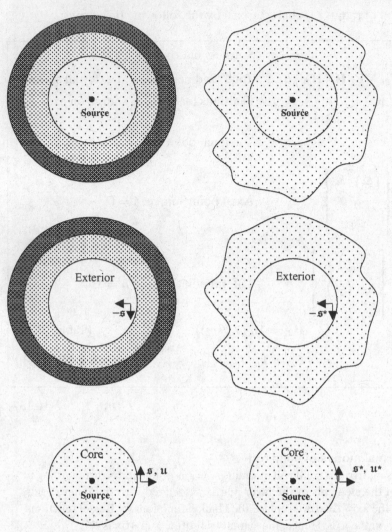

Figure 10.5: Layered solid rod subjected to source on axis (left), and modeling via reference solution for infinite, homogeneous solid (right); see Example 10.10.

with the axis by arguing that the inner radius of the core's cylindrical "layer" is zero, so the *core* of the layered system has the same stiffness matrix as Example 10.9. However, we now have a difficulty in that the source acts directly on the extirpated axis, but we could deal with this situation by means of the rigorous method that we presented in Section 10.3.2d. However, we shall now demonstrate instead an alternative way of solving this problem. The strategy is to define an arbitrary cylindrical reference surface within the core (which could be the external surface itself), evaluate at this location the displacements caused by the source in an *infinite*, homogeneous solid with the same material properties as the core, and apply the substructuring technique used in Example 10.5. With reference to Fig. 10.5, the details are as follows:

Separate the inner and outer systems at the reference surface within the core for both the actual and the infinite solid. For convenience, choose the reference surface to coincide with the core's external interface (i.e., r_N). The source in the *actual* layered medium elicits at r_N displacements and internal stresses $\mathbf{u}_N$, $\mathbf{s}_N$, whereas the same source in the homogeneous, *infinite* medium elicits displacement and internal stresses $\mathbf{u}^*$, $\mathbf{s}^*$. Because the core of the actual and the homogeneous problem are identical and so also is the source, it follows that the difference in displacements at the reference surface is solely the result of the *difference* in internal tractions per radian, i.e.,

$$r_N\left(\mathbf{s}_N - \mathbf{s}^*\right) = \mathbf{K}_{\text{core}}\left(\mathbf{u}_N - \mathbf{u}^*\right) \quad \text{with} \quad \mathbf{K}_{\text{core}} = r_N \mathbf{F}_N \left(\mathbf{H}_N\right)^{-1}$$

(Use Bessel functions J_n.) By taking the difference, we effectively discard the source at the axis. Also, since the exterior region is free from sources, it follows that the stresses for the homogeneous reference problem satisfy the equilibrium condition

$$-r_N \mathbf{s}^* = \mathbf{K}_{ext}\mathbf{u}^* \quad \text{with} \quad \mathbf{K}_{ext}^* = -r_N \mathbf{F}_N^{(2)} \left(\mathbf{H}_N^{(2)}\right)^{-1} \tag{10.140}$$

(Use 2nd Hankel functions.) Hence

$$-r_N \mathbf{s}_N = -\mathbf{K}_{\text{core}}\mathbf{u}_N + \mathbf{K}_{\text{full}}^*\mathbf{u}^* \quad \text{with} \quad \mathbf{K}_{\text{full}}^* = \mathbf{K}_{\text{core}} + \mathbf{K}_{ext}^* \tag{10.141}$$

Finally, the solution to the actual problem follows from

$$\begin{Bmatrix} \begin{bmatrix} \mathbf{K}_{11} & \mathbf{K}_{12} & 0 & \cdots & & 0 \\ \mathbf{K}_{21} & \mathbf{K}_{22} & \mathbf{K}_{23} & \cdots & & 0 \\ 0 & \mathbf{K}_{32} & \mathbf{K}_{33} & \ddots & & \vdots \\ \vdots & \vdots & \ddots & \ddots & \mathbf{K}_{N-1,N} & \\ 0 & 0 & \cdots & \mathbf{K}_{N,N-1} & \mathbf{K}_{NN} \end{bmatrix} \end{Bmatrix} \begin{Bmatrix} \mathbf{u}_1 \\ \mathbf{u}_2 \\ \mathbf{u}_3 \\ \vdots \\ \mathbf{u}_N \end{Bmatrix} = \begin{Bmatrix} 0 \\ 0 \\ 0 \\ \vdots \\ \mathbf{K}_{\text{full}}^*\mathbf{u}^* \end{Bmatrix} \tag{10.142}$$

in which $\mathbf{K}_{NN}$ receives contribution from $\mathbf{K}_{\text{core}}$ as well as the adjoining external layer. The reference displacements $\mathbf{u}^* = \mathbf{u}^*(r_N, k_z, \omega)$ for any type of axial load (a spatially harmonic line load, a line of pressure, dipoles, etc.) can be obtained explicitly from the equations presented in Section 4.9. The only difficulty here is that $\mathbf{K}_{\text{core}}$, which adds to both sides of the equations (i.e., $\mathbf{K}_{NN}$ $\mathbf{K}_{\text{full}}$), has infinite elements at the resonant frequencies of the core with fixed exterior. These are avoided by adding some damping to the system.

10.4 Stiffness matrix method for layered spheres

Consider an elastic spherical layer of arbitrary thickness, which is bounded by external and internal spherical surfaces of radii R_e, R_i, respectively. As we have seen in Section 1.4.3, across any arbitrary spherical surface in the layer there exist internal stresses of the form

$$\bar{\mathbf{s}}_R = \begin{bmatrix} \sigma_R & \sigma_{R\phi} & \sigma_{R\theta} \end{bmatrix}^T = \mathbf{L}_R^T \bar{\mathbf{u}}$$

$$= \left[\mathbf{D}_{RR}\frac{\partial}{\partial R} + \mathbf{D}_{R\phi}\frac{1}{R}\frac{\partial}{\partial \phi} + \mathbf{D}_{R\theta}\frac{1}{R\sin\phi}\frac{\partial}{\partial \theta} + \mathbf{D}_{R1}\frac{1}{R} + \mathbf{D}_{R2}\frac{\cot\phi}{R}\right] \mathbf{T}_n \mathbf{L}_m^n \mathbf{H}_m \mathbf{a}$$

$$\tag{10.143}$$

Also, on each of the external surfaces act external *tractions per unit solid angle*, which we define as

$$\bar{\mathbf{p}}_e = R_e^2\,\bar{\mathbf{s}}_R|_{R=R_e}, \qquad \bar{\mathbf{p}}_i = -R_i^2\,\bar{\mathbf{s}}_R|_{R=R_i} \tag{10.144}$$

Table 10.6. **Stiffness matrices in spherical coordinates**

Note: Stiffness matrices given below are *not* symmetric. See Section 10.4.2 on how to make them symmetric. Also, for greater computational efficiency, solve separately for the SH and SVP degrees of freedom in **K** when solving for $\tilde{\mathbf{u}}$, since they are uncoupled.

For definition of the submatrices below, see Tables 10.7 and 10.8.

R_e = external radius $\qquad R_i$ = internal radius $\qquad R$ = generic radius

$\tilde{\mathbf{p}} = \mathbf{K}\tilde{\mathbf{u}}$ $\qquad\qquad\qquad \mathbf{K} = \mathbf{K}(R_e, R_i, \lambda, \mu, \omega, m)$ $\qquad$ (not a function of n)

$\mathbf{F}_{mj}^{(i)} \equiv \mathbf{F}_m^{(i)}(R_j)$ $\qquad\qquad \mathbf{H}_{mj}^{(i)} \equiv \mathbf{H}_m^{(i)}(R_j)$

Spherical layer: (6×6) $\quad \mathbf{K} = \begin{Bmatrix} \mathbf{K}_{ee} & \mathbf{K}_{ei} \\ \mathbf{K}_{ie} & \mathbf{K}_{ii} \end{Bmatrix}$

$$= \begin{Bmatrix} R_e^2 \mathbf{F}_{me}^{(1)} & R_e^2 \mathbf{F}_{me}^{(2)} \\ -R_i^2 \mathbf{F}_{mi}^{(1)} & -R_i^2 \mathbf{F}_{mi}^{(2)} \end{Bmatrix} \begin{Bmatrix} \mathbf{H}_{me}^{(1)} & \mathbf{H}_{me}^{(2)} \\ \mathbf{H}_{mi}^{(1)} & \mathbf{H}_{mi}^{(2)} \end{Bmatrix}^{-1}$$

Solid sphere: (3×3) $\quad \mathbf{K} = R^2 \mathbf{F}_m \left(\mathbf{H}_m\right)^{-1}$ $\qquad$ Core (use j_m in $\mathbf{F}_m$, $\mathbf{H}_m$)

Cavity: (3×3) $\qquad\quad \mathbf{K} = -R^2 \mathbf{F}_m^{(2)} \left(\mathbf{H}_m^{(2)}\right)^{-1}$ $\quad$ Infinite external space (use $h_m^{(2)}$ for matrices)

Wavenumber–space transform:

$$\tilde{\mathbf{p}}(R, m, n, \omega) = \mathbf{J}^{-1} \int_0^\pi \sin\phi \, \mathbf{L}_m^n \left\{ \int_0^{2\pi} \mathbf{T}_n \, \mathbf{p}(R, \phi, \theta, \omega) \, d\theta \right\} d\phi \qquad \text{traction per steradian}$$

$$\tilde{\mathbf{u}} = \mathbf{K}^{-1}\tilde{\mathbf{p}}$$

$$\mathbf{u}(R, \phi, \theta, \omega) = \sum_{m=0}^{\infty} \sum_{n=0}^{m} \mathbf{T}_n \mathbf{L}_m^n \tilde{\mathbf{u}}(R, m, n, \omega) = \sum_{m=0}^{\infty} \sum_{n=0}^{m} \bar{\mathbf{u}}(R, m, n, \omega)$$

with

$$\mathbf{u} = \begin{Bmatrix} u_R \\ u_\phi \\ u_\theta \end{Bmatrix} = \text{displacements}, \qquad \mathbf{p} = \begin{Bmatrix} p_R \\ p_\phi \\ p_\theta \end{Bmatrix} = \text{tractions per steradian}$$

$$\mathbf{J} = \frac{\pi(1 + \delta_{n0})(n+m)!}{(m+\frac{1}{2})(m-n)!} \begin{Bmatrix} 1 & 0 & 0 \\ 0 & m(m+1) & 0 \\ 0 & 0 & m(m+1) \end{Bmatrix}$$

The negative sign in the second term comes from the fact that external tractions are opposite in direction to the internal stresses at the inner surface. Considering that the wave field in the layer is composed of both outgoing and ingoing waves, and with reference to Tables 10.6, 10.7 and the results in Section 9.3, we write the displacement and traction vectors for specific indices m, n as

$$\bar{\mathbf{u}}(R, \phi, \theta, \omega) = \mathbf{T}_n \mathbf{L}_m^n \left(\mathbf{H}_m^{(1)} \mathbf{a}_1 + \mathbf{H}_m^{(2)} \mathbf{a}_2 \right)$$

$$= \mathbf{T}_n \mathbf{L}_m^n \left\{ \mathbf{H}_m^{(1)} \quad \mathbf{H}_m^{(2)} \right\} \begin{Bmatrix} \mathbf{a}_1 \\ \mathbf{a}_2 \end{Bmatrix} \tag{10.145}$$

Table 10.7. Spheroidal, azimuthal and spherical Bessel matrices

$$\mathbf{L}_m^n = \begin{Bmatrix} P_m^n & 0 & 0 \\ 0 & \dfrac{dP_m^n}{d\phi} & \dfrac{n\,P_m^n}{\sin\phi} \\ 0 & \dfrac{n\,P_m^n}{\sin\phi} & \dfrac{dP_m^n}{d\phi} \end{Bmatrix}, \qquad P_m^n = P_m^n(\phi) = \text{associated Legendre function}$$

$$\mathbf{T}_n = \text{diag}[\cos n\theta \quad \cos n\theta \quad -\sin n\theta\,]$$

or

$$\mathbf{T}_n = \text{diag}[\sin n\theta \quad \sin n\theta \quad \cos n\theta\,]$$

$$\mathbf{H}_m = \begin{Bmatrix} \dfrac{dh_{\mathrm{P}m}}{dz_{\mathrm{P}}} & m(m+1)\dfrac{h_{\mathrm{S}m}}{z_{\mathrm{S}}} & 0 \\ \dfrac{h_{\mathrm{P}m}}{z_{\mathrm{P}}} & \dfrac{1}{z_{\mathrm{S}}}\dfrac{d(z_{\mathrm{S}}h_{\mathrm{S}m})}{dz_{\mathrm{S}}} & 0 \\ 0 & 0 & h_{\mathrm{S}m} \end{Bmatrix}, \qquad \mathbf{F}_m = \begin{Bmatrix} f_{11} & f_{12} & 0 \\ f_{21} & f_{22} & 0 \\ 0 & 0 & f_{33} \end{Bmatrix}$$

$$f_{11} = -k_{\mathrm{P}}\left[(\lambda+2\mu)\,h_{\mathrm{P}m} - \frac{2\mu}{z_{\mathrm{P}}}\left(2h_{\mathrm{P},m+1} + m\,(m-1)\frac{h_{\mathrm{P}m}}{z_{\mathrm{P}}}\right)\right]$$

$$f_{22} = -k_{\mathrm{S}}\left[\mu\,h_{\mathrm{S}m} - \frac{2\mu}{z_{\mathrm{S}}}\left(h_{\mathrm{S},m+1} - (m+1)\,(m-1)\frac{h_{\mathrm{S}m}}{z_{\mathrm{S}}}\right)\right]$$

$$f_{12} = -k_{\mathrm{S}}\,m\,(m+1)\frac{2\mu}{z_{\mathrm{S}}}\left(h_{\mathrm{S},m+1} - (m-1)\frac{h_{\mathrm{S}m}}{z_{\mathrm{S}}}\right)$$

$$f_{21} = -k_{\mathrm{P}}\frac{2\mu}{z_{\mathrm{P}}}\left(h_{\mathrm{P},m+1} - (m-1)\frac{h_{\mathrm{P}m}}{z_{\mathrm{P}}}\right)$$

$$f_{33} = -k_{\mathrm{S}}\mu\left(h_{\mathrm{S},m+1} - (m-1)\frac{h_{\mathrm{S}m}}{z_{\mathrm{S}}}\right)$$

$\mathbf{H}_m,\ \mathbf{F}_m \qquad \rightarrow h_{\mathrm{P}m} = j_m(z_{\mathrm{P}}), \qquad h_{\mathrm{S}m} = j_m(z_{\mathrm{S}})$ (use spherical Bessel functions)

$\mathbf{H}_m^{(1)},\ \mathbf{F}_m^{(1)} \qquad \rightarrow h_{Pm} = h_m^{(1)}(z_{\mathrm{P}}), \qquad h_{Sm} = h_m^{(1)}(z_{\mathrm{S}})$ (use 1st spherical Hankel functions)

$\mathbf{H}_m^{(2)},\ \mathbf{F}_m^{(2)} \qquad \rightarrow h_{Pm} = h_m^{(2)}(z_{\mathrm{P}}), \qquad h_{Sm} = h_m^{(2)}(z_{\mathrm{S}})$ (use 2nd spherical Hankel functions)

$$z_{\mathrm{P}} = k_{\mathrm{P}}R \equiv \Omega_{\mathrm{P}}, \qquad z_{\mathrm{S}} = k_{\mathrm{S}}R \equiv \Omega_{\mathrm{S}}, \qquad k_{\mathrm{P}} = \frac{\omega}{\alpha}, \qquad k_{\mathrm{S}} = \frac{\omega}{\beta}$$

$$\bar{\mathbf{p}}(R,\phi,\theta,\omega) = \mathbf{T}_n\mathbf{L}_m^n\left(\mathbf{F}_m^{(1)}\mathbf{a}_1 + \mathbf{F}_m^{(2)}\mathbf{a}_2\right)$$

$$= \mathbf{T}_n\mathbf{L}_m^n\left\{\mathbf{F}_m^{(1)} \quad \mathbf{F}_m^{(2)}\right\}\begin{Bmatrix} \mathbf{a}_1 \\ \mathbf{a}_2 \end{Bmatrix} \tag{10.146}$$

with subscripted matrices $\mathbf{T}, \mathbf{H}, \mathbf{F}$ defined in Table 10.7, and the overbar being a reminder that this is a particular solution for given m, n. Also, the numbers in parenthesis refer to the kind of spherical Bessel functions used to construct the matrices: $^{(1)}$ refers to $h_m^{(1)}$, and $^{(2)}$ to $h_m^{(2)}$. Alternatively, spherical Bessel functions of the first and second kind can be

Table 10.8. **The spheroidal or co-latitude matrices up to order and rank 3**

$$\mathbf{L}_0^0 = \left\{ \begin{matrix} 1 & & \\ & 0 & \\ & & 0 \end{matrix} \right\}$$

$$\mathbf{L}_1^0 = \left\{ \begin{matrix} \cos\phi & 0 & 0 \\ 0 & -\sin\phi & 0 \\ 0 & 0 & -\sin\phi \end{matrix} \right\}, \qquad \mathbf{L}_1^1 = -\left\{ \begin{matrix} \sin\phi & 0 & 0 \\ 0 & \cos\phi & 1 \\ 0 & 1 & \cos\phi \end{matrix} \right\}$$

$$\mathbf{L}_2^0 = \frac{1}{4} \left\{ \begin{matrix} 3\cos 2\phi + 1 & 0 & 0 \\ 0 & -6\sin 2\phi & 0 \\ 0 & 0 & -6\sin 2\phi \end{matrix} \right\}$$

$$\mathbf{L}_2^1 = -\frac{3}{2} \left\{ \begin{matrix} \sin 2\phi & 0 & 0 \\ 0 & 2\cos 2\phi & 2\cos\phi \\ 0 & 2\cos\phi & 2\cos 2\phi \end{matrix} \right\}$$

$$\mathbf{L}_2^2 = \frac{3}{2} \left\{ \begin{matrix} 1 - \cos 2\phi & 0 & 0 \\ 0 & 2\sin 2\phi & 4\sin\phi \\ 0 & 4\sin\phi & 2\sin 2\phi \end{matrix} \right\}$$

$$\mathbf{L}_3^0 = \frac{1}{8} \left\{ \begin{matrix} 5\cos 3\phi + 3\cos\phi & 0 & 0 \\ 0 & -3(5\sin 3\phi + \sin\phi) & 0 \\ 0 & 0 & -3(5\sin 3\phi + \sin\phi) \end{matrix} \right\}$$

$$\mathbf{L}_3^1 = -\frac{3}{8} \left\{ \begin{matrix} \sin\phi + 5\sin 3\phi & 0 & 0 \\ 0 & \cos\phi + 15\cos 3\phi & 6 + 10\cos 2\phi \\ 0 & 6 + 10\cos 2\phi & \cos\phi + 15\cos 3\phi \end{matrix} \right\}$$

$$\mathbf{L}_3^2 = \frac{15}{4} \left\{ \begin{matrix} \cos\phi - \cos 3\phi & 0 & 0 \\ 0 & 3\sin 3\phi - \sin\phi & 4\sin 2\phi \\ 0 & 4\sin 2\phi & 3\sin 3\phi - \sin\phi \end{matrix} \right\}$$

$$\mathbf{L}_3^3 = -\frac{15}{4} \left\{ \begin{matrix} 3\sin\phi - \sin 3\phi & 0 & 0 \\ 0 & 3(\cos\phi - \cos 3\phi) & 6(1 - \cos 2\phi) \\ 0 & 6(1 - \cos 2\phi) & 3(\cos\phi - \cos 3\phi) \end{matrix} \right\}$$

used. Evaluating these expressions at the inner and outer surfaces, and adding appropriate sub-indices e, i, to identify these surfaces, we can write

$$\left\{ \begin{matrix} \bar{\mathbf{u}}_e \\ \bar{\mathbf{u}}_i \end{matrix} \right\} = \left\{ \begin{matrix} \mathbf{T}_n\,\mathbf{L}_m^n & \mathbf{0} \\ \mathbf{0} & \mathbf{T}_n\,\mathbf{L}_m^n \end{matrix} \right\} \left\{ \begin{matrix} \mathbf{H}_{me}^{(1)} & \mathbf{H}_{me}^{(2)} \\ \mathbf{H}_{mi}^{(1)} & \mathbf{H}_{mi}^{(2)} \end{matrix} \right\} \left\{ \begin{matrix} \mathbf{a}_1 \\ \mathbf{a}_2 \end{matrix} \right\} \tag{10.147}$$

and

$$\left\{ \begin{matrix} \bar{\mathbf{p}}_e \\ \bar{\mathbf{p}}_i \end{matrix} \right\} = \left\{ \begin{matrix} \mathbf{T}_n\,\mathbf{L}_m^n & \mathbf{0} \\ \mathbf{0} & \mathbf{T}_n\,\mathbf{L}_m^n \end{matrix} \right\} \left\{ \begin{matrix} R_e^2\mathbf{F}_{me}^{(1)} & R_e^2\mathbf{F}_{me}^{(2)} \\ -R_i^2\mathbf{F}_{mi}^{(1)} & -R_i^2\mathbf{F}_{mi}^{(2)} \end{matrix} \right\} \left\{ \begin{matrix} \mathbf{a}_1 \\ \mathbf{a}_2 \end{matrix} \right\} \tag{10.148}$$

Next, we define

$$\left\{ \begin{array}{c} \bar{\mathbf{u}}_e \\ \bar{\mathbf{u}}_i \end{array} \right\} = \left\{ \begin{array}{cc} \mathbf{T}_n \mathbf{L}_m^n & \mathbf{0} \\ \mathbf{0} & \mathbf{T}_n \mathbf{L}_m^n \end{array} \right\} \left\{ \begin{array}{c} \tilde{\mathbf{u}}_1 \\ \tilde{\mathbf{u}}_2 \end{array} \right\}, \quad \left\{ \begin{array}{c} \tilde{\mathbf{u}}_e \\ \tilde{\mathbf{u}}_i \end{array} \right\} = \left\{ \begin{array}{cc} \mathbf{H}_{me}^{(1)} & \mathbf{H}_{me}^{(2)} \\ \mathbf{H}_{mi}^{(1)} & \mathbf{H}_{mi}^{(2)} \end{array} \right\} \left\{ \begin{array}{c} \mathbf{a}_1 \\ \mathbf{a}_2 \end{array} \right\} \qquad (10.149)$$

and

$$\left\{ \begin{array}{c} \bar{\mathbf{p}}_e \\ \bar{\mathbf{p}}_i \end{array} \right\} = \left\{ \begin{array}{cc} \mathbf{T}_n \mathbf{L}_m^n & \mathbf{0} \\ \mathbf{0} & \mathbf{T}_n \mathbf{L}_m^n \end{array} \right\} \left\{ \begin{array}{c} \tilde{\mathbf{p}}_e \\ \tilde{\mathbf{p}}_i \end{array} \right\}, \quad \left\{ \begin{array}{c} \tilde{\mathbf{p}}_e \\ \tilde{\mathbf{p}}_i \end{array} \right\} = \left\{ \begin{array}{cc} R_e^2 \mathbf{F}_{me}^{(1)} & R_e^2 \mathbf{F}_{me}^{(2)} \\ -R_i^2 \mathbf{F}_{mi}^{(1)} & -R_i^2 \mathbf{F}_{mi}^{(2)} \end{array} \right\} \left\{ \begin{array}{c} \mathbf{a}_1 \\ \mathbf{a}_2 \end{array} \right\} \qquad (10.150)$$

which express the displacement and load vectors at the two interfaces in terms of local amplitudes, namely the quantities with tilde. Observe that the azimuthal and co-latitude matrices are common to both surfaces, and indeed, to all concentric surfaces in the case of a layered sphere, so they can be factored out for the system as a whole. Eliminating the integration constants between the traction and displacement amplitudes, we obtain

$$\left\{ \begin{array}{c} \tilde{\mathbf{p}}_e \\ \tilde{\mathbf{p}}_i \end{array} \right\} = \left\{ \begin{array}{cc} R_e^2 \mathbf{F}_{me}^{(1)} & R_e^2 \mathbf{F}_{me}^{(2)} \\ -R_i^2 \mathbf{F}_{mi}^{(1)} & -R_i^2 \mathbf{F}_{mi}^{(2)} \end{array} \right\} \left\{ \begin{array}{cc} \mathbf{H}_{me}^{(1)} & \mathbf{H}_{me}^{(2)} \\ \mathbf{H}_{mi}^{(1)} & \mathbf{H}_{mi}^{(2)} \end{array} \right\}^{-1} \left\{ \begin{array}{c} \tilde{\mathbf{u}}_e \\ \tilde{\mathbf{u}}_i \end{array} \right\}$$

$$= \left\{ \begin{array}{cc} \mathbf{K}_{ee} & \mathbf{K}_{ei} \\ \mathbf{K}_{ie} & \mathbf{K}_{ii} \end{array} \right\} \left\{ \begin{array}{c} \tilde{\mathbf{u}}_e \\ \tilde{\mathbf{u}}_i \end{array} \right\} \qquad (10.151)$$

or more compactly

$$\tilde{\mathbf{p}} = \mathbf{K}\tilde{\mathbf{u}} \qquad (10.152)$$

$\mathbf{K}$ is the dynamic stiffness or *impedance matrix*, which relates the amplitudes of the external tractions applied at both surfaces to the amplitudes of displacements observed at these locations. Hence, for given tractions $\tilde{\mathbf{p}}$, we can obtain the displacements $\tilde{\mathbf{u}}$, and from here the actual displacements by appropriate multiplication by the azimuthal and co-latitude matrices.

More generally, in the case of a layered sphere composed of $N-1$ layers whose N interfaces have radii $R_1, R_2, \ldots, R_N$ and the interfaces are numbered from the outside to the inside, we can construct the system impedance matrix by appropriately overlapping the layer impedance matrices. The end result is of the block-tridiagonal form

$$\left\{ \begin{array}{c} \tilde{\mathbf{p}}_1 \\ \tilde{\mathbf{p}}_2 \\ \tilde{\mathbf{p}}_3 \\ \vdots \\ \tilde{\mathbf{p}}_n \end{array} \right\} = \left\{ \begin{array}{cccccc} \mathbf{K}_{11} & \mathbf{K}_{12} & \mathbf{0} & \cdots & & \mathbf{0} \\ \mathbf{K}_{21} & \mathbf{K}_{22} & \mathbf{K}_{23} & \cdots & & \mathbf{0} \\ \mathbf{0} & \mathbf{K}_{32} & \mathbf{K}_{33} & \ddots & & \vdots \\ \vdots & \vdots & \ddots & \ddots & & \mathbf{K}_{n-1,n} \\ \mathbf{0} & \mathbf{0} & \cdots & & \mathbf{K}_{n,n-1} & \mathbf{K}_{nn} \end{array} \right\} \left\{ \begin{array}{c} \tilde{\mathbf{u}}_1 \\ \tilde{\mathbf{u}}_2 \\ \tilde{\mathbf{u}}_3 \\ \vdots \\ \tilde{\mathbf{u}}_n \end{array} \right\} \qquad (10.153)$$

which again allows obtaining the displacement vector amplitudes from the load amplitudes by solving the system of equations, and then applying the azimuthal and co-latitude transformations to obtain the actual displacements.

10.4.1 Properties and use of impedance matrices

The impedance matrix for any layer is a function of the inner and outer radii, the material properties of the layer, the frequency of excitation, and the radial index m. However, it is independent of the azimuthal index n. As we have seen, the 6×6 stiffness matrix for a spherical layer is of the form

$$\mathbf{K} = \left\{ \begin{matrix} \mathbf{K}_{ee} & \mathbf{K}_{ei} \\ \mathbf{K}_{ie} & \mathbf{K}_{ii} \end{matrix} \right\} = \left\{ \begin{matrix} R_e^2 \mathbf{F}_{me}^{(1)} & R_e^2 \mathbf{F}_{me}^{(2)} \\ -R_i^2 \mathbf{F}_{mi}^{(1)} & -R_i^2 \mathbf{F}_{mi}^{(2)} \end{matrix} \right\} \left\{ \begin{matrix} \mathbf{H}_{me}^{(1)} & \mathbf{H}_{me}^{(2)} \\ \mathbf{H}_{mi}^{(1)} & \mathbf{H}_{mi}^{(2)} \end{matrix} \right\}^{-1} \qquad \text{(layer)} \qquad (10.154)$$

In principle, spherical Bessel functions j_m, y_m could be used in place of the spherical Hankel functions $h_m^{(1)}, h_m^{(2)}$ to construct this matrix, but they are not recommended here for *numerical* computation, because they may become quasi-linearly dependent for complex, and especially negative imaginary, arguments.

On the other hand, the 3×3 matrix of a solid sphere of radius R (or of the core of a layered system) *must* be constructed with spherical Bessel functions of the first kind j_m so as to avoid a singularity on the axis:

$$\tilde{\mathbf{u}} = \mathbf{H}_m \mathbf{c}, \qquad \tilde{\mathbf{p}} = R^2 \mathbf{F}_m \mathbf{c}, \qquad \mathbf{K} = R^2 \mathbf{F}_m (\mathbf{H}_m)^{-1} \qquad \text{(core)} \qquad (10.155)$$

By contrast, in the case of a spherical cavity of radius R in an infinite space, the component matrices $\mathbf{H}_m, \mathbf{F}_m$ must be constructed using second spherical Hankel functions $h_m^{(2)}(kR)$, so as to satisfy the radiation condition at infinity. The impedance for the outer region, as seen from the surface of the cavity, is given by a 3×3 matrix:

$$\tilde{\mathbf{u}} = \mathbf{H}_m^{(2)} \mathbf{c}, \qquad \tilde{\mathbf{p}} = -R^2 \mathbf{F}_m^{(2)} \mathbf{c}, \qquad \mathbf{K} = -R^2 \mathbf{F}_m^{(2)} \left(\mathbf{H}_m^{(2)} \right)^{-1} \qquad \text{(infinite exterior)} \quad (10.156)$$

The imaginary part in the resulting impedance matrix represents the radiation damping in the outer region. For a system of layers surrounded by a homogeneous infinite space, the above outer region matrix must be added to the element $\mathbf{K}_{11}$ to complete the global system matrix.

The case $m = 0$ is also special. Here, the co-latitude matrix $\mathbf{L}_0^0$ is zero except for the first diagonal element. Also, the off-diagonal element h_{12} of $\mathbf{H}$ is zero. It follows that each interface is characterized by a single degree of freedom, which is the radial component. Hence, all elements of $\mathbf{H}$ except for h_{11} can be set to zero, and in fact, all other degrees of freedom should be removed before computation.

10.4.2 Asymmetry

As written, the impedance matrices for each layer (and therefore, the system matrix) are not symmetric. If desired, however, they can be brought into a symmetric form by a similarity transformation that, fortunately, depends solely on the index m. For each component matrix, one defines the scaling factor

$$\xi = \frac{1}{\sqrt{m(m+1)}}, \quad m \neq 0 \qquad (10.157)$$

such that

$$
\tilde{\mathbf{F}}_m = \mathbf{Q}_m \, \mathbf{F}_m = \left\{ \begin{array}{ccc} \xi & & \\ & 1 & \\ & & 1 \end{array} \right\} \left\{ \begin{array}{ccc} f_{11} & f_{12} & 0 \\ f_{21} & f_{22} & 0 \\ 0 & 0 & f_{33} \end{array} \right\} \tag{10.158}
$$

$$
\tilde{\mathbf{H}}_m = \mathbf{Q}_m \, \mathbf{H}_m = \left\{ \begin{array}{ccc} \xi & & \\ & 1 & \\ & & 1 \end{array} \right\} \left\{ \begin{array}{ccc} h_{11} & h_{12} & 0 \\ h_{21} & h_{22} & 0 \\ 0 & 0 & h_{33} \end{array} \right\} \tag{10.159}
$$

which implies

$$
\begin{aligned}
\bar{\mathbf{u}} &= \mathbf{T}_n \, \mathbf{L}_m^n \, \mathbf{Q}_m^{-1} \, \mathbf{Q}_m \, \mathbf{H}_m \, \mathbf{c} \\
&= \mathbf{T}_n \, \mathbf{L}_m^n \, \mathbf{Q}_m^{-1} \, \tilde{\mathbf{H}}_m \, \mathbf{c}
\end{aligned} \quad \Rightarrow \quad \bar{\mathbf{u}} = \mathbf{T}_n \, \mathbf{L}_m^n \, \mathbf{Q}_m^{-1} \, \tilde{\mathbf{u}} \tag{10.160}
$$

and

$$
\begin{aligned}
\bar{\mathbf{p}} &= \mathbf{T}_n \, \mathbf{L}_m^n \, \mathbf{Q}_m^{-1} \, \mathbf{Q}_m \, \mathbf{F}_m \, \mathbf{c} \\
&= \mathbf{T}_n \, \mathbf{L}_m^n \, \mathbf{Q}_m^{-1} \, \tilde{\mathbf{F}}_m \, \mathbf{c}
\end{aligned} \quad \Rightarrow \quad \begin{aligned} \bar{\mathbf{p}} &= \mathbf{T}_n \, \mathbf{L}_m^n \, \mathbf{Q}_m^{-1} \, \tilde{\mathbf{p}} \\ &= \mathbf{T}_n \, \mathbf{L}_m^n \, \mathbf{Q}_m^{-1} \, \tilde{\mathbf{K}} \, \tilde{\mathbf{u}} \end{aligned} \tag{10.161}
$$

For example, for an exterior region, the scaled, symmetric impedance matrix would be

$$
\tilde{\mathbf{K}} = -R^2 \tilde{\mathbf{F}}_m \left(\tilde{\mathbf{H}}_m \right)^{-1} \quad \Rightarrow \quad \tilde{\mathbf{K}} = \mathbf{Q}_m \, \mathbf{K} \, \mathbf{Q}_m^{-1} \tag{10.162}
$$

which relates the scaled force and displacement amplitudes vectors at the interface of the outer region. More generally, in the case of a system of layers, we obtain a symmetric global system matrix. However, to avoid proliferation of symbols, we continue labeling the scaled vectors for the system as we did for any individual layer, and write the system equation simply as

$$
\tilde{\mathbf{p}} = \tilde{\mathbf{K}} \, \tilde{\mathbf{u}} \quad \Rightarrow \quad \tilde{\mathbf{u}} = \tilde{\mathbf{K}}^{-1} \, \tilde{\mathbf{p}} \tag{10.163}
$$

In this equation, every third (SH) degree of freedom is uncoupled from the preceding two (SV-P) degrees of freedom, so they should be solved separately. The advantage of using a symmetric stiffness matrix lies in the saving in computational effort, which is a factor of two. For notational transparency, however, we present the ensuing examples in the standard, non-symmetric form.

10.4.3 Expansion of source and displacements into spherical harmonics

The stiffness matrix method described herein is based on a formulation in the frequency–spheroidal-wavenumber domain. Thus, the sources $\mathbf{p} = \mathbf{p}(R, \phi, \theta, \omega)$, if any, must be expressed in terms of spheroidal harmonics of order m,n, i.e., $\tilde{\mathbf{p}} \equiv \tilde{\mathbf{p}}_{mn}(R, \omega)$

Let $\mathbf{p}$ be the 3×1 vector of external tractions per solid angle at a given frequency, which we assume to be defined over a spherical surface of constant radius R. Expansion into spheroidal harmonics yields

$$
\mathbf{p} = \sum_{m=0}^{\infty} \sum_{n=0}^{m} \mathbf{T}_n \, \mathbf{L}_m^n \, \tilde{\mathbf{p}}_{mn} \tag{10.164}
$$

whose inversion is

$$\tilde{\mathbf{p}}_{mn} = \mathbf{J}^{-1} \int_0^\pi \sin\phi \, \mathbf{L}_m^n \left\{ \int_0^{2\pi} \mathbf{T}_n \mathbf{p} \, d\theta \right\} d\phi \tag{10.165}$$

with

$$\mathbf{J} = \frac{\pi(1+\delta_{n0})(n+m)!}{(m+\frac{1}{2})(m-n)!} \left\{ \begin{matrix} 1 & 0 & 0 \\ 0 & m(m+1) & 0 \\ 0 & 0 & m(m+1) \end{matrix} \right\} \tag{10.166}$$

This is based on the orthogonality conditions of the spheroidal and azimuthal matrices,

$$\int_0^\pi \mathbf{L}_m^n \mathbf{L}_m^n \sin\phi \, d\phi = \mathbf{J} \tag{10.167}$$

$$\int_0^{2\pi} \mathbf{T}_n \mathbf{T}_j \, d\theta = \pi \, \delta_{(nj)}(1+\delta_{n0}\delta_{j0}). \tag{10.168}$$

Having found the displacements $\tilde{\mathbf{u}} \equiv \tilde{\mathbf{u}}_{mn}(\omega)$ in the spheroidal domain, the displacements in the physical domain are obtained for each interface as

$$\mathbf{u} = \sum_{m=0}^\infty \sum_{n=0}^m \mathbf{T}_n \mathbf{L}_m^n \tilde{\mathbf{u}}_{mn} \tag{10.169}$$

Application of these equations is demonstrated in Example 10.13.

10.4.4 Rigid body spheroidal modes

An unconstrained, spherical system of finite size admits six rigid body modes that produce no net stresses within the system, namely three translations and three rotations. They are:

Translations

$$U_x = c\,\mathbf{T}_1^{(1)}\,\mathbf{L}_1^1\,\hat{\mathbf{e}}_{12}, \qquad U_y = c\,\mathbf{T}_1^{(2)}\,\mathbf{L}_1^1\,\hat{\mathbf{e}}_{12}, \qquad U_z = c\,\mathbf{T}_0^{(1)}\,\mathbf{L}_1^0\,\hat{\mathbf{e}}_{12} \tag{10.170}$$

Rotations

$$\Omega_x = c\,\mathbf{T}_1^{(2)}\,\mathbf{L}_1^1\,\hat{\mathbf{e}}_3, \qquad \Omega_y = c\,\mathbf{T}_1^{(1)}\,\mathbf{L}_1^1\,\hat{\mathbf{e}}_3, \qquad \Omega_x = c\,\mathbf{T}_0^{(2)}\,\mathbf{L}_1^0\,\hat{\mathbf{e}}_3 \tag{10.171}$$

in which c is an arbitrary constant, and

$$\mathbf{T}_n^{(1)} = \text{diag}[\cos n\theta \quad \cos n\theta \quad -\sin n\theta], \quad \mathbf{T}_n^{(2)} = \text{diag}[\sin n\theta \quad \sin n\theta \quad \cos n\theta]$$

$$\tag{10.172}$$

$$\hat{\mathbf{e}}_{12} = \begin{bmatrix} 1 & 1 & 0 \end{bmatrix}^T, \quad \hat{\mathbf{e}}_3 = \begin{bmatrix} 0 & 0 & 1 \end{bmatrix}^T \tag{10.173}$$

Example 10.11: Free vibration modes of a solid sphere

The dynamic stiffness matrix of the solid sphere is obtained from the preceding by disregarding the terms associated with the inner surface, inasmuch as $R_i = 0$, at the center (i.e., the inner surface shrinks to a point). The result is of the form

$$\tilde{\mathbf{p}} = \mathbf{K}\tilde{\mathbf{u}}, \quad \mathbf{K} = R_e^2 \mathbf{F}_m (\mathbf{H}_m)^{-1} \tag{10.174}$$

in which the 3×3 matrices used to form $\mathbf{K}$ are assembled with spherical functions of the first kind, $\tilde{\mathbf{u}}$ are the displacements on the surface of the sphere, and $\tilde{\mathbf{p}}$ are the external tractions per solid angle. When these tractions vanish, we obtain the free vibration problem $\mathbf{K}\tilde{\mathbf{u}} = \mathbf{0}$. Nontrivial solutions of this problem exist if

$$\det \mathbf{K}(\omega) = 0 \quad \Rightarrow \quad \det \mathbf{F}_m = 0 \tag{10.175}$$

or in full

$$\left.\begin{array}{l} f_{11}\,f_{22} - f_{12}\,f_{21} = 0, \quad m > 0 \\ \qquad\qquad f_{11} = 0, \quad m = 0 \end{array}\right\} \quad \Rightarrow \quad \text{spheroidal modes} \tag{10.176}$$

$$f_{33} = 0, \quad m > 0 \quad \Rightarrow \quad \text{torsional modes} \tag{10.177}$$

the roots of which can be obtained by numerical search for any given parameter m. Having found the frequencies, one can then proceed to find the vibration modes in transformed space by solving the equation $\mathbf{K}\tilde{\mathbf{u}} = \mathbf{0}$ while assigning an arbitrary value to one component of $\tilde{\mathbf{u}}$, say the first. Again, since the torsional (SH) degree of freedom is uncoupled from the spheroidal (SV-P) degrees of freedom, they should be solved separately. Having found the modes in transformed space, the actual modes in physical space are obtained by applying the co-latitude and azimuthal matrices to the vectors thus found. Observe that the natural frequencies do not depend on the azimuthal index n, because $\mathbf{K}$ does not depend on n.

Example 10.12: Free vibration modes of a layered sphere

This is similar to the preceding example. It suffices to set up the system matrix by appropriately overlapping the layer matrices, and then search for the roots of $\det \mathbf{K} = 0$. The inner core, if any, is characterized by the same 3×3 stiffness matrix of the previous example, whereas the outer layers have matrices of size 6×6. In the case of a hollow layered sphere, the core matrix must be left out.

Example 10.13: Response to applied load

We examine next the case of a layered sphere being acted upon by a harmonic load with some known spatial distribution. To keep the presentation simple, we consider the special case of a unit point load in the radial direction at location R_0, ϕ_0, θ_0. This can be expressed as a radial traction per solid angle of the form

$$p_R = \frac{\delta(\phi - \phi_0)\,\delta(\theta - \theta_0)}{4\pi}, \quad \mathbf{p} = p_R \mathbf{e}, \quad \mathbf{e} = \begin{Bmatrix} 1 \\ 0 \\ 0 \end{Bmatrix} \tag{10.178}$$

The load vector in transformed space is then

$$\tilde{\mathbf{p}} = \mathbf{J}^{-1} \int_0^\pi \sin\phi \int_0^{2\pi} \mathbf{L}_m^n \mathbf{T}_n \mathbf{p}\, d\theta\, d\phi = \tfrac{1}{4\pi} \sin\phi_0\, \mathbf{J}^{-1} \mathbf{L}_m^n(\phi_0)\, \mathbf{T}_n(\theta_0)\, \mathbf{e}$$

$$= \frac{(m+\tfrac{1}{2})\,(m-n)!}{4\pi^2(1+\delta_{n0})\,(n+m)!} \sin\phi_0\, P_m^n(\phi_0) \begin{pmatrix} \cos n\theta_0 \\ \sin n\theta_0 \end{pmatrix} \mathbf{e} \tag{10.179}$$

Observe that the load has both symmetric and antisymmetric spheroidal components as indicated by the two alternative variations with the azimuth. Also, notice that the radial load cannot be located at either the north or the south pole, because in such a case $\sin\phi_0 = 0$. This is the result of using herein only spheroidal harmonics of the first kind. Now, we can always assume that the interface at radial distance R (where the load is applied) coincides with one of the interfaces of the layers of the problem at hand, for if this were not the case, it would suffice to subdivide the physical layer in which the load resides into two sub-layers with identical material properties. Thereafter, we augment the vector $\mathbf{e}$ with sufficient number of zeros to match the size of the problem, and proceed to solve for the displacements as

$$\tilde{\mathbf{u}} = \mathbf{K}^{-1}\tilde{\mathbf{p}} \quad \Rightarrow \quad \mathbf{u} = \sum_{m=0}^{\infty} \sum_{n=0}^{m} \mathbf{T}_n \mathbf{L}_m^n \tilde{\mathbf{u}} \tag{10.180}$$

Observe that $\mathbf{K}$ does not depend on n.

11 Basic properties of mathematical functions[1]

11.1 Bessel functions

11.1.1 Differential equation

$$\frac{d^2y}{dx^2} + \frac{1}{r}\frac{dy}{dx} + \left(k^2 - \frac{n^2}{r^2}\right)y = 0 \tag{11.1}$$

This equation has solutions of the form $y = c_1 J_n(kr) + c_2 Y_n(kr)$, with c_1, c_2 arbitrary constants, and $J_n(kr)$, $Y_n(kr)$ the *Bessel functions* of order n and of the first and second kind, respectively (some books write $N_n(kr)$ instead of $Y_n(kr)$ and refer to it as the *Neumann function*). Alternatively, solutions can be written in the form

$$y = c_1 H_n^{(1)}(kr) + c_2 H_n^{(2)}(kr) \tag{11.2}$$

in which $H_n^{(1)} = J_n + i\,Y_n$ and $H_n^{(2)} = J_n - i\,Y_n$ are the *Hankel functions* of the first and second kind (or Bessel functions of the third kind). These functions behave asymptotically as complex exponentials $\exp(\pm ikr)$. Thus, in combination with a harmonic term of the form $\exp(+i\omega t)$, Hankel functions can be interpreted as waves that propagate toward the origin (first kind), or away from it (second kind), so they are often used in wave propagation problems. We refer in the ensuing to *any* of the Bessel functions by means of the symbol $C_n = C_n(z)$, which is shorthand for *cylindrical* functions with a (generally complex) argument $z = kr$.

11.1.2 Recurrence relations

$$\frac{n}{z}C_n = \frac{1}{2}\left(C_{n-1} + C_{n+1}\right) \tag{11.3}$$

$$\frac{dC_n}{dz} = \frac{1}{2}\left(C_{n-1} - C_{n+1}\right) \tag{11.4}$$

---

[1] Extensive tables and properties for mathematical functions can be found in the following two references:
  a) Abramovitz, M. and Stegun, I. A., 1970, *Handbook of mathematical functions*, National Bureau of Standards.
  b) Gradshteyn, I. and Ryzhik, I., 1980, *Table of integrals, series and products*, Academic Press.

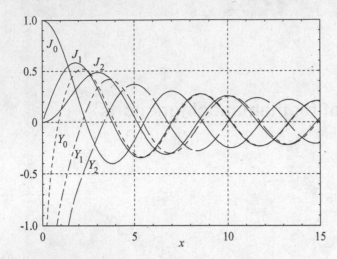

Figure 11.1: Bessel functions $J_n(x)$, $Y_n(x)$.

11.1.3 Derivatives

$$\frac{d^k C_n}{dz^k} = \frac{1}{2^k} \sum_{j=0}^{k} (-1)^j \binom{k}{j} C_{n-k+2j}, \quad k = 0, 1, 2, 3, \ldots \tag{11.5}$$

$$\left(\frac{1}{z}\frac{d}{dz}\right)^k (z^n C_n) = z^{n-k} C_{n-k} \tag{11.6}$$

$$\left(\frac{1}{z}\frac{d}{dz}\right)^k (z^{-n} C_n) = (-1)^k z^{-n-k} C_{n+k} \tag{11.7}$$

11.1.4 Wronskians

$$J_{n+1}(z)\, Y_n(z) - J_n(z)\, Y_{n+1}(z) = \frac{2}{\pi z} \tag{11.8}$$

$$H_n^{(1)}(z)\, H_{n+1}^{(2)} - H_{n+1}^{(1)}\, H_n^{(2)}(z) = \frac{4\,i}{\pi z} \tag{11.9}$$

11.1.5 Orthogonality conditions

First condition

If $J_n(z_{nj}) = 0$ are the zeros of J_n, and $z_{nj} = k_{nj} R$, then

$$\int_0^R J_n(k_{ni} r)\, J_n(k_{nj} r)\, r\, dr = \frac{1}{2} R^2 \left[\frac{d}{dz} J_n(z)\bigg|_{z=k_{nj} R}\right]^2 \delta_{ij}, \quad i, j = 1, 2, \ldots \tag{11.10}$$

When $n \neq 0$, there exists a zero $z_{n0} = 0$ that leads to the trivial condition $0 = 0$, so these must be excluded. To a first approximation, the zeros are given by $z_{nj} = k_{nj} R = \pi\left(j + \frac{1}{2}n - \frac{1}{4}\right)$.

Second condition

If $\frac{d}{dz} J_n(z)\big|_{z'_{nj}} = 0$ are the zeros of the *derivative* of J_n and $z'_{nj} = k'_{nj} R$, then

$$\int_0^R J_n(k'_{ni}r)\, J_n(k'_{nj}r)\, r\, dr = \frac{1}{2} R^2 J_n^2(k'_{ni} R) \left[1 - \left(\frac{n}{k'_{ni} R} \right)^2 \right] \delta_{ij}, \quad i, j = 1, 2, \dots \tag{11.11}$$

The zeros z'_{nj} of $\frac{d}{dz} J_n(z)$ interlace the zeros z_{nj} of $J_n(z)$. When $n \neq 0$, $z'_{nj} \neq 0$. In addition, for $n = 0$, there exists a zero $z'_{00} = 0$ that satisfies

$$\int_0^R J_0^2(0)\, r\, dr = \tfrac{1}{2} R^2 \tag{11.12}$$

11.1.6 Useful integrals

$$\int_0^R J_n(\alpha r)\, J_n(\beta r)\, r\, dr = \frac{R[J_n(\alpha R)\, J_n'(\beta R) - J_n'(\alpha R)\, J_n(\beta R)]}{\alpha^2 - \beta^2}, \quad \alpha \neq \beta \tag{11.13}$$

$$\int_0^R J_n^2(\alpha r)\, r\, dr = \frac{R^2}{2} \left\{ \left(\frac{1}{\alpha} \frac{dJ_n(\alpha r)}{dr} \right)^2 \bigg|_{r=R} + \left[1 - \left(\frac{n}{\alpha R} \right)^2 \right] J_n^2(\alpha R) \right\} \tag{11.14}$$

$$\int_0^R J_n(\alpha r)\, Y_n(\beta r)\, r\, dr = \frac{R[J_n(\alpha R)\, Y_n'(\beta R) - J_n'(\alpha R)\, Y_n(\beta R)] - \frac{2}{\pi} (\alpha/\beta)^n}{\alpha^2 - \beta^2}, \quad \alpha \neq \beta \tag{11.15}$$

$$\int_0^R J_n(\alpha r)\, Y_n(\alpha r)\, r\, dr = \frac{r^2}{2} \left\{ \frac{1}{\alpha^2} \left(\frac{dJ_n(\alpha r)}{dr} \frac{dY_n(\alpha r)}{dr} \right) \right.$$
$$\left. + \left[1 - \left(\frac{n}{\alpha r} \right)^2 \right] J_n(\alpha r)\, Y_n(\alpha r) \right\} \bigg|_{r=R} - \frac{n}{\alpha^2 \pi} \tag{11.16}$$

$$\int_a^b Y_n(\alpha r)\, Y_n(\beta r)\, r\, dr = \frac{r[Y_n(\alpha r)\, Y_n'(\beta r) - Y_n'(\alpha r)\, Y_n(\beta r)]}{\alpha^2 - \beta^2} \bigg|_a^b, \quad \alpha \neq \beta, \ a > 0 \tag{11.17}$$

$$\int_a^b Y_n^2(\alpha r)\, r\, dr = \frac{r^2}{2} \left\{ \left(\frac{1}{\alpha} \frac{dY_n(\alpha r)}{dr} \right)^2 + \left[1 - \left(\frac{n}{\alpha r} \right)^2 \right] Y_n^2(\alpha r) \right\} \bigg|_a^b, \quad \alpha = \beta, \ a > 0 \tag{11.18}$$

11.2 Spherical Bessel functions

11.2.1 Differential equation

$$\frac{\partial^2 y}{\partial R^2} + \frac{2}{R} \frac{\partial y}{\partial R} + \left[k^2 - \frac{m(m+1)}{R^2} \right] y = 0 \tag{11.19}$$

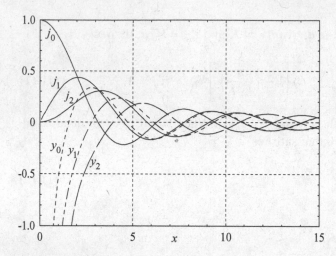

Figure 11.2: Spherical Bessel functions $j_n(x)$, $y_n(x)$.

This equation has solutions of the form $y = c_1\, j_m(kr) + c_2\, y_m(kr)$, with c_1, c_2 arbitrary constants, and $j_m(kr)$, $y_m(kr)$ the *spherical Bessel functions* of order m and of the first and second kind, respectively. Alternatively, solutions can be written in the form

$$y = c_1 h_m^{(1)}(kr) + c_2 h_m^{(2)}(kr) \tag{11.20}$$

in which $h_m^{(1)} = j_m + i\,y_m$ and $h_m^{(2)} = j_m - i\,y_m$ are the *spherical Hankel functions* of the first and second kind (or spherical Bessel functions of the third kind). For integer m, and generally complex argument $z = kr$, these functions relate to the Bessel functions of half-integer order $n = m + \frac{1}{2}$ as

$$j_m(z) = \sqrt{\frac{\pi}{2z}}\, J_{m+1/2}(z), \qquad y_m(z) = \sqrt{\frac{\pi}{2z}}\, Y_{m+1/2}(z) \tag{11.21}$$

11.2.2 Trigonometric representations

The spherical Bessel functions can be represented in terms of elementary functions. The first three spherical Bessel functions of the first, second, and third kinds, with complex argument z are

$$j_0(z) = \frac{\sin z}{z}, \quad j_1(z) = \frac{\sin z}{z^2} - \frac{\cos z}{z}, \quad j_2(z) = \left(\frac{3}{z^3} - \frac{1}{z}\right)\sin z - \frac{3}{z^2}\cos z \tag{11.22}$$

$$y_0(z) = -\frac{\cos z}{z}, \quad y_1(z) = -\frac{\cos z}{z^2} - \frac{\sin z}{z}, \quad y_2(z) = \left(\frac{1}{z} - \frac{3}{z^3}\right)\cos z - \frac{3}{z^2}\sin z \tag{11.23}$$

$$h_0^{(2)}(z) = i\frac{e^{-iz}}{z}, \quad h_1^{(2)}(z) = i\frac{e^{-iz}}{z^2}(1 + iz), \quad h_2^{(2)}(z) = i\frac{e^{-iz}}{z^3}(3 + 3\,i\,z - z^2) \tag{11.24}$$

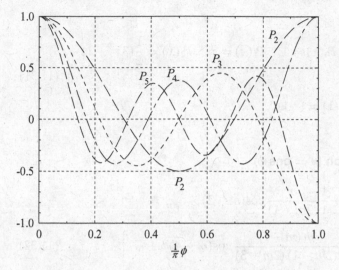

Figure 11.3: Legendre polynomials.

The functions j_m satisfy $j_m(0) = \delta_{m0}$, that is, they vanish at the origin except for j_0. Also, the y_m functions are singular at $z = 0$. More generally,

$$j_m(z) = (-z)^m \left(\frac{1}{z}\frac{d}{dz}\right)^m \frac{\sin z}{z} = (-z)^m \left(\frac{1}{z}\frac{d}{dz}\left(\frac{1}{z}\frac{d}{dz}\cdots\left(\frac{1}{z}\frac{d}{dz}\frac{\sin z}{z}\right)\right)\right), \quad m = 0, 1, 2, \ldots \tag{11.25}$$

$$y_m(z) = -(-z)^m \left(\frac{1}{z}\frac{d}{dz}\right)^m \frac{\cos z}{z} \tag{11.26}$$

11.2.3 Recurrence relations

$s_m(z) =$ *any of the spherical Bessel functions*

$$\frac{2m+1}{z}s_m(z) = s_{m-1}(z) + s_{m+1}(z) \tag{11.27}$$

$$(2m+1)\frac{ds_m(z)}{dz} = ms_{m-1}(z) - (m+1)s_{m+1}(z) \tag{11.28}$$

11.3 Legendre polynomials

11.3.1 Differential equation

$$(1-x^2)\frac{d^2y}{dx^2} - 2x\frac{dy}{dx} + m(m+1)y = 0, \quad m = 0, 1, 2, \ldots, \quad |x| \leq 1 \tag{11.29}$$

For non-negative integer m, the solution to this equation is $y = c_1 P_m + c_2 Q_m$, with c_1, c_2 being arbitrary constants, and $P_m(x)$ the Legendre polynomial of order m. The second solution $Q_m(x)$ to the Legendre differential equation is used much less often. In most applications, $x = \cos\phi$. The trigonometric form of the differential equation is

$$\frac{d^2 P_m}{d\phi^2} + \cot\phi\,\frac{d P_m}{d\phi} + m(m+1)\,P_m = 0 \tag{11.30}$$

11.3.2 Rodrigues's formula

$$P_m(x) = \frac{(-1)^m}{2^m m!} \frac{d^m (1 - x^2)^m}{dx^m}, \quad P_0(x) = 1, \quad P_1(x) = x, \quad P_2(x) = \frac{1}{2}(3x^2 - 1), \ldots$$

$$(11.31)$$

Observe that $P_m(1) = 1$ and $P_m(-1) = (-1)^m$.

11.3.3 Trigonometric expansion ($x = \cos \phi$)

$$P_m(\phi) = 2\frac{(2m-1)!!}{(2m)!!}\left[\cos m\phi + \frac{1}{1}\frac{m}{(2m-1)}\cos(m-2)\phi\right.$$

$$\left. + \frac{1 \times 3}{1 \times 2}\frac{m(m-1)}{(2m-1)(2m-3)}\cos(m-4)\phi + \cdots\right] \quad (11.32)$$

Divide the last term by 2 if m is even! Here, $(2m-1)!! = 1 \times 3 \times 5 \times \cdots \times (2m-1)$ and $(2m)!! = 2 \times 4 \times 6 \times \cdots \times 2m$. The first six Legendre polynomials in trigonometric form are

$$P_0 = 1, \quad P_1 = \cos\phi, \quad P_2 = \frac{1}{4}\left(3\cos 2\phi + 1\right), \quad P_3 = \frac{1}{8}\left(5\cos 3\phi + 3\cos\phi\right) \quad (11.33)$$

$$P_4 = \frac{1}{64}\left(35\cos 4\phi + 20\cos 2\phi + 9\right), \quad P_5 = \frac{1}{128}\left(63\cos 5\phi + 35\cos 3\phi + 30\cos\phi\right)$$

$$(11.34)$$

11.3.4 Recurrence relations

$$(m+1)P_{m+1} = (2m+1)x\, P_m - m\, P_{m-1} \quad \text{(simplest way to find } P_m\text{)} \quad (11.35)$$

$$(1 - x^2)\frac{dP_m}{dx} = m\left(P_{m-1} - x\, P_m\right) = \frac{m(m+1)}{2m+1}\left(P_{m-1} - P_{m+1}\right) \quad (11.36)$$

$$\sin\phi\frac{dP_m}{d\phi} = m\left(\cos\phi\, P_m - P_{m-1}\right) = \frac{m(m+1)}{2m+1}\left(P_{m+1} - P_{m-1}\right) \quad (11.37)$$

11.3.5 Orthogonality condition

$$\int_{-1}^{+1} P_m(x)\, P_n(x)\, dx = \frac{2}{2m+1}\delta_{mn} \quad (11.38)$$

In its trigonometric form, with $x = \cos\phi$, the orthogonality condition is

$$\int_0^\pi P_m(\cos\phi)\, P_n(\cos\phi)\,\sin\phi\, d\phi = \frac{2}{2m+1}\delta_{mn} \quad (11.39)$$

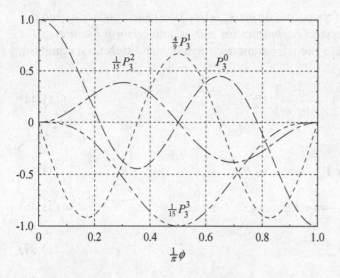

Figure 11.4: Associated Legendre functions.

11.3.6 Expansion in Legendre series

For any function $f(x)$ defined in the interval $-1 \le x \le 1$ that satisfies the so-called Dirichlet conditions (most practical functions do), one can express it in a Legendre series of the form

$$f(x) = \sum_{m=0}^{\infty} c_m P_m(x), \quad c_m = \left(m + \tfrac{1}{2}\right) \int_{-1}^{+1} f(x)\, P_m(x)\, dx \tag{11.40}$$

$$g(\phi) = \sum_{m=0}^{\infty} d_m P_m(\phi), \quad d_m = \left(m + \tfrac{1}{2}\right) \int_{0}^{\pi} g(\phi)\, P_m(\phi)\, \sin\phi\, d\phi \tag{11.41}$$

When the series is truncated at some fixed m, the resulting mth degree polynomial provides the best fit to $f(x)$ in the least squares sense.

11.4 Associated Legendre functions (spheroidal harmonics)

11.4.1 Differential equation

$$(1 - x^2)\frac{d^2 y}{dx^2} - 2x\frac{dy}{dx} + \left[m(m+1) - \frac{n^2}{1 - x^2}\right] y = 0$$

$$m = 0, 1, 2, \ldots, \quad n = 0, 1, 2, \ldots, \quad n \le m, \quad |x| \le 1 \tag{11.42}$$

The solution to this differential equation is $y = c_1\, P_m^n(x) + c_2\, Q_m^n(x)$, with c_1, c_2 arbitrary constants, and P_m^n, Q_m^n the associated Legendre functions or *spheroidal harmonics* of the first and second kind, respectively. The first of these functions is related to the Legendre polynomials as

$$P_m^n(x) = (-1)^n (1 - x^2)^{n/2} \frac{d^n P_m(x)}{dx^n} \tag{11.43}$$

Recall that P_m is a polynomial of order m, so it has no derivatives higher than the mth, which implies $n \le m$. Also, $P_m^0 = P_m$.

The related functions $\cos n\theta\ P_m^n(\cos\phi)$ and $\sin n\theta\ P_m^n(\cos\phi)$ are called *surface harmonics* of the first kind – in particular, *tesseral harmonics* for $m > n$, and *sectoral harmonics* for $m = n$. Setting $x = \cos\phi$, we obtain the trigonometric form of the differential equation, which is

$$\frac{d^2 P_m^n}{d\phi^2} + \cot\phi\frac{d P_m^n}{d\phi} + \left[m(m+1) - \frac{n^2}{\sin^2\phi}\right] P_m^n = 0 \tag{11.44}$$

11.4.2 Recurrence relations

$$(m - n + 1)P_{m+1}^n = (2m+1)\,x\,P_m^n - (m+n)\,P_{m-1}^n \tag{11.45}$$

$$(1 - x^2)\frac{d P_m^n}{dx} = (m+n)\,P_{m-1}^n - mx\,P_m^n \tag{11.46}$$

$$\frac{d P_m^n}{d\phi} = m\cot\phi\,P_m^n - \frac{m+n}{\sin\phi}\,P_{m-1}^n, \tag{11.47}$$

$$\frac{d P_m^n}{d\phi} = \frac{m-n+1}{\sin\phi}\,P_{m+1}^n - (m+1)\cot\phi\,P_m^n$$

List of spheroidal harmonics P_m^n up to third degree and order

$n\backslash m$	0	1	2	3
0	$P_0^0 = 1$	$P_1^0 = \cos\phi$	$P_2^0 = \frac{1}{4}(1+3\cos 2\phi)$	$P_3^0 = \frac{1}{8}(5\cos 3\phi + 3\cos\phi)$
1	—	$P_1^1 = -\sin\phi$	$P_2^1 = -\frac{3}{2}\sin 2\phi$	$P_3^1 = -\frac{3}{8}(\sin\phi + 5\sin 3\phi)$
2	—	—	$P_2^2 = \frac{3}{2}(1-\cos 2\phi)$	$P_3^2 = \frac{15}{4}(\cos\phi - \cos 3\phi)$
3	$-/+$	—	—	$P_3^3 = -\frac{15}{4}(3\sin\phi - \sin 3\phi)$

11.4.3 Orthogonality conditions

$$\int_{-1}^{+1} P_m^n(x)\,P_k^n(x)\,dx = \frac{(n+m)!}{(m+\frac{1}{2})(m-n)!}\delta_{mk} \qquad \text{(varying degree)} \tag{11.48}$$

$$\int_{-1}^{+1}(1-x^2)^{-1}P_m^n(x)\,P_m^k(x)\,dx = \begin{cases} \dfrac{(n+m)!}{n(m-n)!} & 0 < n = k \leq m \\ 0 & n \neq k \end{cases} \qquad \text{(varying order)} \tag{11.49}$$

In their trigonometric form, these orthogonality conditions are

$$\int_0^\pi P_m^n(\phi)\,P_k^n(\phi)\,\sin\phi\,d\phi = \frac{(n+m)!}{(m+\frac{1}{2})(m-n)!}\delta_{mk} \qquad \text{(varying degree)} \tag{11.50}$$

$$\int_0^\pi P_m^n(\phi)\,P_m^k(\phi)\,d\phi = \begin{cases} \dfrac{(n+m)!}{n(m-n)!} & 0 < n = k \leq m \\ 0 & n \neq k \end{cases} \qquad \text{(varying order)} \tag{11.51}$$

11.4.4 Orthogonality of co-latitude matrix

$$
\mathbf{L}_m^n = \left\{ \begin{array}{ccc} P_m^n & 0 & 0 \\[2mm] 0 & \dfrac{d P_m^n}{d\phi} & \dfrac{n\, P_m^n}{\sin \phi} \\[4mm] 0 & \dfrac{n\, P_m^n}{\sin \phi} & \dfrac{d P_m^n}{d\phi} \end{array} \right\}
\tag{11.52}
$$

(used in problems formulated in spherical coordinates).

$$
\int_0^\pi \mathbf{L}_m^n \, \mathbf{L}_k^n \, \sin \phi \, d\phi = \delta_{mk} \frac{(n+m)!}{(m+\frac{1}{2})\,(m-n)!} \, \mathrm{diag}\{ 1 \quad m(m+1) \quad m(m+1) \}
\tag{11.53}
$$

This is based on the integral

$$
\int_0^\pi \left[\frac{d P_m^n}{d\phi} \frac{d P_k^n}{d\phi} + \frac{n^2}{\sin^2 \phi} P_m^n P_k^n \right] \sin \phi \, d\phi = \frac{m(m+1)}{(m+\frac{1}{2})} \frac{(n+m)!}{(m-n)!} \delta_{mk}
\tag{11.54}
$$

which can be proved integrating the first term by parts, using the differential equation for the associated Legendre functions, and applying the orthogonality condition. Also,

$$
n \int_0^\pi \left[\frac{d P_m^n}{d\phi} P_k^n + P_m^n \frac{d P_k^n}{d\phi} \right] d\phi = n\, P_m^n P_k^n \big|_0^\pi = 0
\tag{11.55}
$$

11.4.5 Expansion of arbitrary function in spheroidal harmonics

$$
f(x) = \sum_{m=0}^\infty c_m P_m^n(x), \quad c_m = \left(m+\tfrac{1}{2}\right) \frac{(m-n)!}{(m+n)!} \int_{-1}^{+1} f(x)\, P_k^n(x)\, dx
\tag{11.56}
$$

$$
g(\phi) = \sum_{m=0}^\infty d_m P_m^n(\cos \phi), \quad d_m = \left(m+\tfrac{1}{2}\right) \frac{(m-n)!}{(m+n)!} \int_0^\pi g(\phi)\, P_k^n(\phi)\, \sin \phi \, d\phi
\tag{11.57}
$$

11.4.6 Leibniz rule for the derivative of a product of two functions

$$
\frac{d^n\,[f(x)\, g(x)]}{dx^n} = f g^{(n)} + \frac{n}{1} f' g^{(n-1)} + \frac{n(n-1)}{1 \times 2} f' g^{(n-1)}
$$
$$
+ \frac{n(n-1)(n-3)}{1 \times 2 \times 3} f'' g^{(n-2)} + \cdots + f^{(n)} g
\tag{11.58}
$$

This rule is often useful in the context of Legendre functions, for example when $f(x) = (1 - x^2)^k$. Observe that the coefficients are the same as those of the binomial expansion.

12 Brief listing of integral transforms

12.1 Fourier transforms

For any sufficiently well-behaved function $f(t)$ there exists a Fourier transform $F(\omega)$ such that

$$f(t) = \frac{1}{2\pi} \int_{-\infty}^{+\infty} F(\omega)\, e^{i\omega t}\, d\omega \quad \leftrightarrow \quad F(\omega) = \int_{-\infty}^{+\infty} f(t)\, e^{-i\omega t}\, dt$$

$$\frac{\partial^n f(t)}{\partial t^n} \quad \leftrightarrow \quad (i\omega)^n F(\omega)$$

$$t^n f(t) \quad \leftrightarrow \quad i^n \frac{\partial^n F(\omega)}{\partial \omega^n}$$

provided that $f(t)$ and/or $F(\omega)$ decay sufficiently fast with t or ω, that is, that they have a finite number of discontinuities and satisfy the so-called *Dirichlet condition* $\int_{-\infty}^{+\infty} |f(t)|\, dt \neq \infty$. Although the derivatives of such functions do not necessarily satisfy these conditions – and hence the integrals do not converge – the above rules can still be used to obtain formal transforms to the derivatives of f or F in the sense of distributions (or singularity functions). Such transforms are identified in the listings below with an asterisk.

a) Transforms in time–frequency

$$f(t) = \frac{1}{2\pi} \int_{-\infty}^{+\infty} F(\omega)\, e^{i\omega t}\, d\omega \quad \leftrightarrow \quad F(\omega) = \int_{-\infty}^{+\infty} f(t)\, e^{-i\omega t}\, dt$$

$\delta(t)$	$\underset{*}{\leftrightarrow}$	1
$\dfrac{\mathcal{H}(t - t_S)}{\sqrt{t^2 - t_S^2}}$	$\leftrightarrow$	$-\dfrac{\pi\, i}{2} H_0^{(2)}(\omega t_S)$
$\sqrt{t^2 - t_S^2}\; \mathcal{H}(t - t_S)$	$\underset{*}{\leftrightarrow}$	$\dfrac{\pi\, i}{2} \dfrac{t_S}{\omega} H_1^{(2)}(\omega t_S)$
$\dfrac{\mathcal{H}(t - t_S)}{\left(t^2 - t_S^2\right)^{3/2}}$	$\underset{*}{\leftrightarrow}$	$\dfrac{\pi\, i}{2} \dfrac{\omega}{t_S} H_1^{(2)}(\omega t_S)$
$\dfrac{t}{t_S^2} \left[\mathcal{H}(t - t_P) - \mathcal{H}(t - t_S) \right]$	$\leftrightarrow$	$e^{-i\omega t} \dfrac{1 + i\omega t}{\omega t_S^2} \Big|_{t_P}^{t_S}$

b) Wavenumber–space integrals

From Gradshteyn and Ryzhik (Chapter 11, footnote 1), p. 1151, eq. 17.34.4, we have

$$\int_0^{+\infty} \frac{\cos kx \, dx}{(x^2 + a^2)^{\nu + 1/2}} = \sqrt{\pi} \left(\frac{k}{2a}\right)^\nu \frac{K_\nu(ka)}{\Gamma(\nu + \frac{1}{2})}, \qquad \mathrm{Re}\,\nu > -\tfrac{1}{2}, \quad \mathrm{Re}\,a > 0, \quad k > 0$$

We use this expression to derive some of the results given below, using symmetry and antisymmetry considerations, together with the relations

$$K_\nu(z) = \begin{cases} \frac{1}{2}\pi \mathrm{i}^{\nu+1} H_\nu^{(1)}(\mathrm{i}z) & -\pi < \arg z \le \frac{1}{2}\pi \\ \frac{1}{2}\pi(-\mathrm{i})^{\nu+1} H_\nu^{(2)}(-\mathrm{i}z) & -\frac{1}{2}\pi < \arg z \le \pi \end{cases} \qquad \text{(Abramovitz \& Stegun, eq. 9.6.4b)}$$

and

$$\Gamma(n + \tfrac{1}{2}) = \frac{1 \times 3 \times 5 \times \cdots \times (2n-1)}{2^n} \Gamma(\tfrac{1}{2}) = \frac{(2n)!}{2^{2n} n!} \sqrt{\pi} \quad \text{(Abramovitz \& Stegun, eq. 6.1.12)}$$

Starting from the above, we obtain

$f(x) = \dfrac{1}{2\pi} \displaystyle\int_{-\infty}^{+\infty} F(k)\, e^{\mp ikx} dk$	$\leftrightarrow$	$F(k) = \displaystyle\int_{-\infty}^{+\infty} f(x)\, e^{\pm ikx} dx$	
$\dfrac{1}{(x^2 + a^2)^{\nu + 1/2}}$	$\leftrightarrow$	$2\sqrt{\pi} \left(\dfrac{|k|}{2a}\right)^\nu \dfrac{K_\nu(|k|a)}{\Gamma(\nu + \frac{1}{2})}$	$\mathrm{Re}\,a > 0, \mathrm{Re}\,\nu > -\frac{1}{2}$
$\dfrac{1}{(x^2 + a^2)^{n + 1/2}}$	$\leftrightarrow$	$\dfrac{2^{n+1} n!}{(2n)!} \left(\dfrac{|k|}{a}\right)^n K_n(|k|a)$	$\mathrm{Re}\,a > 0$
$\dfrac{1}{\sqrt{x^2 + a^2}}$	$\leftrightarrow$	$2 K_0(|k|a)$	$\mathrm{Re}\,a > 0$
$\dfrac{1}{2\mathrm{i}} \dfrac{\Gamma(\frac{1}{2})}{\Gamma(\nu + \frac{1}{2})} \left(\dfrac{-|x|}{2k_0}\right)^\nu H_\nu^{(2)}(k_0|x|)$	$\leftrightarrow$	$\dfrac{1}{(k^2 - k_0^2)^{\nu + 1/2}}$	$\mathrm{Im}\,k_0 < 0, \mathrm{Re}\,\nu > -\frac{1}{2}$
$\dfrac{1}{2\mathrm{i}} \dfrac{2^n n!}{(2n)!} \left(\dfrac{-|x|}{k_0}\right)^n H_n^{(2)}(k_0|x|)$	$\leftrightarrow$	$\dfrac{1}{(k^2 - k_0^2)^{n + 1/2}}$	$\mathrm{Im}\,k_0 < 0$
$\dfrac{1}{2\mathrm{i}} H_0^{(2)}(k_0|x|)$	$\leftrightarrow$	$\dfrac{1}{\sqrt{k^2 - k_0^2}}$	$\mathrm{Im}\,k_0 < 0$

Additional formulas may be obtained by contour integration and derivation under the integral sign with respect to parameter. Integrals that violate the Dirichlet conditions are valid only in the sense of distributions. We assume throughout that $\mathrm{Im}(k_0) \le 0$. Integrals for $\mathrm{Im}(k_0) > 0$ can be obtained by conjugation. Results for $\mathrm{Im}(k_0) = 0$ are principal value (PV).

$$\frac{1}{2\pi} \int_{-\infty}^{\infty} \frac{e^{\pm ikx}}{k^2 - k_0^2} \, dk = \begin{cases} -\dfrac{\mathrm{i}}{2k_0} e^{-\mathrm{i}k_0|x|} & \mathrm{Im}\,k_0 < 0 \\ -\dfrac{1}{2k_0} \sin k_0|x| & \mathrm{Im}\,k_0 = 0 \end{cases}$$

$$\frac{1}{2\pi} \int_{-\infty}^{\infty} \frac{k e^{\pm ikx}}{k^2 - k_0^2} \, dk = \begin{cases} \pm \dfrac{\mathrm{i}}{2} e^{-\mathrm{i}k_0|x|} \mathrm{sgn}(x) & \mathrm{Im}\,k_0 < 0 \\ \pm \dfrac{\mathrm{i}}{2} \cos k_0 x \, \mathrm{sgn}(x) & \mathrm{Im}\,k_0 = 0 \end{cases}$$

$$\frac{1}{2\pi} \int_{-\infty}^{\infty} \frac{e^{\pm ikx}}{k\left(k^2 - k_0^2\right)}\, dk = \begin{cases} \pm \dfrac{i}{2k_0^2}\left(e^{-ik_0|x|} - 2\right)\mathrm{sgn}(x) & \text{Im}\, k_0 < 0 \\[3mm] \pm \dfrac{i}{2k_0^2}\left(\cos k_0 x - 2\right)\mathrm{sgn}(x) & \text{Im}\, k_0 = 0 \end{cases}$$

$$\frac{1}{2\pi} \int_{-\infty}^{\infty} \frac{e^{\pm ikx}}{\left(k^2 - k_0^2\right)^2}\, dk = \begin{cases} -\dfrac{1}{4k_0^3}\left(k_0\,|x| - i\right)e^{-ik_0|x|} & \text{Im}\, k_0 < 0 \\[3mm] -\dfrac{1}{4k_0^2}\,|x|\,\cos k_0 x & \text{Im}\, k_0 = 0 \end{cases}$$

$$\frac{1}{2\pi} \int_{-\infty}^{\infty} \frac{k\,e^{\pm ikx}}{\left(k^2 - k_0^2\right)^2}\, dk = \begin{cases} \pm \dfrac{x}{4k_0}\,e^{-ik_0|x|} & \text{Im}\, k_0 < 0 \\[3mm] \pm \dfrac{i}{4k_0^2}\left(\cos k_0 x - k_0 x\,\sin k_0 x\right)\mathrm{sgn}\, x & \text{Im}\, k_0 = 0 \end{cases}$$

$$\frac{1}{2\pi} \int_{-\infty}^{\infty} \frac{k^2\,e^{\pm ikx}}{\left(k^2 - k_0^2\right)^2}\, dk = \begin{cases} \dfrac{1}{4ik_0}\left(1 - ik_0\,|x|\right)e^{-ik_0|x|} & \text{Im}\, k_0 < 0 \\[3mm] \dfrac{|x|}{4}\,\cos k_0 x & \text{Im}\, k_0 = 0 \end{cases}$$

$$\frac{1}{2\pi} \int_{-\infty}^{\infty} \frac{e^{\pm ikx}}{\left(k^2 - k_0^2\right)^{1/2}}\, dk = \begin{cases} \dfrac{1}{2i}\,H_0^{(2)}(k_0\,|x|) & \text{Im}\, k_0 < 0 \\[3mm] \dfrac{1}{i}\,J_0(k_0\,|x|) & \text{Im}\, k_0 = 0 \end{cases}$$

$$\frac{1}{2\pi} \int_{-\infty}^{\infty} \frac{k\,e^{\pm ikx}}{\left(k^2 - k_0^2\right)^{1/2}}\, dk = \begin{cases} \pm \tfrac{1}{2}k_0\,H_1^{(2)}(k_0\,|x|)\,\mathrm{sgn}(x) & \text{Im}\, k_0 < 0 \\[3mm] \pm k_0\,J_1(k_0\,|x|)\,\mathrm{sgn}(x) & \text{Im}\, k_0 = 0 \end{cases}$$

$$\frac{1}{2\pi} \int_{-\infty}^{\infty} \frac{e^{\pm ikx}}{\left(k^2 - k_0^2\right)^{3/2}}\, dk = \begin{cases} \dfrac{i\,|x|}{2k_0}\,H_1^{(2)}(k_0\,|x|) & \text{Im}\, k_0 < 0 \\[3mm] \dfrac{i\,|x|}{k_0}\,J_1(k_0\,|x|) & \text{Im}\, k_0 = 0 \end{cases}$$

$$\frac{1}{2\pi} \int_{-\infty}^{\infty} \frac{k\,e^{\pm ikx}}{\left(k^2 - k_0^2\right)^{3/2}}\, dk = \begin{cases} \pm \tfrac{1}{2}x\,H_0^{(2)}(k_0\,|x|) & \text{Im}\, k_0 < 0 \\[3mm] \pm x\,J_0(k_0\,|x|) & \text{Im}\, k_0 = 0 \end{cases}$$

12.2 Hankel transforms

The Hankel transform is defined by the integral

$$F_n(k) = \int_0^{\infty} r^{\nu}\, f(r)\, J_n(kr)\, dr$$

in which ν is defined in various references as either $1/2$ or 1. Since the Bessel function behaves asymptotically as a harmonic function while its amplitude decays as $r^{-1/2}$, the Hankel transform exists only if the integral $\int_0^{\infty} \left| r^{\nu - 1/2} f(r)\right|\, dr$ exists (i.e., if the *Dirichlet condition* is satisfied), in which case the transform is self-reciprocating, namely

$$f(r) = \int_0^\infty k^\nu \, F_n(k) \, J_n(kr) \, dk$$

We list below some Hankel transforms using $\nu = 1$.

$f(r) = \int_0^\infty F_n(k) \, J_n(kr) \, k \, dk$		$F_n(k) = \int_0^\infty f(r) \, J_n(kr) \, r \, dr$
$\dfrac{\delta(r)}{r}$	$\underset{n=0}{\longleftrightarrow}$	1
$\dfrac{\delta(r-a)}{r}$	$\underset{n}{\longleftrightarrow}$	$J_n(ka)$
$e^{-\alpha r^2}$	$\underset{n=0}{\longleftrightarrow}$	$\dfrac{1}{2\alpha} e^{-k^2/4\alpha}$

$$\int_0^\infty \frac{k}{k^2 - k_0^2} J_\nu(kR) J_\nu(kr) \, dk = \begin{cases} \dfrac{\pi}{2i} J_\nu(k_0 r) \, H_\nu^{(2)}(k_0 R), & r \le R \\[2ex] \dfrac{\pi}{2i} J_\nu(k_0 R) \, H_\nu^{(2)}(k_0 r), & R \le r \end{cases} \qquad \text{Im}(k_0) < 0$$

$$\int_0^\infty \frac{1}{k^2 - k_0^2} J_1(kR) J_0(kr) \, dk = \begin{cases} \dfrac{\pi}{2i k_0} J_0(k_0 r) \, H_1^{(2)}(k_0 R) - \dfrac{1}{Rk_0^2}, & r \le R \\[2ex] \dfrac{\pi}{2i k_0} J_1(k_0 R) \, H_0^{(2)}(k_0 r), & R \le r \end{cases} \qquad \text{Im}(k_0) < 0$$

$$\int_0^\infty \frac{1}{k(k^2 - k_0^2)} J_1(kR) J_1(kr) \, dk = \begin{cases} \dfrac{\pi}{2i k_0^2} J_1(k_0 r) \, H_1^{(2)}(k_0 R) - \dfrac{r}{2Rk_0^2}, & r \le R \\[2ex] \dfrac{\pi}{2i k_0^2} J_1(k_0 R) \, H_1^{(2)}(k_0 r) - \dfrac{R}{2r \, k_0^2}, & R \le r \end{cases} \qquad \text{Im}(k_0) < 0$$

12.3 Spherical Hankel transforms

Starting from the expression $j_n(kr) = \sqrt{\pi/2kr} \, J_{n+1/2}(kr)$, which relates the spherical Bessel function of the first kind to the half-integer order conventional Bessel function, we conclude that an appropriate spherical Hankel transform is

$$f(r) = \int_0^\infty k^{\nu+\frac{1}{2}} F_n(k) \, j_n(kr) \, dk \quad \Leftrightarrow \quad F_n(k) = \int_0^\infty r^{\nu+\frac{1}{2}} f(r) \, j_n(kr) \, dr$$

provided that the integrals $\int_0^\infty |r^\nu f(r)| \, dr$ and/or $\int_0^\infty |k^\nu F(k)| \, dk$ exist. A convenient value is $\nu = 1/2$, in which case

$$f(r) = \int_0^\infty k \, F_n(k) \, j_n(kr) \, dk \quad \Leftrightarrow \quad F_n(k) = \int_0^\infty r \, f(r) \, j_n(kr) \, dr$$

13 MATLAB programs

The following Matlab programs are listed herein:

SH2D_FULL	SH line load in full space
SVP2D_FULL	SVP line load in full space
BLAST2D	Blast line source in full space
CAVITY2D	2-D cavity in full space
POINT_FULL	Point source in full 3-D space
TORSION_FULL	Torsional point source in full space
CAVITY3D	3-D cavity in full space
SH2D_HALF	SH line source in half-space
GARVIN	Blast line source in half-space
LAMB2D	SVP line source in half-space
LAMB3D	Point source in half-space (horizontal and vertical)
TORSION_HALF	Torsional point source in half-space
SH_PLATE	SH line load in homogeneous plate
SH_STRATUM	SH line source in homogeneous stratum
SVP_PLATE	SVP line source in plate with mixed boundary conditions
SPHEROIDAL	Spheroidal and torsional modes of homogeneous sphere

In addition, the following functions are called by some of the programs

EASYPLOT	Writes text files for plotting (most programs)
CISIB	Sine and cosine integrals (cavity 2-D and 3-D)
ELLIPINT3	Elliptic integrals (Lamb 3-D)

Note: Some programs have same-named functions embedded within, but they are not identical to each other, even if functionally related. Thus, they are not interchangeable.

```
function [] = SH2D_Full()

% SH line source in a full, homogeneous 2D space

% Default data
cs = 1;                   % Shear wave velocity
rho = 1                   % Mass density
np = 200;                 % Number of time and/or frequency intervals
tmax = 10;               % Maximum dimensionless time for plotting
wmax = 10;               % maximum dimensionless frequency for plotting (rad/s)
mu = rho*cs^2;          % Shear modulus
dw = wmax/np;
dt = tmax/np;
W = [dw:dw:wmax];
G = -i*0.25*besselh(0,2,W)/mu;

tit = 'Full-space displacement due to SH line load';
titx = 'Dimensionless frequency w*r/Cs = kr';
plot (W,real(G));
hold on;
plot (W,imag(G),'r');
grid on;
title(tit);
xlabel(titx);
pause;
hold off;

% Frequency domain
W2 = W.^0.5;
W = [0, W];
G = [0, G.*W2];
plot (W,real(G));
hold on;
plot (W,imag(G),'r');
tit = 'Full-space displacement due to SH line load * sqrt(kr)';
title(tit);
xlabel(titx);
grid on;
pause;
hold off;
EasyPlot('SH1-FD.ezp', tit, titx, W, G, 'c');

% Time domain
T1 = [0 1 1];
T2 = [1+dt:dt:tmax];
Uyy = 2*pi*mu*sqrt(T2.^2-1);
Uyy = [0 0 inf 1./Uyy];
Uyy(3) = Uyy(4);   % avoid singularity and discontinuity in plot
T = [T1, T2];
plot (T, Uyy);
```

```
tit = 'Full-space displacement due to SH line load';
title(tit);
titx = 'Dimensionless time, t*Cs/r';
xlabel(titx);
grid on;
pause;
EasyPlot('SH1-TD.ezp', tit, titx, T, Uyy, 'r');
close;
return
```

```
function [] = SVP2D_Full(x, z, poi)

  % SVP line source in a full 2D space
  % Arguments:
  %       x, z = coordinates of receiver
  %       pois = Poisson's ratio
  % Note: Uxz = Uzx, Gxz = Gzx
  % To obtain the components in cylindrical coordinates,
  % simply run with x=1,z=0, and then x=0,z=1

  % Default data
  rho = 1;        % mass density
  cs = 1;         % shear wave velocity
  nt = 500;       % number of time points
  tmax = 5;       % maximum time
  nf = 500;       % number of frequency points
  wmax = 20;      % max. frequency

  mu = rho*cs^2;    % Shear modulus
  alfa = sqrt((1-2*poi)/(2-2*poi));       % cs/cp
  r = sqrt(x^2+z^2);
  theta = atan2(z, x);

  % Time domain solution
  dt = tmax/nt;
  T = [0:nt]*dt;
  T2 = T.^2;
  U = impulse_response(r, theta, cs, mu, alfa, T2);
  T = T*cs/r;    % dimensionless time
  plot (T, U(1,:));
  tit = sprintf(...
    'Uxx at x=%5.3f, z=%5.3f due to SVP line load in full space', x, z);
  titx = 'Dimensionless time, t*Cs/r';
  grid on;
  title(tit);
  xlabel(titx);
  EasyPlot('SVP_TDxx.ezp', tit, titx, T, U(1,:), 'r');
  pause;
  plot (T, U(2,:));
  tit = sprintf(...
    'Uxz at x=%5.3f, z=%5.3f due to SVP line load in full space', x, z);
  grid on;
  title(tit);
  xlabel(titx);
  EasyPlot('SVP_TDxz.ezp', tit, titx, T, U(2,:), 'r');
  pause;
  plot (T, U(3,:));
  tit = sprintf(...
    'Uzz at x=%5.3f, z=%5.3f due to SVP line load in full space', x, z);
  grid on;
```

```
    title(tit);
    xlabel(titx);
    EasyPlot('SVP_TDzz.ezp', tit, titx, T, U(3,:), 'r');
    pause;

    % Frequency domain solution
    dw = wmax/nf;
    W = [dw:dw:wmax];
    H = Green(r, theta, cs, mu, alfa, W);
    W = W*r/cs;   % dimensionless frequency
    plot (W, real(H(1,:)));
    hold on;
    plot (W, imag(H(1,:)),'r');
    tit = sprintf(...
       'Gxx at x=%5.3f, z=%5.3f due to SVP line load in full space', x, z);
    titx = 'Dimensionless frequency w*r/Cs = k*r';
    grid on;
    title(tit);
    xlabel(titx);
    EasyPlot('SVP_FDxx.ezp', tit, titx, W, H(1,:), 'c');
    pause;
    hold off;
    plot (W, real(H(2,:)));
    hold on;
    plot (W, imag(H(2,:)),'r');
    tit = sprintf(...
       'Gxz at x=%5.3f, z=%5.3f due to SVP line load in full space', x, z);
    grid on;
    title(tit);
    xlabel(titx);
    EasyPlot('SVP_FDxz.ezp', tit, titx, W, H(2,:), 'c');
    pause;
    hold off;
    plot (W, real(H(3,:)));
    hold on;
    plot (W, imag(H(3,:)),'r');
    tit = sprintf(...
       'Gzz at x=%5.3f, z=%5.3f due to SVP line load in full space', x, z);
    grid on;
    title(tit);
    xlabel(titx);
    EasyPlot('SVP_FDzz.ezp', tit, titx, W, H(3,:), 'c');
    pause;
    hold off;
    close all;
    return

    function [G] = Green(r, theta, cs, mu, alfa, W)
    % Green's functions for SVP line load in full space
    Ws = W*r/cs;
```

```
    Wp = alfa*Ws;
    psi = (besselh(1,2,Ws)-alfa*besselh(1,2,Wp))./Ws - besselh(0,2,Ws);
    chi = alfa^2*besselh(2,2,Wp)- besselh(2,2,Ws);
    c = cos(theta);
    s = sin(theta);
    f = 0.25*i/mu;
    G = f*[psi+chi*c^2; chi*c*s; psi+chi*s^2];   % [gxx; gxz=gzx; gzz]
    % G = f*[psi+chi; psi];   % radial & tangential components
    % Radial changes as cos(theta), tangential as -sin(theta)
    return

    function [U] = impulse_response(r, theta, cs, mu, alfa, T2)
    % Impulse response functions for SVP line load in full space
    ts = r/cs;
    tp = alfa*ts;
    ts2 = ts^2;
    tp2 = tp^2;
    k = find(tp2<T2);
    if isempty(k)
      PSI = zeros(size(T2));
      CHI = PSI;
    else
      k = k(1);
      k1 = k-1;
      n = length(T2);
      S = sqrt((T2(k:n)-tp2));
      PSI = [zeros(1,k1),-S/ts2];   % yes, we divide by ts2, not tp2
      CHI = [zeros(1,k1),alfa^2./S]-2*PSI;
      if tp2==T2(k1), CHI(k1)=inf; end
      k = find(ts2<T2);
      if ~isempty(k)
        k = k(1);
        k1 = k-1;
        n = length(T2);
        S = sqrt((T2(k:n)-ts2));
        PSI = PSI + [zeros(1,k1),S/ts2+1./S];
        CHI = CHI - [zeros(1,k1),2*S/ts2+1./S];
        if tp2==T2(k1), PSI(k1)=inf; CHI(k1)=inf; end
        end
    end
    c = cos(theta);
    s = sin(theta);
    f = 0.5/pi/mu;
    U = f*[PSI+CHI*c^2; CHI*c*s; PSI+CHI*s^2];   % [uxx; uxz=uzx; uzz]
    % U = f*[PSI+CHI; PSI];   % radial & tangential components
    % Radial changes as cos(theta), tangential as -sin(theta)
    return
```

```
function [] = Blast2D(pois)

    % Computes the plane-strain response elicited by an
    % SV-P line blast source in a full, homogeneous 2D space
    % Arguments:
    %   pois = Poisson's ratio

    % Default data
    nt = 300;       % Number of time intervals
    nf = 200;       % Number of frequency intervals
    cs = 1;         % Shear wave velocity
    rho = 1;        % Mass density
    r = 1;          % epicentral distance
    tmax = 5;       % Maximum time for plotting
    wmax = 20;      % maximum frequency for plotting (rad/s)

    mu = rho*cs^2;       % Shear modulus
    cp = cs*sqrt((2-2*pois)/(1-2*pois)); % P-wave velocity
    a = cs/cp;
    a2 = a^2;

    % Frequency domain
    f = 0.25/mu/cp;
    dw = wmax/nf;
    Ws = [dw:dw:wmax]*r/cs;   % dimensionless frequency for S waves
    Wp = a*Ws;          % dimensionless frequency for P waves
    H1p = besselh(1,2,Wp);
    Gr = f*Wp.*H1p;     % Green's function for radial displacement
    tit = 'Radial displacement in full space due to line blast load';
    plot (Ws,real(Gr));
    hold on;
    plot (Ws,imag(Gr),'r');
    grid on;
    title(tit);
    titx = 'Dimensionless frequency w*r/Cs = k*r';
    xlabel(titx);
    pause;
    hold off;
    EasyPlot('B2D_FDr.ezp', tit, titx, Ws, Gr, 'c');
    clear G* H*

    % Time domain
    f = 1/(2*pi*mu*r);
    tp = r/cp;      % arrival time of P waves
    ts = r/cs;      % arrival time of SV waves
    ts2 = ts^2;
    tp2 = tp^2;
    dt = tmax/nt;
    T = [0:dt:tmax]+eps;
    P = sqrt(T.^2-tp2);
```

```
Ur = f*T.*real(1./P);
T = T*cs/r;      % dimensionless time
plot (T, Ur);
tit = 'Radial displacement in full space due to line blast load';
title(tit);
titx = 'Dimensionless time, t*Cs/r';
xlabel(titx);
grid on;
pause;
EasyPlot('B2D_TDr.ezp', tit, titx, T, Ur, 'r');
close;
return
```

```
function [] = Cavity2D(r, r0, pois)

% Cylindrical (2-D) cavity in a full, homogeneous space
% subjected to a harmonic/impulsive pressure
% The pressure is scaled so that p*pi*r0^2=1
%
% Arguments (in consistent units)
%   r   = distance to receiver
%   r0   = radius of cavity
%   pois = Poisson's ratio
%
% Returned transfer functions are   g*mu*r    vs. w*r/Cp
% Returned time histories are        u*rho*r^2 vs. t*Cp/r
% in which mu is the shear modulus, rho is the mass density,
% and Cp is the P-wave velocity

% Basic data
Cs = 1;   % S-wave velocity (the results do not depend on this value)
a2 = (2-2*pois)/(1-2*pois);
a = sqrt(a2);      % Cp/Cs
Cp = a*Cs;       % P-wave velocity


rr = r0/r;
nf = 1024;
wmax = 2*pi*Cp/r0;
dw = wmax/nf;
w = [dw:dw:wmax]; % true frequency
w0 = w*r0/Cp;
w1 = w*r/Cp;       % dimensionless frequency

wmax = wmax*r/Cp; % max. dimensionless freq.
dw = dw*r/Cp;   % dimensionless freq. step

% Frequency domain solution
f1 = r/r0/pi;       % factor for cavity
f2 = 0.25/i;       % factor for line of pressure
h0 = besselh(0,2,w0);
h1 = besselh(1,2,w0);
y = 2*h1-a2*w0.*h0;
h1 = besselh(1,2,w1);
H = [0.5/pi, f1*h1./y];  % transfer function for cylindrical cavity
G = [0.5/pi, f2*w1.*h1]; % transfer function for line of pressure
w1 = [0, w1];       % frequency vector
plot(w1,real(H));
hold on;
plot(w1,imag(H),'r');
%plot(w1,real(G),':');  % Compare with TF for line of pressure
%plot(w1,imag(G),'r:');
tit = sprintf( ...
'T. F. for pressure in cylindrical cavity, r0=%5.4f, \\nu=%5.2f', rr, pois);
```

```
titx = 'Frequency (rad/s)';
title (tit);
xlabel(titx);
grid on;
%axis([0 wmax -1 1]);
pause;
hold off;
close;
EasyPlot('Cav2D-FD.ezp', tit, titx, w1, [H;G], 'c');

% Response at receiver in the time domain
t0 = 1-rr;              % dimensionless arrival time
H = H.*exp(i*w1*t0);    % delay TF by arrival time
tmax = 2;               % maximum dimensionless (delayed) time
nt = 100;               % No. of time steps
dt = tmax/nt;           % time step
T = [0:nt]*dt;          % dimensionless, delayed time axis
U = zeros(size(T));
for j=1:length(T)  % Fourier transform by direct integration (no FFT here)
  t = T(j);             % Advantage: can use arbitrary size & No. of time steps
  U(j) = sum(real(H.*exp(i*w1*t)));
end
U = U*dw*a/pi;

f = 2/pi;
fac = 0.5/a/pi/rr^1.5;
u = fac*(1-f*cisib(wmax*T));   % tail
U = U+u;
T = [0 t0 T+t0];    % shift delayed time axis by wave arrival time
U = [0 0 U];
plot(T,U);       % plot response
grid on;
tit = sprintf(...
'Cylindrical cavity under pressure, response at r/r0=%5.2f, \\nu=%5.2f',...
  1/rr, pois);
title(tit);
titx = 'Time (s)';
xlabel(titx);
pause;
close;
EasyPlot('Cav2D-TD.ezp', tit, titx, T, U, 'r');
return
```

```
function [] = Point_Full(pois)

% Computes the response elicited by a point load
% in a full, 3D homogeneous space
% Input arguments:
%   pois = Poisson's ratio

% Basic data
nt = 400;          % Number of time intervals
nf = 200;          % Number of frequency intervals
cs = 1;            % Shear wave velocity
rho = 1;           % Mass density
mu = rho*cs^2;     % Shear modulus
r = 1;             % epicentral distance
tmax = 2;          % Maximum dimensionless time for plotting
wmax = 20;         % maximum dimensionless frequency for plotting (rad/s)

% Frequency domain
dw = wmax/nf;
dt = tmax/nt;
cp = cs*sqrt((2-2*pois)/(1-2*pois));   % P-wave velocity
a = cs/cp;
a2 = a^2;

Ws = [dw:dw:wmax];   % Dimensionless frequency for S waves
Wp = a*Ws;           % Dimensionless frequency for P waves
Es = exp(-i*Ws);
Ep = a2*exp(-i*Wp);

Fp = Ep.*(i+1./Wp)./Wp;
Fs = Es.*(i+1./Ws)./Ws;

% Displacements in spherical coordinates due to point load in x direction
PSI = Es-Fs+Fp;          % Rigid-body component
CHI = Ep-Es-3*(Fp-Fs);   % Distortional component
clear E* F*

tit = sprintf('\\psi in full 3D space, \\nu=%5.2f',pois);
plot (Ws,real(PSI));
hold on;
plot (Ws,imag(PSI),'r');
grid on;
title(tit);
titx = 'Dimensionless frequency w*r/Cs';
xlabel(titx);
ylabel('\psi');
pause;
EasyPlot('PSI-3D-FD.ezp', tit, titx, Ws, PSI, 'c');
hold off;
tit = sprintf('\\chi in full 3D space, \\nu=%5.2f',pois);
plot (Ws,real(CHI));
```

```
hold on;
plot (Ws,imag(CHI),'r');
grid on;
title(tit);
titx = 'Dimensionless frequency w*r/Cs';
xlabel(titx);
ylabel('\chi');
hold off;
EasyPlot('CHI-3D-FD.ezp', tit, titx, Ws, CHI, 'c');
pause;

% Time domain
clear G*
tp = r/cp;          % arrival time of P waves
ts = r/cs;          % arrival time of SV waves
ts2 = ts^2;
T = [0:dt:tmax];
Tp = T.*(T>=tp)/ts2;
Ts = T.*(T>=ts)/ts2;
PSI = Ts-Tp;
CHI = -3*PSI;
jp = ccil(tp/dt)+1;
js = floor(ts/dt);
amax = 10*max(abs(PSI));   % to simulate a Dirac delta
mm = length(PSI);
PSI = [0,0,0,PSI(jp),PSI(jp:js),PSI(js),amax,0,0];
CHI = [0,0,amax,CHI(jp),CHI(jp:js),CHI(js),- amax,0,0];
tp1 = 0.5*(T(jp-1)+T(jp));
ts1 = 0.5*(T(js)+T(js+1));
T = [0,tp1,tp1,tp1,T(jp:js),ts1,ts1,ts1,tmax];
plot (T, PSI);
tit = sprintf('\\psi in full 3D space, \\nu=%5.2f',pois);
title(tit);
titx = 'Dimensionless time, t*Cs/r';
xlabel(titx);
grid on;
axis([0 tmax -1.5 0.5]);
EasyPlot('PSI-3D-TD.ezp', tit, titx, T, PSI, 'r');
pause;

plot (T, CHI);
tit = sprintf('\\chi in full 3D space, \\nu=%5.2f',pois);
title(tit);
titx = 'Dimensionless time, t*Cs/r';
xlabel(titx);
axis([0 tmax -0.5 5]);
grid on;
EasyPlot('CHI-3D-TD.ezp', tit, titx, T, CHI, 'r');
pause;
close;
return
```

```
function Torsion_Full(x,y,z,td,rho,cs)
  % Computes the response of a full 3D space to a torsional
  % point source with time variation (pi/2/td)*sin(pi*t/td), 0<=t<=td
  % (i.e. a sine pulse with unit area, or unit impulse)
  % The torsional moment is applied at the origin and has vertical axis
  % Arguments
  %   x, y, z      = coordinates of receiver
  %   td           = duration of pulse
  %   rho          = mass density
  %   cs           = shear wave velocity

  mu = rho*cs^2;       % shear modulus
  nt = 200;            % Number of points
  dt = td/nt;          % time step
  r = sqrt(x^2 + y^2);
  R = sqrt (x^2 + y^2 + z^2);
  phi = r/R;           % cos of vertical angle
  ap = pi/2/td;        % amplitude of sine pulse
      ap = 1;
  A = ap*phi/8/pi/mu;  % Scaling factor for response
  ta = R/cs;           % arrival time
  ang = pi/td;
  a = A/R^2;
  b = ang*A/cs/R;
  T = [0:dt:td];
  tau = ang*T;
  v = a*sin(tau)+b*cos(tau);
  T = [0,ta,T+ta,ta+td,ta+2*td];
  v = [0,0,v,0,0];
  % plot actual response vs. actual time
  plot(T,v)
  tit = sprintf(...
  'Response at (x,y,z)=(%5.2f,%5.2f,%5.2f) due to torsional sine pulse',...
  x,y,z);
  xlabel('Time');
  title(tit)
  grid on;
  pause;
  % plot dimensionless response & time
  tau = T*cs/R;
  v = (mu*R^2)*v;
  plot(tau,v);
  grid on;
  tit = sprintf(...
  'Response at (x/R,y/R,z/R)=(%5.2f,%5.2f,%5.2f) due to torsional sine
  pulse',... x/R, y/R, z/R);
  ylabel('v*mu*R^2');
  xlabel('Dimensionless time t*Cs/R');
  title(tit);
```

```
pause;
close
% Make file for plotting with EasyPlot
fout = fopen ('tors.ezp','w');
fprintf (fout, '/et g "%s"\n', tit);
fprintf (fout, '/et x "t*Cs/R"\n');     .
fprintf (fout, '/og on\n');
fprintf (fout, '/sd off\n');
fprintf (fout, '/sm off\n');
fprintf (fout, '%15.5e%15.5e\n', [tau;v]);
fclose (fout);
return
```

```
function [] = Cavity3D(r, pois)

% Spherical (3-D) cavity in a full, homogeneous space
% subjected to a harmonic/impulsive pressure p
% Arguments:
%    r = R/R0 = (distance to receiver)/(radius of cavity)
%    pois = Poisson's ratio
%
% This program assumes p0=1 (unit pressure), and r0=1 (radius of cavity)

% Basic parameters
a2 = (2-2*pois)/(1-2*pois);
a = sqrt(a2);         % Cp/Cs
R0 = 1;               % Radius of cavity
R = r*R0;             % Receiver
rho = 1;              % mass density
Cs = 1;               % S-wave velocity
Cp = a*Cs;            % P-wave velocity
mu = rho*Cs^2;        % shear modulus

% Frequency and time vectors
nf = 1024;
wmax = 16*pi;
dw = wmax/nf;
w = [dw:dw:wmax];
tmax = 6;             % maximum dimensionless time
nt = 200;             % No. of time steps
dt = tmax/nt;         % time step
T = [0:nt]*dt;        % dimensionless, delayed time (= t*Cp/R)

% Response in frequency domain
fac = 0.25*R0/mu;   % assuming p=1 here
w0 = w*R0/Cp;
w1 = w0*r;
G0 = 1+i*w0;
G1 = 1+i*w1;
G = fac*(G0./(G0-0.25*a2*w0.^2));   % Cavity wall at R0
plot(w,real(G));
hold on;
plot(w,imag(G),'r');
grid on;
tit = sprintf(...
   'T.F. for pressure in spherical cavity, R/R0=%5.4f, \\nu=%5.2f',...
     r, pois);
titx = 'Frequency (rad/s)';
title (tit);
xlabel(titx);
H = G.*exp(-i*(r-1)*w0).*G1./G0/r^2;   %Receiver at R
plot(w,real(H),':');
plot(w,imag(H),'r:');
```

```
grid on;
hold off;
pause;
fout = fopen('Cav3D-FD.ezp', 'w');
fprintf (fout, '/et g "%s"\n', tit);
fprintf (fout, '/et x "%s"\n', titx);
fprintf (fout, '/og on\n');
f0 = 0.75*mu/R0/pi;   % factor for unit strength pressure
f1 = f0*r^2;
fprintf (fout, '%15.5f %15.5f %15.5f\n', [w1;f1*real(H);f1*imag(H)]);
fprintf (fout, '%s \n', '//nc');
fprintf (fout, '%15.5f %15.5f %15.5f\n', [w0;f0*real(G);f0*imag(G)]);
fclose (fout);

% a) Time domain, response at wall by numerical Fourier transform
U = zeros(size(T));
u0 = 0.5*fac;
for j=1:length(T)  % Fourier transform by direct integration (no FFT here)
  t = T(j);        % Advantage: can use arbitrary size & No. of time steps
   U(j) = u0+sum(real(G.*exp(i*w*t)));
end
U = U*dw/pi;
tail = (1-2/pi*cisib(wmax*T))*2*fac*Cs^2/Cp/R0; %tail of Fourier integral
U = U+tail;
plot(T,U);          % plot response by FT
grid on;
% b) Time domain, response at wall by exact formula
fac = sqrt((1-pois)/2)/rho/Cs;
wn = 2*Cs/R0;          % natural frequency
xi = Cs/Cp;            % damping
wd = wn*sqrt(1-xi^2);   % damped frequency
b = xi*wn;
phi = asin(pois/(1-pois));
U = fac*exp(-b*T).*cos(wd*T-phi);   % Exact response
hold on;
plot(T,U,'r:');
tit = sprintf(...
   'Cylindrical cavity under pressure, response at wall, \\nu=%5.2f',...
     pois);
title(tit);
titx = 'Time (s)';
xlabel(titx);
pause;
% Create EasyPlot file
fout = fopen('Cav3D-TD0.ezp', 'w');
fprintf (fout, '/et g "%s"\n', tit);
fprintf (fout, '/et x "%s"\n', titx);
fprintf (fout, '/og on\n');
f0 = 0.75*rho*Cs*R0^3/pi;   % factor for unit strength pressure
```

```
f1 = f0*r^3;
fprintf (fout, '%15.5f %15.5f\n', [T;U*f0]);
fclose (fout);


% c) Response at receiver by numerical Fourier transform
hold off;
t0 = r-1;          % dimensionless arrival time
H = H.*exp(i*w0*t0);   % delay TF by arrival time
V = zeros(size(T));
u0 = 0.5*fac/r^2;
for j=1:length(T)   % Fourier transform by direct integration (no FFT here)
  t = T(j);          % Advantage: can use arbitrary size & No. of time steps
  V(j) = u0+sum(real(H.*exp(i*w*t)));
end
V = V*dw/pi;
V = V+tail/r;
T1 = [0 t0 T+t0];      % shift delayed time by wave arrival time
V = [0 0 V];
plot(T1,V);              % plot response by FT
grid on;
% d) Response at receiver by exact formula
fac = sqrt((1-pois)/2)/rho/Cs;
V = fac*exp(-b*T).*(cos(wd*T-phi)-(1-1/r)*sin(wd*T))/r;   % Exact response
V = [0 0 V];
hold on;
plot(T1,V,'r:');
tit = sprintf(...
  'Cylindrical cavity under pressure, response at r/r0=%5.2f, \\nu=%5.2f',...
r, pois);
title(tit);
titx = 'Time (s)';
xlabel(titx);
pause;
close;
% Create EasyPlot file
fout = fopen('Cav3D-TD1.ezp', 'w');
fprintf (fout, '/et g "%s"\n', tit);
fprintf (fout, '/et x "%s"\n', titx);
fprintf (fout, '/og on\n');
fprintf (fout, '%15.5f %15.5f\n', [T1;V*f1]);
fclose (fout);
return
```

```
    function [] = SH2D_Half(xs, zs, xr, zr)

% Computes displacements due to SH line source in a homogeneous half-space
% The source and the receiver can be placed anywhere
%
% Input arguments:
%   xs, zs  = Coordinates of source
%   xr, zr  = Coordinates of receiver
%
%   The origin of coordinates is taken at the free surface.
%   The vertical axis may be taken either up or down. Hence, zs and zr
%     must have the same sign (either both positive or both negative)

if (zs>0 & zr<0) | (zs<0 & zr>0)
   'Error, zs and zr in call to SH2 must have the same sign!'
   return
end
% Basic data
cs = 1;            % Shear wave velocity
rho = 1;           % Mass density
mu = rho*cs^2;     % Shear modulus
tmax = 10;         % Maximum dimensionless time for plotting
wmax = 10;         % maximum dimensionless frequency for plotting (rad/s)
np = 500;          % Number of time and/or frequency intervals
dw = wmax/np;
dt = tmax/np;
dx = xr-xs;
dz1 = zr-zs;
dz2 = zr+zs;
r1 = sqrt(dx^2+dz1^2);   % source-receiver distance
r2 = sqrt(dx^2+dz2^2);   % image source - receiver distance
r = r2/r1;
if r<1, r=1/r; end

W = [dw:dw:wmax];
G = -i*0.25*(besselh(0,2,W)+besselh(0,2,W*r))/mu;

tit = sprintf(...
'Half-space SH displacement, Source at (%5.2f,%5.2f), Receiver at (%5.2f,%5.2f)',...
   xs, zs, xr, zr);
plot (W,real(G));
hold on;
plot (W,imag(G),'r');
grid on;
title(tit);
titx = 'Dimensionless frequency w*r/Cs';
xlabel(titx);
pause;
hold off;
% Create EasyPlot file
```

```
fout = fopen('SH2-FD.ezp', 'w');
fprintf (fout, '/et g "%s"\n', tit);
fprintf (fout, '/et x "%s"\n', titx);
fprintf (fout, '/sm off\n');
fprintf (fout, '/sd off\n');
fprintf (fout, '/og on\n');
fprintf (fout, '%15.5f %15.5f %15.5f\n', [W;real(G);imag(G)]);
fclose (fout);

f = 1/(2*pi*mu);
T1 = [0 1 1];
T2 = [1+dt:dt:tmax];
Uyy = f./sqrt(T2.^2-1)+real(f./sqrt((T2.^2-r^2)));
Uyy = [0 0 inf Uyy];
Uyy(3) = Uyy(4);   % avoid singularity & discontinuity in plot
T = [T1, T2];
plot (T, Uyy);
titx = 'Dimensionless time, t*Cs/r';
xlabel(titx);
title(tit);
grid on;
pause;
% Create EasyPlot file
fout = fopen('SH2-TD.ezp', 'w');
fprintf (fout, '/et g "%s"\n', tit);
fprintf (fout, '/et x "%s"\n', titx);
fprintf (fout, '/sm off\n');
fprintf (fout, '/sd off\n');
fprintf (fout, '/og on\n');
fprintf (fout, '%15.5f %15.5f\n', [T;Uyy]);
fclose (fout);
close;
```

```
function [T, Ux, Uz] = Garvin(x,z,pois);

% Garvin's line blast problem, plane strain (2-D)
% Step blast line load applied at depth z below surface of elastic half-space
%
% Written by Eduardo Kausel, MIT, Room 1-271, Cambridge, MA
%
% [T, Ux, Uz] = garvin(x,z,pois)
% Input arguments:
%     x = range of receiver on surface (z = 0)
%     z = depth of source at x = 0
%     pois = Poisson's ratio
% Output arguments
%     T = Dimensionless time vector, tau = t*Cs/r
%     Ux = Horizontal displacement at surface
%     Uz = Vertical displacement at surface
% Unit soil properties are assumed (Cs=1, rho=1)
%
% Sign convention:
%   x from left to right, z=0 at the surface, z points up.
%   Displacements are positive up and to the right.
%   If z > 0 ==> an upper half-space is assumed, z=0 is the lower boundary
%   If z < 0 ==> a lower half-space is assumed, z=0 is the upper boundary

% Reference:
%   W. W. Garvin, Exact transient solution of the buried line source problem,
%               Proceedings of the Royal Society of London, Series A
%               Vol. 234, No. 1199, March 1956, pp. 528-541

% Basic data
cs = 1;                 % Shear wave velocity
rho = 1;                % Mass density
np = 200;               % Number of time intervals
tmax = 2.;              % Max. time for plotting

mu = rho*cs^2;          % Shear modulus
a2 = (1-2*pois)/(2-2*pois);
a = sqrt(a2);           % Cs/Cp
dt = (tmax-a)/np;       % time step

r = sqrt(x^2+z^2);      % Source-receiver distance
c = abs(z)/r;           % direction cosine w.r.t. z axis
s = abs(x)/r;           %      "       "    "            x axis
theta = asin(s);        % Source-receiver angle w.r.t. vertical
fac = 1/(pi*mu*r);      % scaling factor for displacements
T0 = [0 a];             % time interval before arrival of P waves
Ux0 = [0, 0];
Uz0 = [0, 0];
T = [a+dt:dt:tmax];     % Time vector
T2 = T.^2;
```

```
T1 = conj(sqrt(T2-a2));      % make the imaginary part negative
q1 = c*T1+i*s*T;             % Complex horizontal slowness, P waves
p1 = c*T+i*s*T1;             %     "  vertical      "    "    "
Q1 = q1.^2;
s1 = sqrt(Q1+1);
S1 = 2*Q1+1;
R1 = S1.^2 - 4*Q1.*p1.*s1;   % Rayleigh function, P waves
D1 = p1./T1./R1;             % Derivative of q1 divided by R1
Ux1 = 2*fac*sign(x)*imag(q1.*s1.*D1);
Uz1 = -fac*sign(z)*real(S1.*D1);
T = [T0,T];
Ux = [Ux0,Ux1];
Uz = [Uz0,Uz1];

% Plot results
plot (T,r*Ux);
tmax = max(T);
tit = sprintf(...
  'Horizontal displacement due to line blast load, \\nu=%5.2f',pois);
title (tit);
grid on;
%axis([0 tmax -2 2]);
pause;
% Create EasyPlot file
fout = fopen('garvin-x.ezp', 'w');
fprintf (fout, '/et g "%s"\n', tit);
fprintf (fout, '/og on\n');
fprintf (fout, '/sd off\n');
fprintf (fout, '/sm off\n');
fprintf (fout, '%15.5f %15.5f\n', [T;r*Ux]);
fclose (fout);

plot (T,r*Uz);
tit = sprintf(...
  'Vertical displacement due to line blast load, \\nu=%5.2f',pois);
title (tit);
grid on;
%axis([0 tmax -2 2]);
pause;
% Create EasyPlot file
fout = fopen('garvin-z.ezp', 'w');
fprintf (fout, '/et g "%s"\n', tit);
fprintf (fout, '/og on\n');
fprintf (fout, '/sd off\n');
fprintf (fout, '/sm off\n');
fprintf (fout, '%15.5f %15.5f\n', [T;r*Uz]);
fclose (fout);
close;
```

```
function [T, Uxx, Uxz, Uzz] = lamb2D(x,z,pois)

% Lamb's problem in plane strain (2-D)
% Impulsive line load applied onto the surface of an elastic
% half-space and receiver at depth z
%
% To obtain the solution for an interior source and receiver at the
% surface, exchange the coupling terms, and reverse their signs.
%
% Written by Eduardo Kausel, MIT, Room 1-271, Cambridge, MA
%
% [T, Uxx, Uxz, Uzz] = lamb2(x,z,pois)
% Input arguments:
%       x, z = coordinates of receiver relative to source at (0,0)
%        pois = Poisson's ratio
% Output arguments
%       T = Dimensionless time vector, tau = t*Cs/r
%       Uxx = Horizontal displacement caused by a horizontal load
%       Uxz = Horizontal displacement caused by a vertical load
%       Uzx = Vertical displacement caused by a horizontal load
%       Uzz = Vertical displacement caused by a vertical load
%
% Unit soil properties are assumed (Cs=1, rho=1)
%
% Sign convention:
%   x from left to right, z=0 at the surface, z points up.
%   Displacements are positive up and to the right.
%   If z > 0 ==> an upper half-space is assumed, z=0 is the lower boundary
%            Vertical impulse at z=0 is compressive, i.e. up)
%   If z < 0 ==> a lower half-space is assumed, z=0 is the upper boundary
%            Vertical impulse at z=0 is tensile, i.e. up)
%   If z = 0 ==> a lower half-space is assumed, z=0 is the upper boundary
%            both the load and the displacements are at the surface
%
% References:
%   Walter L. Pilant, Elastic Waves in the Earth, Elsevier, 1979
%   Eringen & Suhubi, Elastodynamics, Vol. 2, page 615, eq. 7.16.5,
%              Pages 606/607, eqs. 7.14.22 and 7.14.27
%              Page 612, eq. 7.15.9, page 617, eqs. 7.16.8-10

% Basic data
cs = 1;             % Shear wave velocity
rho = 1;            % Mass density
np = 200;           % Number of time intervals
tmax = 2.;          % Max. time for plotting
mang = 0.4999*pi;   % Max. angle for interior solution

mu = rho*cs^2;      % Shear modulus
a2 = (1-2*pois)/(2-2*pois);
```

```
a = sqrt(a2);              % Cs/Cp
dt = (tmax-a)/np;          % time step


r = sqrt(x^2+z^2);         % Source-receiver distance
c = abs(z)/r;              % direction cosine w.r.t. z axis
s = abs(x)/r;              %     "     "     "   x axis
theta = asin(s);           % Source-receiver angle w.r.t. vertical
crang = asin(a);           % Critical angle w.r.t. vertical
fac = cs/(pi*mu*r);        % scaling factor for displacements


% t < tp=a (two points suffice)
T0 = [0 a];
Uxx0 = [0, 0];
Uxz0 = [0, 0];
Uzx0 = [0, 0];
Uzz0 = [0, 0];
T = [a+dt:dt:tmax];        % Time vector
T2 = T.^2;
jl = length(T);
if theta > mang
  % Displacements on surface
  T1 = [a:dt:1];           % interval from tp to ts
  T2 = [1+dt:dt:tmax];     % t > ts=1
  % a=tp <= t <= ts=1
  t2 = T1.^2;
  p = t2-a2;
  s = 1-t2;
  q = 2*t2-1;
  d = q.^4 + 16*p.*s.*t2.^2;
  p = sqrt(p);
  Uxx1 = 4*fac*t2.*s.*p./d;
  Uzz1 = -fac*q.^2.*p./d;
  s = sqrt(s);
  Uxz1 = 2*fac*T1.*q.*s.*p./d;
  % t > ts=1
  t2 = T2.^2;
  p = sqrt(t2-a2);
  s = sqrt(t2-1);
  q = 2*t2-1;
  d = q.^2-4*t2.*p.*s;
  Uxx2 = -fac*s./d;
  Uzz2 = -fac*p./d;
  Uxz2 = zeros(size(Uxx2));
  cr = cs*(0.874+(0.197-(0.056+0.0276*pois)*pois)*pois);
  tr = cs/cr;     % dimensionless arrival time of R waves
  xr = (cr/cs)^2;
  wr = 0.25*fac*pi*(1-0.5*xr)^3/(1-0.5*xr^2+0.125*xr^3-a2);
  jr = floor((tr-a)/dt+1);
```

```
    % Combine solutions
    Uxx = [Uxx1, Uxx2];
    Uzz = [Uzz1, Uzz2];
    Uxz = [Uxz1, Uxz2];
    Uzx = -Uxz;
    T = [T1 T2];
elseif theta < 1e-3
    % Displacements on epicentral line
    D1 = (T2-a2+0.5).^2 - T.*(T2-a2).*sqrt(T2-a2+1);
    Uxx = -T.*sqrt(T2-a2).*sqrt(T2-a2+1)./D1;
    Uzz = T2.*(T2-a2+0.5)./sqrt(T2-a2)./D1;
    js = 1+floor((1-a)/dt+1); % first element after arrival of S waves
    T2 = T2(js:jl);
    T1 = T(js:jl);
    D2 = (T2-0.5).^2 - T1.*(T2-1).*sqrt(T2+a2-1);
    Uxx(js:jl) = Uxx(js:jl) + T2.*(T2-0.5)./D2./sqrt(T2-1);
    Uzz(js:jl) = Uzz(js:jl) - T1.*sqrt(T2-1).*sqrt(T2-1+a2)./D2;
    clear T1;
    f = 0.5*fac;
    Uxx = f*Uxx;
    Uzz = f*Uzz;
    Uxz = zeros(size(T));
    Uzx = Uxz;
else
    % Displacements in the interior
    T1 = conj(sqrt(T2-a2)); % make the imaginary part negative
    T2 = conj(sqrt(T2-1));
    q1 = c*T1+i*s*T;   % Complex slowness, P waves (Cagniard-De Hoop path)
    q2 = c*T2+i*s*T;   %    "           "    S  "
    p1 = c*T+i*s*T1;
    p2 = c*T+i*s*T2;
    Q1 = q1.^2;
    Q2 = q2.^2;
    s1 = sqrt(Q1+1);
    s2 = sqrt(Q2+a2);
    S1 = 2*Q1+1;
    S2 = 2*Q2+1;
    R1 = S1.^2 - 4*Q1.*p1.*s1;   % Rayleigh function, P waves
    R2 = S2.^2 - 4*Q2.*p2.*s2;   % Rayleigh function, S waves
    D1 = p1./T1./R1;     % Derivative of q1 divided by R1
    D2 = p2./T2./R2;     %    "      q2  "    "    R2
    % Check critical angle
    if (theta > crang)
        tcrit = cos(theta-crang);
    else
        tcrit = 1;
    end
    % Apply Heaviside step function
    k = floor((tcrit-a)/dt+1);
```

```
      if k >=1; D2(1:k)=0; end
      S1 = S1.*D1;
      S2 = S2.*D2;
      D1 = 2*D1.*s1;
      D2 = 2*D2.*s2;
    % Displacements due to impulsive line load
      sgn = fac*sign(x)*sign(z);
      Uxx = fac*real(p2.*S2-Q1.*D1);
      Uzz = fac*real(p1.*S1-Q2.*D2);
      Uzx = sgn*imag(q1.*p1.*D1-q2.*S2);
      Uxz = sgn*imag(q1.*S1-q2.*p2.*D2);
end
T = [T0,T];
Uxx = [Uxx0,Uxx];
Uzz = [Uzz0,Uzz];
Uxz = [Uxz0,Uxz];
Uzx = [Uzx0,Uzx];

% Plot results
plot (T,Uxx);
tmax = max(T);
tit = sprintf(...
'Horizontal displacement due to horizontal line load, \\nu=%5.2f',...
pois);
titx = 'Time';
title (tit);
grid on;
%axis([0 tmax -2 2]);
pause;

plot (T,Uzz);
tit = sprintf((...
'Vertical displacement due to vertical line load, \\nu=%5.2f',pois);
title (tit);
grid on;
%axis([0 tmax -2 2]);
pause;

plot (T,Uzx);
tit = sprintf((...
'Vertical displacement due to horizontal line load, \\nu=%5.2f',pois);
title (tit);
grid on;
%axis([0 tmax -0.5 0.5]);
if theta > mang
    hold on
    plot ([tr tr], [0 -wr]);
end
hold off;
pause;
```

```
plot (T,Uxz);
tit = sprintf(...
'Horizontal displacement due to vertical line load, \\nu=%5.2f',pois);
title (tit);
grid on;
%axis([0 tmax -0.5 0.5]);
if theta > mang
   hold on
   plot ([tr tr], [0 wr]);
end
pause;
hold off;
pfac = rho*cs*r;
EasyPlot('Lamb2Dxx.ezp', tit, titx, T, pfac*Uxx, 'r');
EasyPlot('Lamb2Dzz.ezp', tit, titx, T, pfac*Uzz, 'r');
if z~=0
  % reverse sign & coupling term for source at depth, receiver at surface
  EasyPlot('Lamb2Dxz.ezp', tit, titx, T, -pfac*Uzx, 'r');
  EasyPlot('Lamb2Dzx.ezp', tit, titx, T, -pfac*Uxz, 'r');
else
  % source+ receiver are at the surface of a lower half-space
  EasyPlot('Lamb2Dxz.ezp', tit, titx, T, pfac*Uxz, 'r');
  EasyPlot('Lamb2Dzx.ezp', tit, titx, T, pfac*Uzx, 'r');
end
close;
```

```
function [T, Uxx, Utx, Uzz, Urz] = Lamb3D(r, pois)

% Point loads suddenly applied onto the surface of a lower elastic half-space.
% Time variation of load is a unit step function (Heaviside), which is
% applied at the origin on the surface (x=0, z=0).
% Both horizontal and vertical loads are considered.
% Unit soil properties are assumed (Cs=1, rho=1)
%
% Written by Eduardo Kausel, MIT, Room 1-271, Cambridge, MA
%
% [T, Uxx, Utx, Uzz, Urz] = point3D(r, pois)
% Input arguments:
%      r = range of receiver relative to source at (0,0)
%      pois = Poisson's ratio
% Output arguments (cylindrical coordinates)
%      T = Dimensionless time vector, tau = t*Cs/r
%      Urx = Radial displacement caused by a horizontal load
%           Varies as cos(theta) with azimuth
%      Utx = Tangential displacement caused by a horizontal load
%           Varies as -sin(theta) with azimuth
%      Uzz = Vertical displacement caused by a vertical load
%           (no variation with azimuth)
%      Urz = Radial displacement caused by a vertical load
%           (no variation with azimuth)
%      Uzx = Vertical displacement caused by a horizontal load
%            Not returned, because Uzx = -Urz, except that
%            Uzx varies as cos(theta) (reciprocity)
%
% Sign convention:
% Vertical load and vertical displacements at the surface point up
% Response given in terms of dimensionless time tau=t*Cs/r
%
% Note:
% Solution for vertical loads for any Poisson's ratio, but coupling term
% Urz available only up to Poisson's ratio 0.2631. This is the value
% for which the two false roots of the Rayleigh equation turn complex.
% References:
%   a) Vertical loads:
%   Eringen and Suhubi, Elastodynamics, Vol. II, 748-750, pois<0.26
%   Mooney, 1974, BSSA, V. 64, No.2, pp. 473-491, pois > 0.2631, but
%     vertical component only
%   b) Horizontal loads: (pois = 0.25 only)
%     Chao, C. C. Dynamical response of an elastic halfspace to
%     tangential surface loading, Journal of Applied Mechanics,
%     Vol 27, September 1960, pp 559-567

mu = 1;      % Shear modulus
Nt = 400;
tmax = 2;      % Maximum time for plotting (t=1 => arrival of S waves)
fac = 1/16/pi/r;
```

```
% Vertical displacements due to vertical step load
f = (1-pois)*fac;
a2 = (1-2*pois)./(2 - 2*pois);
b2 = 1-a2;
p = [-16*(1-a2), 8*(3-2*a2), -8, 1];
x = sort(roots(p));
x1 = x(1);
x2 = x(2);
x3 = x(3);
n1 = b2/(a2-x1);
if pois==0
  x2 = a2;
  n2 = inf;
else
  n2 = b2/(a2-x2);
end
n3 = b2/(a2-x3);
tp = sqrt(a2);           % Arrival time of P waves
ts = 1;                  %     "       "    " S  "
tr = sqrt(x3);           %     "       "    " R  "
dt = (ts-tp)/Nt;
T0 = [0 tp];
U0 = [0 0];
T1 = [tp+dt:dt:ts];
t2 = T1.^2;
if (imag(x1)==0)
  A1 = (1-2*x1)^2*sqrt(a2-x1)/((x1-x2)*(x1-x3));
  A2 = (1-2*x2)^2*sqrt(a2-x2)/((x2-x1)*(x2-x3));
  A3 = (1-2*x3)^2*sqrt(x3-a2)/((x3-x1)*(x3-x2));
  U1 = f*(4-A1./sqrt(t2-x1)-A2./sqrt(t2-x2)-A3./sqrt(x3-t2));
else
  D = t2-a2;
  Z = (a2-x1)./D;
  Q = 1+2*Z+2*sqrt((Z+1).*Z);
  for k=1:length(Q)
    if abs(Q(k))>1, Q(k)=1/Q(k); end
  end
  Q = (1./Q - Q).*D;
  A1 = (1-2*x1)^2*(a2-x1)/((x1-x2)*(x1-x3));
  Q = real(A1./Q);
  A3 = (1-2*x3)^2*sqrt(x3-a2)/real((x3-x1)*(x3-x2));
  U1 = f*(4 - 8*Q - A3./sqrt(x3-t2));
end
T2 = [ts+dt:dt:tr-dt];
t2 = T2.^2;
U2 = 2*f*(4 - A3./sqrt(x3-t2));
y = U2(length(U2));   % save last point
Uzz = 8*f;
T3 = [tr];
```

```matlab
U3 = [-sign(A3)*inf];
T4 = [tr+dt:dt:tmax];
U4 = Uzz*ones(size(T4));
U4(1) = y;
T = [T0, T1, T2, T3, T4];
Uzz = [U0, U1, U2, U3, U4];
plot(T, Uzz);
grid on;
%axis ([0 tmax -0.1 0.2]);
tit = sprintf(...
  'Vertical displacement due to vertical point (step) load, \\nu=%5.3f'...,
    pois);
title (tit);
titx = 'Dimensionless time';
xlabel(titx);
EasyPlot('Lamb3Dzz.ezp', tit, titx, T, Uzz, 'r');
pause;
hold off;

% Radial displacements due to vertical load
fac = 1/(8*pi*pi*r);
if (imag(x1)==0)
  t2 = T1.^2;
  k2 = (t2-a2)/b2;          % = k^2
  B1 = ellipint3(90,n1*k2,k2)*(1-2*x1)*(1-x1)/(x1-x2)/(x1-x3);
  B2 = ellipint3(90,n2*k2,k2)*(1-2*x2)*(1-x2)/(x2-x1)/(x2-x3);
  B3 = ellipint3(90,n3*k2,k2)*(1-2*x3)*(1-x3)/(x3-x1)/(x3-x2);
  U1 = 2*ellipint3(90,0,k2) - B1 - B2 - B3;
  f = fac/sqrt(b2^3);
  U1 = f*U1.*T1;
  t2 = T2.^2;
  k2 = b2./(t2-a2);   % inverse of k^2
  B1 = ellipint3(90,n1,k2)*(1-2*x1)*(1-x1)/(x1-x2)/(x1-x3);
  B2 = ellipint3(90,n2,k2)*(1-2*x2)*(1-x2)/(x2-x1)/(x2-x3);
  B3 = ellipint3(90,n3,k2)*(1-2*x3)*(1-x3)/(x3-x1)/(x3-x2);
  U2 = 2*ellipint3(90,0,k2) - B1 - B2 - B3;
  U2 = f*U2.*T2.*sqrt(k2);
  U3 = U2(length(U2));   % save last point
  t2 = T4.^2;
  C = (2*x3-1)^3/(1-4*x3+8*b2*x3^3);
  U4 = 2*pi*fac*C*T4./sqrt(t2-x3);
  Urz = [U0, U1, U2, U3, U4];
  plot (T,Urz);
  grid on;
  axis ([0 tmax -0.2 0.6]);
  tit = sprintf(...
  'Radial displacement due to vertical point (step) load, \\nu=%5.3f',...
    pois);
  title (tit);
```

```
    xlabel(titx);
    EasyPlot('Lamb3Drz.ezp', tit, titx, T, Urz, 'r');
    pause;
else
    % THIS PART NOT YET FUNCTIONAL FOR Urz
    Warning =...
    'Sorry: solution for Urz not available for Poisson' 'ratio > 0.263'
    break
end

% Horizontal step load (Chao's solution)
% Note: Chao's figures are upside down, and multiplied by pi (but not here)
% Radial displacements
fac = 0.5/pi/mu/r;
c1 = 0.75*sqrt(3);
c2 = 0.125*sqrt(6*sqrt(3)+10);
c3 = 0.125*sqrt(6*sqrt(3)-10);
x1 = 0.25;
x2 = 0.25*(3-sqrt(3));
x3 = 0.25*(3+sqrt(3));
t2 = T1.^2;
U1 = fac*t2.*(c1./sqrt(t2-x1)-c2./sqrt(t2-x2)-c3./sqrt(x3-t2));
t2 = T2.^2;
U2 = fac*(1 -2*c3*t2./sqrt(x3-t2));
U4 = fac*ones(size(T4));
T = [T0, T1, T2, T4];
Urx = [U0, U1, U2, U4];
plot (T,Urx);
grid on;
tit = 'Radial displacement due to horizontal point (step) load, \\nu=0.25';
title (tit);
xlabel(titx);
EasyPlot('Chao3Drx.ezp', tit, titx, T, Urx, 'r');
pause;

% Tangential displacements
fac = -0.375/pi/mu/r;
t2 = T1.^2;
f = 8/3;
U1 = 0.5*fac*(1- f*(c1*sqrt(t2-x1)-c2*sqrt(t2-x2)+c3*sqrt(x3-t2)));
t2 = T2.^2;
U2 = fac*(1 - f*c3*sqrt(x3-t2));
U4 = fac*ones(size(T4));
Utx = [U0, U1, U2, U4];
plot (T,Utx);
grid on;
tit = 'Tangential displacement due to horizontal point (step) load, \\nu=0.25';
title (tit);
xlabel(titx);
```

```
EasyPlot('Chao3Dtx.ezp', tit, titx, T, Utx, 'r');
pause;

% Chao's functions on the axis (point load, displ. under load.)
fac = 0.5/pi/mu/r;
t = T1;
t2 = t.^2;
f = (2*t2+1/3).^2-4*t.*(t2-1/3).*sqrt(t2+2/3);
f = t.*(t2-1/3).*sqrt(t2+2/3)./f;
U1 = -fac*f;
t = [T2,T4];
t2 = t.^2;
f = (2*t2+1/3).^2-4*t.*(t2-1/3).*sqrt(t2+2/3);
f = t.*(t2-1/3).*sqrt(t2+2/3)./f;
g = (2*t2-1).^2-4*t.*(t2-1).*sqrt(t2-2/3);
g = t2.*(t2-1).*(2*t.*sqrt(t2-2/3)-2*t2+1)./g;
U2 = fac*(0.5*(t2+1)-f+g);
T = [T0,T1,t];
U = [U0,U1,U2];
plot (T,U);
grid on;
tit = 'Displacement at axis under horizontal point (step) load, \\nu=0.25';
title(tit);
xlabel(titx);
EasyPlot('Chao3Daxis.ezp', tit, titx, T, U, 'r');
pause;
close all;
```

```
function [] = Torsion_Half(r, z, z0)

% Torsional point source with vertical axis acting within an
% elastic half-space at depth zo. The source has a time
% variation of the form f(t) = sin(pi*t/td), 0<=t<=td
% Origin is at surface
% Arguments:
%   r, z  = coordinates of receiver (horizontal range and depth)
%   z0    = depth of source

if (z>0&z0<0) | (z<0&z0<0)
   disp('Error: z and z0 must have the same sign');
   return
end
mu = 1;        % Shear modulus
rho = 1;       % mass density
cs = 1;        % shear wave velocity
td = 1.0;      % duration of pulse

fac = 1/8/pi/mu;
R1 = sqrt(r^2+(z-z0)^2);
R2 = sqrt(r^2+(z+z0)^2);
S1 = fac*r/R1^3;   % fac*sin(phi-1)/R1^2
S2 = fac*r/R2^3;

% Time domain solution
m = 200;       % Number of intervals in pulse

dt = td/m;
t1 = R1/cs;
t2 = R2/cs;
nt = m;
tarr = 2*max(t1,t2);  % latest arrival
while nt*dt<tarr, nt=2*nt; end
T = [0:nt]*dt;
U = zeros(size(T));
Tp = pi/m*[0:m];
M = sin(Tp);
dM = pi/td/cs*cos(Tp);
dM(1) = 0.5*dM(1);     % smooth out discontinuity for FFT
dM(m+1) = 0.5*dM(m+1);
k1 = floor(t1/dt)+1;
k2 = floor(t2/dt)+1;
U(k1:k1+m) = S1*(M+R1*dM);
U(k2:k2+m) = U(k2:k2+m)+S2*(M+R2*dM);
plot (T, U);
tit = sprintf(...
   'Response at r=%5.3f, z=%5.3f due to torsional point load at z=%5.3f',...
      r, z, z0);
titx = 'Time';
```

```
        grid on;
        title(tit);
        xlabel(titx);
        pause;
        EasyPlot('TorHalf_TD.ezp', tit, titx, T, U, 'r');

        % Transfer function by brute force Fourier Transform
        % Reason: Only a dense set of low frequencies is of interest
        % (difficult and computationally expensive to accomplish with FFT)
        nf = 200;
        fmax = 2;
        df = fmax/nf;
        f = [0:df:fmax];
        Tp = dt*[0:m];
        U = U(k1:k2+m);
        T = T(k1:k2+m);
        H = zeros(1,nf+1);
        for j=1:nf+1
          w = i*2*pi*f(j);
          Uf = sum(U.*exp(-w*T));
          Um = sum(M.*exp(-w*Tp));
          H(j) = Uf/Um;
        end
        plot (f, real(H));
        hold on;
        plot (f, imag(H),'r');
        tit = 'Transfer function by FT ';
        title (tit);
        titx = 'Frequency';
        xlabel(titx);
        grid on;
        pause;
        hold off;
        A1 = abs(H);
        plot(f, A1);
        grid on;
        title ('Abs. Val. transfer function by FT ');
        xlabel(titx);
        pause;
        hold off;
        ang1 = unwrap(angle(H));
        plot(f, ang1);
        grid on;
        title ('Angle transfer function by FT ');
        xlabel(titx);
        pause;

        % Transfer function by frequency domain solution
        W1 = i*2*pi*R1/cs*f;
```

```
W2 = i*2*pi*R2/cs*f;
%H = S1*(1+W1).*exp(-W1)+S2*(1+W2).*exp(-W2);
H = S1*(1+W1).*exp(-W1)+S2*(1+W2).*exp(-W2);
plot (f, real(H));
hold on;
plot(f,imag(H),'r');
tit = sprintf(...
    'Transfer function at r=%5.3f, z=%5.3f due to torsional load at z=%5.3f',...
      r, z, z0);
title (tit);
xlabel(titx);
grid on;
pause;
hold off;
EasyPlot('TorHalf_FD.ezp', tit, titx, f, H, 'c');
A2 = abs(H);
plot(f, A2);
grid on;
tit = 'Abs. Val. transfer function, direct';
title (tit);
xlabel(titx);
pause;
ang2 = unwrap(angle(H));
plot(f, ang2);
grid on;
title ('Angle transfer function, direct');
xlabel(titx);
pause;
plot(f, A1./A2);
grid on;
title ('A1/A2');
xlabel(titx);
pause;
ang = (ang2-ang1)/pi;
plot(f, ang);
xlabel(titx);
grid on;
title ('Diff. Angle / pi');
xlabel(titx);
pause;
close all;
return
```

```
      function [] = SH_Plate(x, z, z0)

% SH line source at elevation zo above the origin in an elastic plate.
% Plate has unit thickness and material properties
% Origin is at the bottom surface, z is positive up
% Arguments:
%   x, z  = coordinates of receiver
%   z0  = elevation of source

% Default data
mu = 1;        % Shear modulus
rho = 1;       % mass density
cs = 1;        % shear wave velocity
h = 1;         % plate thickness
xi = 0.0001;   % material damping
N = 10;        % parameter for max. time and max. frequency

if z>h | z0>h
  disp('Error, source/receiver z must be <= than thickness of
    plate');
  return
end
% Time domain solution via method of images
% Choose number of images ni=N such that other images arrive after
tmax
tfund = 2*h/cs;     % fundamental period of plate
ffund = 1./tfund;   % fundamental frequency (Hz)
fac = 1/2/pi/mu;
tmax = N*tfund;
nt = 400;
dt = tmax/nt;
cs2 = cs^2;
x2 = x^2;
dz = z-z0;
sz = z+z0;
h2 = 2*h;
T = [0:nt]*dt;
T2 = T.^2;
U = zeros(size(T));
for j=-N:N
  t1 = (x2+(dz-j*h2)^2)/cs2;
  t2 = (x2+(sz-j*h2)^2)/cs2;
  U = U + response(T2, t1) + response(T2, t2);
end
U = fac*U;
plot (T, U);
tit = sprintf(...
  'Impulse response at x=%5.3f, z=%5.3f due to SH line load in plate at z=%5.3f',...
    x, z, z0);
grid on;
```

```
title(tit);
titx = 'Time';
xlabel(titx);
pause;
EasyPlot('SH_plate_TD.ezp', tit, titx, T, U, 'r');


% Frequency domain solution via method of images
ni = 20;      % Number of images
fac = 0.25/mu/i;
fmax = N*ffund;
nf = 600;
df = fmax/nf;
F = [df:df:fmax];
W = 2*pi*F;
H = zeros(size(F));
for j=-ni:ni
  t1 = sqrt(x2+(dz-j*h2)^2)/cs;
  t2 = sqrt(x2+(sz-j*h2)^2)/cs;
  H = H + besselh(0,2,t1*W) + besselh(0,2,t2*W);
end
H = fac*H;
plot (F, real(H));
hold on;
plot (F, imag(H),'r');
tit = sprintf(...
'Green function at x=%5.3f, z=%5.3f due to SH line load in plate at z=%5.3f',...
  x, z, z0);
titx = 'Frequency (rad/s)';
grid on;
title(tit);
xlabel(titx);
pause;
hold off;
EasyPlot('SH_plate_FD1.ezp', tit, titx, F, H, 'c');


% Frequency domain solution via normal modes
nm = 20;           % number of normal modes
fac = 1/mu/i;
tit = sprintf(...
'Green function at x=%5.3f, z=%5.3f due to SH line load in plate at z=%5.3f',...
  x, z, z0);
x = abs(x/h);
z = z/h;
z0 = z0/h;
W = W*h*(1-i*xi)/cs;
H = 0.5*exp(-i*x*W)./W;
W2 = W.^2;
for j=1:nm
  j1 = pi*j;
```

```
  K = sqrt(W2-j1^2);
  H = H + cos(j1*z)*cos(j1*z0)*exp(-i*x*K)./K;
end
H = fac*H;
plot (F, real(H));
hold on;
plot (F, imag(H),'r');
grid on;
title(tit);
xlabel(titx);
pause;
hold off;
EasyPlot('SH_plate_FD2.ezp', tit, titx, F, H, 'c');
close all;
return

function [U] = response(T2, t2)
k = find(t2<T2);
if isempty(k)
  U = zeros(size(T2));
else
  k = k(1);
  k1 = k-1;
  n = length(T2);
  U = [zeros(1,k1), 1./real(sqrt((T2(k:n)-t2)))];
  if t2==T2(k1), U(k1)=inf; end
end
return
```

```
    function [] = SH_Stratum(x, z, z0)

% SH line source at elevation zo above the origin in an elastic stratum
% Stratum has unit thickness and material properties
% Origin is at the bottom surface, z is positive up
% Arguments:
%   x, z  = coordinates of receiver
%   z0    = elevation of source
% Method of images used for both the time domain and the frequency domain
% In addition, the frequency domain solution is determined again with
% the normal modes method (gives more accurate answers).

mu = 1;            % Shear modulus
rho = 1;           % mass density
cs = 1;            % shear wave velocity
h = 1;             % plate thickness
xi = 0.0001;       % material damping
N = 10;            % parameter for max. time and max. frequency
if z>h | z0>h
  disp('Error, source/receiver z must be <= than thickness of plate');
  return
end


% Time domain solution via method of images
% Choose number of images ni=N such that other images arrive after tmax
tfund = 2*h/cs;    % fundamental period of plate
ffund = 1./tfund;  % fundamental frequency (Hz)
fac = 1/2/pi/mu;
tmax = N*tfund;
nt = 400;
dt = tmax/nt;
cs2 = cs^2;
x2 = x^2;
dz = z-z0;
sz = z+z0;
h2 = 2*h;
T = [0:nt]*dt;
T2 = T.^2;
U = zeros(size(T));
for j=-N:N
  t1 = (x2+(dz-j*h2)^2)/cs2;
  t2 = (x2+(sz-j*h2)^2)/cs2;
  U = U + (response(T2, t1) - response(T2, t2))*(-1)^j;
end
U = fac*U;
plot (T, U);
tit = sprintf(...
  'Impulse response at x=%5.3f, z=%5.3f due to SH line load in stratum at z=%5.3f',...
    x, z, z0);
titx = 'Time';
```

```
xlabel(titx);
grid on;
title(tit);
pause;
EasyPlot('SH_stratum_TD.ezp', tit, titx, T, U, 'r');

% Frequency domain solution via method of images
ni = 100;      % Number of images
fac = 0.25/mu/i;
fmax = N*ffund;
nf = 600;
df = fmax/nf;
F = [df:df:fmax];
W = 2*pi*F;
H = zeros(size(F));
for j=-ni:ni
   t1 = sqrt(x2+(dz-j*h2)^2)/cs;
   t2 = sqrt(x2+(sz-j*h2)^2)/cs;
   H = H + (besselh(0,2,t1*W) - besselh(0,2,t2*W))*(-1)^j;
end
H = fac*H;
plot (F, real(H));
hold on;
plot (F, imag(H),'r');
tit = sprintf(...
   'Green function at x=%5.3f, z=%5.3f due to SH load in stratum at z=%5.3f',...
      x, z, z0);
grid on;
title(tit);
titx = 'Frequency (rad/s)';
xlabel(titx);
pause;
hold off;
EasyPlot('SH_stratum_FD1.ezp', tit, titx, F, H, 'c');


% Frequency domain solution via normal modes
nm = 20;      % number of normal modes
fac = 1/mu/i;
tit = sprintf(...
   'Green function at x=%5.3f, z=%5.3f due to SH load in stratum at z=%5.3f',...
      x, z, z0);
x = abs(x/h);
z = z/h;
z0 = z0/h;
W = W*h*(1-i*xi)/cs;
H = zeros(size(W));
W2 = W.^2;
for j=1:nm
   j1 = pi*(j-0.5);
```

```
          K = sqrt(W2-j1^2);
          H = H + sin(j1*z)*sin(j1*z0)*exp(-i*x*K)./K;
       end
       H = fac*H;
       plot (F, real(H));
       hold on;
       plot (F, imag(H),'r');
       grid on;
       title(tit);
       xlabel(titx);
       pause;
       hold off;
       EasyPlot('SH_stratum_FD2.ezp', tit, titx, F, H, 'c');
       close all;
       return


       function [U] = response(T2, t2)
       k = find(t2<T2);
       if isempty(k)
         U = zeros(size(T2));
       else
         k = k(1);
         k1 = k-1;
         n = length(T2);
         U = [zeros(1,k1), 1./real(sqrt((T2(k:n)-t2)))];
         if t2==T2(k1), U(k1)=inf; end
       end
       return
```

```
function [] = SVP_Plate(x, z, z0, poi, BC1, BC2)

% SVP line source at elevation zo in an elastic plate
% with mixed boundary conditions
% Plate has unit thickness and material properties
% Origin is at the bottom surface, z is positive up
% Arguments:
%   x, z  = coordinates of receiver
%   z0   = elevation of source
%   pois  = Poisson's ratio
%   BC1   = Boundary condition at bottom ('FC' or 'CF')
%   BC2   = Boundary condition at top    ('FC' or 'CF')
% Method of images used for both the time domain and the frequency domain
% In addition, the frequency domain solution is determined again with
% the normal modes method (Gives more accurate answers, and it is MUCH faster).

% Default data
mu = 1;          % Shear modulus
rho = 1;         % mass density
cs = 1;          % shear wave velocity
h = 1;           % plate thickness
xi = 0.0001;     % material damping
N = 10;          % parameter for max. time and max. frequency

if z>h | z0>h
   disp('Error, source/receiver z must be <= than thickness of plate');
   return
end

alfa = sqrt((1-2*poi)/(2-2*poi));  % cs/cp
tfund = 2*h*alfa/cs;   % fundamental period of plate (dilatation)
ffund = 1./tfund;      % fundamental frequency (Hz)
x2 = x^2;
dz = z-z0;
sz = z+z0;
h2 = 2*h;
BC1 = upper(BC1);
BC2 = upper(BC2);
BC = strcat(BC1,BC2);  % boundary conditions

% Method of images
% Parameters for time domain
%tmax = N*tfund;  % choose ni=N such that other images arrive after tmax
tmax = 10;     %%%%%%%%%%%%%
nt = 400;
dt = tmax/nt;
cs2 = cs^2;
T = [0:nt]*dt;
T2 = T.^2;
n = length(T);
```

```
U = zeros(4,n);
for j=-N:N
   z1 = dz-j*h2;
   z2 = sz-j*h2;
   r1 = sqrt(x2+z1^2);
   r2 = sqrt(x2+z2^2);
   U1 = impulse_response(r1, x, z1, mu, cs, alfa, T2);
   U2 = impulse_response(r2, x, z2, mu, cs, alfa, T2);
   if BC=='FCFC'
      U = U+[U1(1:2,:)+U2(1:2,:);U1(2:3,:)-U2(2:3,:)];
   elseif BC=='CFCF'
      U = U+[U1(1:2,:)-U2(1:2,:);U1(2:3,:)+U2(2:3,:)];
   elseif BC=='FCCF'
      U = U+(-1)^j*[U1(1:2,:)+U2(1:2,:);U1(2:3,:)-U2(2:3,:)];
   elseif BC=='CFFC'
      U = U+(-1)^j*[U1(1:2,:)-U2(1:2,:);U1(2:3,:)+U2(2:3,:)];
   else
      disp('Not a valid boundary condition');
      return
   end
end
tit = sprintf(...
   'Uxx at x=%5.3f, z=%5.3f due to SVP line load in plate at z=%5.3f',...
      x, z, z0);
titx = 'Time';
direc = ['xx', 'zx', 'xz', 'zz'];
fname = 'Uxx_SVP_plate_TD.ezp';
for j=1:4
   j2 = 2*j;
   j1 = j2-1;
   tit(2:3) = direc(j1:j2);
   fname(2:3) = direc(j1:j2);
   plot (T, U(j,:));
   grid on;
   title(tit);
   xlabel(titx);
   pause;
   EasyPlot(fname, tit, titx, T, U(j,:), 'r');
end

% Frequency domain solution via method of images
% Parameters frequency domain
ni = 50;      % Number of images frequency domain
fmax = 5;
nf = 500;
df = fmax/nf;
F = [df:df:fmax];
W = 2*pi*F;
n = length(W);
```

```
H = zeros(4,n);
for j=-ni:ni
  z1 = dz-j*h2;
  z2 = sz-j*h2;
  r1 = sqrt(x2+z1^2);
  r2 = sqrt(x2+z2^2);
  H1 = Green(r1, x, z1, mu, cs, alfa, W);
  H2 = Green(r2, x, z2, mu, cs, alfa, W);
  if BC=='FCFC'
    H = H+[H1(1:2,:)+H2(1:2,:);H1(2:3,:)-H2(2:3,:)];
  elseif BC=='CFCF'
    H = H+[H1(1:2,:)-H2(1:2,:);H1(2:3,:)+H2(2:3,:)];
  elseif BC=='FCCF'
    H = H+(-1)^j*[H1(1:2,:)+H2(1:2,:);H1(2:3,:)-H2(2:3,:)];
  elseif BC=='CFFC'
    H = H+(-1)^j*[H1(1:2,:)-H2(1:2,:);H1(2:3,:)+H2(2:3,:)];
  else
    disp('Not a valid boundary condition');
    return
  end
end
tit = sprintf(...
  'Hxx at x=%5.3f, z=%5.3f due to SVP line load in plate at z=%5.3f',...
    x, z, z0);
direc = ['xx', 'zx', 'xz', 'zz'];
fname = 'Hxx_SVP_plate_FD1.ezp';
titx = 'Frequency (rad/s)';
for j=1:4
  j2 = 2*j;
  j1 = j2-1;
  tit(2:3) = direc(j1:j2);
  fname(2:3) = direc(j1:j2);
  plot (F, real(H(j,:)));
  hold on;
  plot (F, imag(H(j,:)), 'r');
  hold off;
  grid on;
  title(tit);
  xlabel(titx);
  pause;
  EasyPlot(fname, tit, titx, F, H(j,:), 'c');
end
clear U H T

% Frequency domain solution via normal modes
nm = 20;   % number of normal modes
tit = sprintf(...
    'Green function at x=%5.3f, z=%5.3f due to SVP load in plate at z=%5.3f',...
      x, z, z0);
```

```
   sgn = sign(x);
   x = abs(x/h);
   z = z/h;
   z0 = z0/h;
   Ws = W*h*(1-i*xi)/cs;
   Wp = alfa*Ws;
   Ws2 = Ws.^2;
   Wp2 = Wp.^2;
   H = zeros(4,length(Ws));
   if BC=='FCFC'
      H(1,:) = 0.5*Wp.*exp(-i*x*Wp);
   elseif BC=='CFCF'
      H(4,:) = 0.5*Ws.*exp(-i*x*Ws);
   end
   for j=1:nm
      if BC1==BC2
         j1 = pi*j;
      else
         j1 = pi*(j-0.5);
      end
      j2 = j1^2;
      Ks = sqrt(Ws2-j2);
      Kp = sqrt(Wp2-j2);
      Es = exp(-i*x*Ks);
      Ep = exp(-i*x*Kp);
      sr = sin(j1*z);
      ss = sin(j1*z0);
      cr = cos(j1*z);
      cs = cos(j1*z0);
      H1 = Kp.*Ep+j2*Es./Ks;
      H2 = j1*(Ep-Es);
      H4 = Ks.*Es+j2*Ep./Kp;
      if BC1=='FC'
         H(1,:) = H(1,:) + cs*cr*H1;
         H(2,:) = H(2,:) - cs*sr*H2;
         H(3,:) = H(3,:) + ss*cr*H2;
         H(4,:) = H(4,:) + ss*sr*H4;
      elseif BC1=='CF'
         H(1,:) = H(1,:) + ss*sr*H1;
         H(2,:) = H(2,:) + ss*cr*H2;
         H(3,:) = H(3,:) - cs*sr*H2;
         H(4,:) = H(4,:) + cs*cr*H4;
      end
   end
   clear H1 H2 H3
   tit = sprintf(...
      'Hxx at x=%5.3f, z=%5.3f due to SVP line load in plate at z=%5.3f',...
         x, z, z0);
   direc = ['xx', 'zx', 'xz', 'zz'];
```

```
fname = 'Hxx_SVP_plate_FD2.ezp';
f = [-i, sgn, sgn, -i]/mu;
for j=1:4
   j2 = 2*j;
   j1 = j2-1;
   tit(2:3) = direc(j1:j2);
   fname(2:3) = direc(j1:j2);
   H(j,:) = f(j)*H(j,:)./Ws2;
   plot (F, real(H(j,:)));
   hold on;
   plot (F, imag(H(j,:)), 'r');
   hold off;
   grid on;
   title(tit);
   xlabel(titx);
   pause;
   EasyPlot(fname, tit, titx, F, H(j,:), 'c');
end
close all;
return


function [G] = Green(r, x, z, mu, cs, alfa, W)
% Returns the Green's functions for an SVP line load in a full space
% r=source receiver distance, c=cos(theta), s=sintheta), alfa=cs/cp
ts = r/cs;
Ws = ts*W;
Wp = alfa*Ws;
psi = (besselh(1,2,Ws)-alfa*besselh(1,2,Wp))./Ws - besselh(0,2,Ws);
chi = alfa^2*besselh(2,2,Wp)- besselh(2,2,Ws);
f = 0.25*i/mu;
c = x/r;      % cos(theta)
s = z/r;      % sin(theta)
G = f*[psi+chi*c^2; chi*c*s; psi+chi*s^2];   % [gxx; gzx; gzz]
return


function [U] = impulse_response(r, x, z, mu, cs, alfa, T2)
% Returns the impulse response functions for an SVP line load in a
full space
ts = r/cs;
tp = alfa*ts;
tp2 = tp^2;
k = find(tp2<T2);
if isempty(k)
   U = zeros(3,length(T2));
   return
else
   ts2 = ts^2;
   k = k(1);
   k1 = k-1;
```

```
        n = length(T2);
        S = sqrt((T2(k:n)-tp2));
        PSI = [zeros(1,k1),-S/ts2];   % yes, we must divide by ts2, not tp2
        CHI = [zeros(1,k1),alfa^2./S]-2*PSI;
        if tp2==T2(k), CHI(k1)=inf; end
        k = find(ts2<T2);
        if ~isempty(k)
          k = k(1);
          k1 = k-1;
          n = length(T2);
          S = sqrt((T2(k:n)-ts2));
          PSI = PSI + [zeros(1,k1),S/ts2+1./S];
          CHI = CHI - [zeros(1,k1),2*S/ts2+1./S];
          if tp2==T2(k), PSI(k1)=inf; CHI(k1)=inf; end
        end
    end
    f = 1/2/pi/mu;
    c = x/r;   % cos(theta)
    s = z/r;   % sin(theta)
    U = f*[PSI+CHI*c^2; CHI*c*s; PSI+CHI*s^2];   % [uxx; uzx; uzz]
    return
```

```
function [ztors, zspher] = spheroidal (pois, m, wmin, wmax)

% Computes the torsional and spheroidal modes of a solid sphere, and
% plots the two determinant functions whose zero crossings define the roots
% Assumes a sphere of radius R=1 and shear wave velocity Cs=1
% Note: the modes are independent of the azimuthal index n
%
% Input arguments:
%     pois  = Poisson's ratio
%     m     = radial index
%     wmin  = minimum frequency for search
%     wmax  = maximum frequency for search
% Output arguments:
%     ztors   = dimensionless frequencies, torsional modes
%     zspher  = dimensionless frequencies, spheroidal modes
% Actual frequencies (rad/s) are obtained multiplying by Cs/R
% where Cs is the shear wave velocity, and R is the radius
%
% For enhanced accuracy, it uses the Bessel functions of half order
% instead of the actual spherical Bessel functions. Reason:
% The factors sqrt(pi/2z) cancel out in the determinant equations
%
% Makes first a rough search in the search interval to determine approximate
% values of the roots, then refines the results using the bisection method
%
% Results agree with Eringen & Suhubi, Elastodynamics, Vol. II, pp. 814,818

dz = 0.1;      % initial search step
if wmin<dz
  zs=[dz:dz:wmax];
else
  zs=[wmin:dz:wmax];
end


tit = sprintf('Torsional modes for m =%d', m);
titx = 'Mode number';
fname = sprintf('Tors%d.ezp',m);
if m>0
  f = dettors(m,zs);
  ztors = zerox(f,dz,zs(1));
  nz = length(ztors);
  for k=1:nz
  z1 = ztors(k)-dz;
  z2 = ztors(k)+dz;
  ztors(k) = bisect2('dettors', m, z1, z2, 1.e-5,pois);
  end
  plot(zs,f);
  axis([0 wmax -1 1]);
  grid on;
  pause;
```

```
    close;
    EasyPlot(fname, tit, titx, [1:nz], ztors, 'r')
  else
    disp('No torsional modes exist for m=0');
    ztors = [];
  end

  tit = sprintf('Spheroidal modes for m =%d', m);
  fname = sprintf('Spher%d.ezp',m);
  f = detspher(m,zs,pois);
  plot(zs,f);
  grid on;
  axis([0 wmax -1 1]);
  pause;
  close;
  zspher= zerox(f,dz,zs(1));
  nz = length(zspher);
  for k=1:nz
    z1 = zspher(k)-dz;
    z2 = zspher(k)+dz;
    z = bisect2('detspher', m, z1, z2, 1.e-5,pois);
    if z~=[], zspher(k) = z; end
  end
  EasyPlot(fname, tit, titx, [1:nz], zspher, 'r')
  return

  function [f] = dettors(m,zs,pois)
  jsm = besselj(m+0.5,zs);
  jsm1 = besselj(m-0.5,zs);
  f = jsm1-(m+2)*jsm./zs;
  return

  function [f] = detspher(m,zs,pois)
  cm = 2*(1-pois)/(1-2*pois);   % Constrained modulus
  a=sqrt(1/cm);   % Cs/Cp
  zp = a*zs;
  jpm = besselj(m+0.5,zp);
  jpm1 = besselj(m-0.5,zp);
  if m>0
    jsm = besselj(m+0.5,zs);
    jsm1 = besselj(m-0.5,zs);
    f11 = -cm*jpm- 2*(2*jpm1-(m+1)*(m+2)*jpm./zp)./zp;
    f12 = 2*m*(m+1)*(jsm1-(m+2)*jsm./zs)./zs;
    f21 = 2*(jpm1-(m+2)*jpm./zp)./zp;
    f22 = -jsm-2*(jsm1-m*(m+2)*jsm./zs)./zs;
    f = f11.*f22-f12.*f21;
  else
    f = -cm*jpm- 4*jpm1./zp;
```

```
      end
      return

      function center = bisect2(f, m, left, right, tolerance, pois)
      % Finds the root of a function f in interval [left, right]
      % where a root is known to exist. Uses the bisection method
      fleft = feval(f,m,left,pois);
      fright = feval(f,m,right,pois);
      if (fleft == 0)
        center = left;
        return
      elseif fright == 0
        center = right;
        return
      elseif fleft* fright > 0
        center = [];
        return
      end
      while (right-left) > tolerance
        center = (left + right)/ 2;
        if feval(f,m,left,pois)* feval(f,m,center,pois) <= 0
          right = center;
        else
          left = center;
        end
      end
      return

      function z = zerox(a, dz, z1)
      % Finds the zero crossings of a vector
      n = length(a);
      nz = 0;
      z = [];
      if (n < 2) return; end
      j = 1;
      while j<n
        sg1 = sign(a(j));
        while (sg1 == 0)
          nz = nz+1;
          z = [z,j];
          j = j + 1;
          sg1 = sign(a(j));
        end
        j = j + 1;
        while j<=n & sign(a(j))==sg1
          j = j+1;
        end
        if j<=n
```

```
                    % found a root
                    nz = nz+1;
                    dj = abs(a(j))/(abs(a(j-1))+abs(a(j)));
                    z = [z,j-dj];
                end
            end
            z = (z-1)*dz+z1;
            return
```

```
function EasyPlot(fname, tit, titx, T, U, flag)
            % Creates files for display with the EasyPlot program
            fout = fopen (fname, 'w');
            fprintf (fout, '/et g "%s"\n', tit);
            fprintf (fout, '/et x "%s"\n', titx);
            fprintf (fout, '/og on\n');
            fprintf (fout, '/sd off\n');
            fprintf (fout, '/sm off\n');
            [nr,nc] = size(U);
            for n=1:nr
              fprintf (fout, '//nc\n');
              if flag=='r'
              fprintf (fout, '%15.5f %15.5f\n', [T;U(n,:)]);
              elseif flag=='c'
              fprintf (fout, ''%15.5f %15.5f %15.5f\n',
            [T;real(U(n,:));imag(U(n,:))]);
              else
              'Not a valid option in EasyPlot';
              end
            end
            fclose (fout);
            return
```

```
function [si] = cisib(x)
    %      =================================================
    %      Purpose: Compute cosine and sine integrals
    %            Si(x) and Ci(x)
    %      Input: x --- Argument of Ci(x) and Si(x)
    %      Output: CI --- Ci(x)
    %                  SI --- Si(x)
    %      =================================================
    %%%%%%%%%%%%%%%%%%%%%%%%%%%%%%% added stuff
    n = length(x);
    n1 = 1;
    while (n1<n & x(n1)<=0)
       x(n1) = 0;
       n1 = n1+1;
    end
    n3 = n1;
    while (n3<n & x(n3)<=1)
       n3 = n3+1;
    end
    %%%%%%%%%%%%%%%%%%%%%%%%%%%%%%%%%%%%
    if (n1>1)
       n0 = n1-1;
       ci(1:n0) = -1.0d+300;
       si(1:n0) = 0.0d0;
    end
    if (n3>n1)
       n2 = n3-1;
       xx = x(n1:n2);
       x2 = xx.^2;
       ci(n1:n2) = (((((-3.0d-8*x2+3.10d-6).*x2-2.3148d-4).*x2+...
       1.041667d-2).*x2-0.25).*x2+0.577215665d0+log(xx);
       si(n1:n2) = (((((3.1d-7*x2-2.834d-5).*x2+...
       1.66667d-003).*x2-5.555556d-002).*x2+1.0).*xx;
    end
    if (n3<=n)
       xx = x(n3:n);
       x2 = xx.^2;
       fx = (((((x2+38.027264d0).*x2+265.187033d0).*x2+...
       335.67732d0).*x2+38.102495d0)./(((((x2 +40.021433d0).*x2+...
       322.624911d0).*x2+570.23628d0).*x2+157.105423d0);
       gx = (((((x2+42.242855d0).*x2+302.757865d0).
    *x2+352.018498d0).*x2+...
       21.821899d0)./(((((x2 +48.196927d0).*x2+...
       482.485984d0).*x2+1114.978885d0).*x2+449.690326d0)./xx;
       ci(n3:n) = fx.*sin(xx)./xx-gx.*cos(xx)./xx;
       si(n3:n) = 1.570796327d0-fx.*cos(xx)./xx-gx.*sin(xx)./xx;
    end
```

```
function [el3] = ellipint3(phi,N,M)

%    [EL3] = ELLIPINT3 (phi,N,M) returns the elliptic integral of
%           the third kind, evaluated for each value of N, M
%           Can also be used to obtain the elliptic integral
%           of the first kind by setting N=0.
%    Arguments: phi - Upper limit of integration (in degrees)
%              M   - Modulus  (some authors use k=sqrt(m))
%                  M can be a scalar or vector
%              N --- parameter (some authors use c=-n)
%                  N can be also be scalar or vector, but if
%                  the latter, it must agree in size with M
%    Definition: If n, m are elements of N, M, then
%
%
%              phi
%    el3 = integ |   [dt/((1+n*sin^2(t))*sqrt(1-m*sin^2(t)))
%              0
%
%    Observe that m = k^2 is the square of the argument
%    used by some authors for the elliptic integrals
%    Method: 10-point Gauss-Legendre quadrature
if (phi < 0)
   'Error, first argument in ellipint3 cannot be negative'
   el3 = 0;
   return
end
if length(N) ==1
   N = N*ones(size(M));
elseif length(N) ~= length(M)
   'Error, wrong size of second argument in ellipint3.'
   'Should be size(N)=1 or size(N)=size(M)'
   el3 = 0;
   return
end
tol = 1-1d-8;
ang = phi*pi/180;
psi = ang/2;
t = [.9931285991850949,.9639719272779138,.9122344282513259,...
     .8391169718222188,.7463319064601508,.6360536807265150,...
     .5108670019508271,.3737060887154195,.2277858511416451,...
     .7652652113349734d-1];
w = [.1761400713915212d-1,.4060142980038694d-1, .6267204833410907d-1,
     .8327674157670475d-1,.1019301198172404,.1181945319615184,...
     .1316886384491766,.1420961093183820,.1491729864726037,...
     .1527533871307258];
t1 = psi*(1+t);
t2 = psi*(1-t);
s1 = sin(t1).^2;
s2 = sin(t2).^2;
el3 = zeros(size(M));
```

```
              s = sin(ang)^2;
          for j=1:length(M)
            k2 = M(j);
            n = N(j);
            flag = 1;
            if phi<=90   % assuming phi is in degrees here
              if k2*s>=tol | -n*s >= tol
                 flag = 0;
              end
            elseif k2 >= tol | -n >=tol
              flag = 0;
            end
            if flag
              f1 = 1./((1+n*s1).*sqrt(1-k2*s1));
              f2 = 1./((1+n*s2).*sqrt(1-k2*s2));
              el3(j) = (f1+f2)*w';
            else
              el3(j) = inf;
            end
          end
          el3 = psi*el3;
          return
```